Polymer Rheology: Theory and Practice

Polymer Rheology: Theory and Practice

Yuri G. Yanovsky

Institute of Applied Mechanics,
Russian Academy of Sciences,
Moscow, Russia

SPRINGER-SCIENCE+BUSINESS MEDIA, B.V

First edition 1993

Originally published by Chapman & Hall in 1993
Softcover reprint of the hardcover 1st edition 1993

Typeset in 10/12pt Times by Interprint Ltd, Malta.

ISBN 978-94-010-4938-2

A catalogue record for this book is available from the British Library

Library of Congress Cataloging-in-Publication data

Yanovsky, Yuri G.
Polymer rheology: theory and practice / Yuri G. Yanovsky.
p. cm.
Includes bibliographical references and index.
ISBN 978-94-010-4938-2 ISBN 978-94-011-2116-3 (eBook)
DOI 10.1007/978-94-011-2116-3
1. Polymers—Rheology. I. Title.
TA455.P58Y36 1993
620.1'920423—dc20 92-24454
CIP

Preface

The present book is devoted to a rapidly developing field of science which studies the behavior of viscoelastic materials under the influence of deformation—the rheology of polymers.

Rheology has long been treated as the theoretical foundation of polymer processing, and from this standpoint it is difficult to overestimate its importance in practice. Rheology plays an important role in developing our ideas on the nature of viscoelastic behavior in connection with the structural features of polymers and composites based on them. This expands the possibilities of employing rheological methods to characterize a variety of materials and greatly magnifies the interest in this field of research.

The rheological properties of polymer systems are studied experimentally, chiefly under conditions of shear and tensile strains. One explanation is that many aspects of polymer material processing are associated with the stretching of melts or a combination of shear and tensile strains.

In scientific investigations, either periodic or continuous conditions of shear deformation are employed. Each mode provides widespread information. In periodic deformation, most attention is generally given to conditions with low deformation amplitudes that do not alter the structure of the polymer system during an experiment (the region of linear deformation conditions). Here the viscoelastic parameters are generally determined with respect to the frequency. Continuous deformation involves considerable strains, and may be attended by significant reversible and irreversible changes in the structure of a polymer. Of major importance in these cases is the appraisal of the parameters characterizing both steady flow conditions in closed channels and conditions causing a loss of stability of the flow. The latter is manifest in the form of separation of the material from the channel walls, distortion of the shape of the streams flowing out of the channels, and even their disintegration into parts. A comparison of the parameters determined in low-amplitude periodic deformation and in continuous deformation enables one to compare the properties of polymer systems with changed and unchanged structures.

Analysis of the rheological properties of a system in tension can show the behavior of a polymer system within a very broad range of magnitude of the strains. The latter include regions of both linear and nonlinear deformation, and pre-ultimate values up to failure of a system. Here one can trace the relation between the viscoelastic, stress–strain, and relaxation characteristics.

The book contains sections concerning modern approaches to the theoretical, model, and experimental descriptions of the viscoelastic behavior of individual polymers, blends of polymers and copolymers, and filled compositions, and also to problems of the relation between the rheological, physicomechanical, and stress–strain properties. They are discussed, in particular, with a view to the achievements of Russian scientists during the last few years. The selected chapter topics reflect the various aspects of the scientific and applied problems of rheology.

Since the rheological behavior of individual fluid polymers has been analyzed in detail in many monographs, in the following these matters are discussed only superficially. The characteristics and properties of individual polymers and systems with a polymer matrix differing in chemical structure, molecular weight, inherent viscosity, melting point, etc., are not compared in detail where this is not required for elucidating the examples. The customary description of the specific results would be cumbersome and of only partial significance. (References to the original literature are an important source of specific information.)

I have also assumed that the reader is acquainted with the fundamentals of rheological science.

In view of the limited volume of the book, I have focused attention on the aspects that are most urgent today, namely the general laws of rheological behavior of composite systems, including blends of polymers, copolymers, and filled compositions. It is exactly the structural features of these objects that ensure their useful properties as materials intended for engineering purposes. Blends, copolymers, and composites are reinforced materials because many factors improving their resistance to failure are realized in them. Although it is difficult to describe the strengthening processes with the aid of a unified theory, one can follow the path of model ideas and construct theoretical substantiations of definite experimental facts.

The book consists of five chapters.

The first chapter is devoted mainly to discussing modern theoretical approaches and also methods for the analytical and numerical calculation of the viscoelastic characteristics of polymer systems (individual polymers, blends, filled polymers) under various deformation conditions. The appendices, which include prototypes of programs for imitation (method of Brownian dynamics) and geometric (method of finite elements) modeling and the test calculations, illustrate the possibilities of the approaches and methods being discussed and the expediency of employing them.

The second chapter deals with the deformation and relaxation behavior of flowing polymer systems within broad temperature and rate (or frequency) ranges covering various physical states of a system. Attention is given mainly to the critical parameters of deformation of fluid polymer systems. They include, on the one hand, the critical stresses and strains determining the conditions of the transitions of a polymer system from the fluid to the rubbery and leather-like states; and, on the other hand, the parameters characterizing the initial state of a system, i.e., those corresponding to the zero values of strains and rates of deformation, determined by the initial viscosity of the polymers. The relation between these two groups of limiting parameters is traced for the example of the dependence of service life on the initial viscosity of a system, its molecular weight and temperature.

The relation between the viscoelastic and relaxation properties measured in the linear and nonlinear deformation regions, on the one hand, and the stress–strain properties, on the other, is analyzed, as well as approaches to determining the relaxation characteristics in the nonlinear deformation region up to failure of a system.

The discussion of the viscoelastic properties of polymer blends and copolymers (Chapters 3 and 4) reveals the general principles of their rheological behavior. Polymer blending is in many ways similar to copolymerization as a way of achieving required properties by combining various chemical structures. The only difference is that blending reaches this goal physically, and not chemically (as for copolymerization).

The features of a number of typical blends exhibiting the properties of various classes of these compositions—compatible (miscible), limitedly compatible, and incompatible—are set out, as well as the features of random and block copolymers. The advantages of the model approach when describing the viscoelastic behavior of the above compositions within a wide range of changes in the composition of the components, their molecular weights, and the nature of the polymers, are discussed.

Copolymers are a structural form of macromolecules having a number of merits and shortcomings in comparison with mechanical mixtures. In some cases, copolymers (block copolymers in particular) exhibit unique possibilities that cannot be realized for other polymer materials. In others, they have no advantages.

The fifth chapter analyzes the rheological behavior of filled polymers using as examples both model systems, viz. filled linear flexible-chain polymers with a narrow molecular weight distribution including fillers of various activities, and samples of commercial filled polymers. The possible relation between the viscoelastic characteristics of filler polymer melts and parameters showing the physicomechanical and stress–strain properties of compositions based on them in the solid state is dealt with. Also discussed are the relaxation properties of filled polymers within a broad range of temperatures covering various physical states of a material—glassy, rubbery, fluid—and also the relaxation process due to the mobility of certain kinetic units.

In writing the book, I did everything possible to combine the individual problems instead of compiling it from separate treatises. At the same time, each chapter is relatively independent to avoid sending the reader frequently to other chapters.

Every book is intended for a definite circle of readers. I hope the present one will be helpful to specialists and scientists in the field of the mechanics of non-Newtonian liquids, and to specialists in the production and processing of polymer materials, both to those with experience and to beginners in these fields. Although the book has not been conceived as a textbook, it will nevertheless be an excellent aid for senior students and postgraduates who are interested in polymer rheology.

I wish to thank everyone who participated in discussing the separate chapters of the book. I am especially grateful to Professor Dr V.N. Pokrovsky, Dr E.K. Borisenkova, Dr V.U. Novikov and Dr Yu.K. Kokorin for their aid in working on the separate chapters, and also to G.L. Leib for his creative approach to the translation of sections of the book.

I also appreciate highly the support and collaboration by the Department of Mechanics and the Institute of Applied Mechanics of the Russian Academy of Sciences, and the aid of the technical staff.

Yuri Yanovsky

Contents

Chapter 1

Some Theoretical and Numerical Approaches to Describing the Viscoelastic Properties of Polymer Systems

1.1 INTRODUCTION

Polymers are widely used in practice and have a number of unique rheological and physicomechanical properties [1–5]. They differ from other substances in the size of their molecules, which are correctly called macromolecules. At temperatures above the glass transition temperature or melting point, a polymer system (concentrated solution or melt of a polymer) can be considered as a viscoelastic liquid whose behavior is determined by a system of weakly bound macromolecules. When a system is excited (mechanically or by heat), the macromolecules easily change their neighbors, but the intactness of each of them is not disturbed.

The structure of polymer liquids is more complicated than that of solids and low-molecular-weight liquids and, at the same time, has features of both, namely, order of the atoms is observed within one macromolecule and chaos in the position of the centers of mass of individual macromolecules and their parts. The mechanical reaction of polymer systems combines the elasticity of a solid and the fluidity of a liquid. Their behavior is defined as viscoelastic [1–5]. This property is one of the manifestations of slow relaxation processes (with a relaxation time up to one second and more). It is associated primarily with the relaxation of an individual macromolecule in a system [2, 5].

The diverse anomalies in viscoelastic behavior are also connected with the features of behavior of long macromolecules. The experimental facts, widely known in recent years, have been supplemented with new ones, e.g. the inflection in the terminal zone on the plot of the modulus of storage vs the frequency [6, 7] or the inflection and extrema in the region of medium frequencies on the frequency dependence of the modulus of the dynamo-optical coefficient and phase angle, respectively, in oscillating birefringence [8]. These features, associated with superslow relaxation processes, had been observed previously in various situations but had not been explained authentically.

The last few decades have seen active searches for a concept from whose standpoint one could understand and describe the nonequilibrium phenomena in polymers unambiguously.

A large number of phenomenological rheological relations have been proposed to describe the behavior of polymer liquids [9, 10]. However, a phenomenological description does not disclose the microscopic nature of the parameters in the relations and their connection to such practically important characteristics as the concentration of a polymer, its molecular weight, and polydispersity. Attempts to explain the observed properties on undilute systems from the viewpoint of the molecular theory of dilute solutions of polymers in a viscous liquid [11–18] were not successful. This was due in part to the fact that until recently the question of the microstructure of such liquids remained open [1, 2] (though the concept of entanglements of macromolecules with one another in undiluted systems had been developed for many years, apparently beginning with the work of Busse [16]). It was not clear whether the polymer chains in them are entangled or separated. Experimentalists gave preference to the model of separated chains. This gave birth to a number of theories [1, 11] in which intermolecular interaction consisted in amplifying the hydrodynamic interaction through the viscous medium or, actually, in a growth in the external friction of a macromolecule.

The answer to the question of the microstructure of undiluted polymer liquids was found to depend on the dimension of space. For two-dimensional space, preference was given [2] to the model of separated chains. For the conventional three-dimensional situation, on the other hand, the results of neutron-scattering experiments [17, 18] favored the entangled chain model. Moreover, these experiments confirmed Flory's important conclusion [19] that every chain in a dense system is ideal and Gaussian.

In a dense system, the volume interactions of a geometric nature or, in other words, the topological entanglements of the macromolecules whose influence did not consist only in a simple increase in the external friction of a chain, determined the need to construct a molecular theory of undiluted polymer liquids with consideration of the microstructure. Definite achievements were reached here with respect to the equilibrium properties [2, 11]. However, the description of the nonequilibrium relaxation phenomena involved substantial difficulties.

Attempts to consider the mutual impermeability of macromolecules with the aid of an approach [20–23] using directly the topological description of the entanglements of the macromolecules in a system led to substantial mathematical difficulties and did not result in noticeable progress in constructing a theory.

1.2 SINGLE MOLECULE APPROACH IN THE THEORY OF VISCOELASTICITY OF POLYMER CONCENTRATED SOLUTIONS AND MELTS

The considerable progress achieved in recent years in the field of the theoretical physics of polymers has also determined the achievements in the structural theory of polymer concentrated solutions, melts, and blends.

We find that the single molecule approach holds the most prospects and meanwhile has yielded the most results [22–29].

An adequately formulated equation of the dynamics of a macromolecule is the foundation on which a structural theory of equilibrium and nonequilibrium relaxation phenomena can be constructed in a single molecule approximation for undiluted polymer systems with entanglements. The correctness of the postulated ideas on the nature of its thermal motion is determined by how well the proposed model of the dynamics describes the entire set of available experimental facts.

The single molecule approach considers the motion of only one arbitrarily singled out macromolecule schematized by a model instead of that of all the mutually entangled chains. The influence of the entanglements and surroundings—other chains and a solvent, if present—on the dynamics of the singled-out macromolecule is then taken into account in an effective way employing the parameters of the theory. The latter proceeds from a more detailed treatment of the interactions between chains as, for instance, was done for the first time by Bueche [30, 31].

The possibility of introducing the single molecule approach is based in essence on an important experimental fact. The quantities determining the behavior of a polymer system depend in a clear and unambiguous way on the length of a macromolecule. Indeed, with slow motion or in the region of low frequencies, the macroscopic properties of polymer systems depend in a universal way on the individual characteristic of a macromolecule—its molecular weight. It is just this circumstance that justified going over from a multichain problem to the single molecule approach.

An important merit of the single molecule scheme is that here undiluted polymer systems can be treated as diluted suspensions of noninteracting test chains, and the available achievements of the theory of dilute solutions and suspensions (see, for instance, [32, 33]) can be used to calculate the macroscopic properties.

The choice of a specific macromolecule model in the single molecule approach is not significant and is mainly determined by the temporal and spatial scales in which the theory attempts to describe the behavior of the system and also by mathematical considerations. Conversely, the choice of the way of schematizing the surroundings and their influence via the entanglement on the singled out macromolecule is fundamentally important. The differences in the ways of considering the surroundings give birth to different variants of the single molecule approach.

For example, attempts to schematize the surroundings of a test macromolecule of a viscous liquid without memory were not successful primarily because here the relaxation properties (aftereffects) of the polymer medium were not considered.

The basic problem in constructing the theory of undiluted linear polymers in a single molecule approximation thus consists in the adequate consideration of the influence of the surroundings and entanglements on the dynamics of a test macromolecule.

At about the same time, two approaches were formulated, namely two equations of motion of a polymer chain solving the problem in fundamentally different ways [25, 28, 29]. One of these equations of dynamics [25], summarizing, as it were, the development of the concept of a primitive chain [20], a chain in a 'tube', and de Gennes's reptation model [24], forms the content of the model of a primitive chain advanced by Doi and Edwards.

In the reptation or tubular approach, the influence of the surroundings is imitated by introducing a 'tube' of a certain diameter in which the test macromolecule moves. The macromolecule cannot intersect the walls of the tube, and large-scale movements of the center of mass of the chain take place by small-scale movements of its segments along the tube. The dynamics of this process, for instance in the model of a primitive chain, are assumed to be such [25] that only the 'head' particle in 'forward' motion occupies its new position by chance. It adds a new section to the tube, whereas the remaining particles can determinately pass over only to the site of their nearest neighbor (predecessor). The corresponding section at the opposite end of the tube vanishes simultaneously: the tube is continuously being renewed.

These illustrative ideas of the dynamics of macromolecules, which generalize de Gennes's reptation model [24], can be strictly formulated mathematically in the form of an equation of motion [25]. The latter is then the basis for calculating the physical quantities of interest.

The approach based on tubular models played a definite, positive role in the development of ideas on the dynamics of a macromolecule in a dense system. For example, consideration of the diffusion of a long macromolecule among its neighbors on the basis of the dynamics of a polymer chain restricted by a tube made it possible to describe the dependence of the diffusivity on the molecular weight, $D \propto M^{-2}$, observed in some experiments [2, 34–44]. This probably played a substantial role in the subsequent broad dissemination and theoretical and experimental studies of the reptation tubular approach. It was soon revealed, however, that such an approach cannot be used to describe many experimental facts [2, 5–8, 11, 42, 45–47], including deviations from the law $D \propto M^{-2}$. It also fails to explain superslow relaxation processes.

An alternative theory [28, 29] was based on a fundamentally different equation of motion, and already the first corollaries following from this equation of dynamics have shown that it describes the viscoelastic properties of linear polymers with entanglements very well. The calculated [29] frequency dependences of the storage G' and loss G'' moduli for a monodisperse system coincided with the experimentally obtained ones within a broad range of frequencies [7]. The relaxation spectrum of the system consisted of long and short relaxation times divided into groups, which is typical of undiluted polymers [1, 11]. Moreover, a shoulder was predicted on linear sections of the plots of G' vs frequency ω in the terminal zone at some values of the theoretical parameters. Such a shoulder was later found experimentally [6, 7] for melts of polybutadienes and was associated [7] with the superslow relaxation times. However, the origin and nature of this shoulder for monodisperse systems have not been explained [6, 29].

The equation of the dynamics of the approach described by Pokrovsky and Volkov [28, 29] was employed successfully for constructing a rheological determining relation and description from common standpoints of the linear and nonlinear effects of viscoelasticity in dense systems [7, 48–58]. The equation of the dynamics underlay the development of a theory for a whole group of phenomena, viz. viscoelasticity [57, 58], diffusion [53] and optical anisotropy [55], and also the processes of light and neutron scattering in dense systems; i.e., in the aggregate a theory of undiluted solutions of linear polymers with entanglements was constructed. We shall consider some of the results obtained in the cited publications and some of their corollaries in greater detail.

1.2.1 Equation of Dynamics of a Macromolecule

1.2.1.1 Dynamics of a Macromolecule without Entanglements

To construct an equation of dynamics, one must formulate the basic equations determining the dynamics of a single macromolecule in a viscous liquid when there are no entanglements and the features of the changes in these conditions when going over to a concentrated solution or melt of noticeably entangled long macromolecules. The latter allows us to construct an equation of motion of a macromolecule in an undilute system with entanglements in the single molecule approximation, and also the correlation functions of the normal coordinates and their derivatives needed to calculate the required physical quantities.

In a dilute Θ solution [59] at a concentration of $C < C^*$ (C^* is the critical concentration), the macromolecules are isolated from one another and there are no entanglements between them. But interactions of a thermodynamic nature are also absent [59]. At the same time within the confines of each individual macromolecule, bulk interactions of a topological nature or self-entanglements are possible [11, 60].

The model of Gaussian subchains was employed as the model of a macromolecule [61]. It proved its merits when dealing with slow motion. In this model, a polymer chain is schematized by a set of $(N+1) \gg 1$ Brownian particles or beads of a subchain bound by harmonic forces. They represent a sufficiently large number of units of a real macromolecule. The distance b between the subchain ends obeys Gaussian statistics.

As regards the interaction of such a macromolecule model with its surroundings (solvent and other macromolecules), the simplest situation is observed in dilute solutions of polymers in viscous Newtonian liquid. Here a Brownian particle is surrounded by a solvent—a liquid. The relaxation times and, accordingly, the aftereffects (memory effects) of such a liquid may be ignored. The coefficient of friction of the subchains ξ is assumed to be constant and localized at the points of their connection [13, 14].

The system of equations of motion for all the particles of one chain in the very simple case of a Rouse model (no interaction of the excluded volume, hydrodynamic and internal friction) in a linear approximation with respect to the gradient of the medium velocity is written in the form [3, 10]

$$m\ddot{r}_i^\alpha(t) = -\xi[\dot{r}_i^\alpha(t) - v_{ij}(t)r_j^\alpha(t)] - 2T\mu A_{\alpha\beta}r_i^\beta(t) + \Phi_i^\alpha(t) \tag{1.1}$$

Here the Greek superscripts number the particles and vary from 0 to N; the subscripts are their Cartesian components and take on values of 1, 2, 3; m is the weight of a subchain; $\vec{r}^\alpha$ is its position; a dot over a letter signifies the derivative with respect to the time t; $v_{ij} = \delta v_i/\delta x_j$ is the tensor of the unperturbed gradients of the medium velocity; $A_{\alpha\beta}$ is the matrix of elastic interaction of pair bonds by the chain random force Φ_i^α. We shall also use the following notation for a symmetrized and antisymmetrized velocity gradient, respectively:

$$\gamma_{ij} = \tfrac{1}{2}(v_{ij} + v_{ji});\ \omega_{ij} = \tfrac{1}{2}(v_{ij} - v_{ji})$$

The procuring of specific results with the aid of eqn (1.1) is facilitated by the introduction of normal coordinates in which the equations divide. Here the matrix of elastic pairwise interactions of adjacent beads along a chain is diagonalized with the aid of an orthogonal matrix. Hence in normal coordinates determined by the formulae

$$\rho_i^\gamma = R_{\beta\gamma}r_i^\beta,\ r_i^\alpha = R_{\alpha\gamma}\rho_i^\gamma \tag{1.2}$$

where $R_{\beta\gamma}$ and $R_{\alpha\gamma}$ are linear transformations and ρ_i^γ is the normal coordinate with number γ, each of the $(N+1)$ equations of dynamics (1.1) has the form

$$m\ddot{\rho}_i^\alpha(t) = -\xi(\dot{\rho}_i^\alpha - v_{ij}\rho_j^\alpha) - 2T\mu\lambda_\alpha\rho_i^\alpha + R_{\alpha\beta}^{-1}\Phi_i^\beta(t) \tag{1.3}$$

where λ_α is the proper moment of matrix $A_{\alpha\gamma}$, and describes motion of only one mode.

The first term from the right is the force of hydrodynamic resistance to the carrying along of a particle in a flow with an unperturbed velocity gradient v_{ij}. The next term describes the elastic pairwise interaction of neighboring particles along the chain [13]. The elastic constant is written in the form $2T\mu$, where T is the temperature of the system, and $\mu = 3N/2R^2$. For an ideal chain, the mean square distance between its ends is $R^2 = Nb^2$ [59]. The last term at the right in eqn (1.3) is the random

Brownian force that is distributed by a Gaussian law. Equation (1.3) leads to the only discrete set of relaxation times of a macromolecule (the Rouse relaxation times [12]):

$$\tau_\alpha^R = \frac{\xi}{4T\mu\lambda_\alpha} = \frac{\tau^*}{\alpha^2} \tag{1.4}$$

where α is the number of mode, with the maximum characteristic time $\tau^* \equiv \tau_1^R$.

The additional consideration of an effect in eqn (1.3) results in a perturbed set of eigenvalues of λ_α^1 and, accordingly, in perturbed relaxation times τ_α^1 [1, 11]. In the long run, the physical quantities of interest will be expressed in terms of these quantities. However, one of the generalizations of the equation of dynamics (1.3) considered the excluded volume V and the hydrodynamic interaction h of various parts of a macromolecule [62], and another considered the internal viscosity γ [14].

Hence, in the considered approximation, the dynamics of a macromolecule in a viscous liquid without entanglements are determined by the parameters τ^*, γ, V, and h, i.e. by the maximum relaxation time, coefficient of internal viscosity, excluded volume, and hydrodynamic interaction, respectively.

1.2.1.2 Dynamics of a Macromolecule with Entanglements

When constructing an equation of motion of a macromolecule in an undiluted system, the following were considered.

(1) The ideal nature of the macromolecules in a melt and a sufficiently concentrated solution was adopted. This made it possible to disregard the interactions of the excluded volume for these systems ($V=0$) and consider the macromolecules to be under Θ conditions [17–19].

(2) The hydrodynamic entrainment in an undiluted system is realized already not so much via the solvent as via the entanglements [1, 2, 63]. Consequently, the role of the hydrodynamic interaction of individual subchains with one another becomes insignificant: $h=0$.

(3) In the macroscopic deformation of an undiluted polymer liquid, both a separately taken (test) macromolecule and the macromolecules of its surroundings are deformed. Hence, this process, like the transmission of hydrodynamic entrainment, is of a cooperative nature. In such a medium, the concept of internal

viscosity associated with deformation of exactly the test polymer chain loses its previous meaning and acquires a qualitatively new content.

(4) In an undiluted system surrounding a test chain, the medium has pronounced relaxation properties or, in other words, a memory [11]. The latter, being a reliably established experimental fact [1–5], enables one to consider the influence of the surroundings on the dynamics of a test macromolecule in a fundamentally different way than that proposed by de Gennes, Doi, Edwards, *et al.* [2, 5, 11, 22–24].

The single molecule approach undertaken by Pokrovsky and Volkov [28, 29] presumes that the medium surrounding a test chain (formed by similar macromolecules joined to one another and, possibly, a solvent) is schematized by an apparent continuous viscoelastic medium with a single relaxation time τ. This is extremely important in the theory and signifies that the motion of a macromolecule at a specific moment of time must take the entire history of the system into account. Since a test chain interacts with its continuous surroundings by friction, this is considered in the equation of dynamics by the appropriate choice of the dissipative properties of a macromolecule in which entanglements are considered effectively.

With a view to these remarks and the notation adopted above, the equations of motion of a macromolecule in a moving medium will be written as follows [28, 29]:

$$m\ddot{r}_i^\alpha(t) = -\int_0^\infty B_{\alpha\beta}(s)(\dot{r}_i^\beta - v_{ij}r_j^\beta)_{t-s}\,\mathrm{d}s - \int_0^\infty G_{\alpha\beta}(s)(\dot{r}_i^\beta - \omega_{ij}r_j^\beta)_{t-s}\,\mathrm{d}s$$
$$-2T\mu A_{\alpha\beta}r_i^\beta(t) + \phi_i^\alpha(t) \tag{1.5}$$

where s is the current (real) time.

The first term on the right-hand side of eqn (1.5) describes the forces of hydrodynamic resistance to the entrainment of the α-particle in its motion relative to the medium and other particles of the chain. The second term describes the forces of internal (see the foregoing remark) viscosity in the deformation of the macromolecule. B and G are friction nuclei.

Further manipulations with eqn (1.5) presume a transition to normal coordinates in which the matrices of B, G, and A are diagonalized simultaneously. In the general case, taking into account the hydrodynamic interaction, internal friction, excluded volume, screening and other effects, it is quite difficult to write down an equation of the dynamics of a macromolecule. For this reason, with a view to the features of the system being considered (strong interaction along a chain and weak interaction between the macromolecules), the normal coordinates can be established in two stages. In the first stage, the dynamics of one macromolecule are determined in the surroundings of the others whose influence is considered not only in terms of the averaged characteristics. In the second stage, the normal coordinates of an individual macromolecule are determined. (For greater detail see [52].)

Equation (1.5) of the dynamics of a macromolecule in a dense system and a linear approximation with respect to the velocity gradients of the medium in normal coordinates (see eqn (1.2)) is written in the form

$$m\ddot{\rho}_i^\alpha(t) = -\int_0^\infty \beta_\alpha(s)(\dot{\rho}_i^\alpha - v_{ij}\dot{\rho}_j^\alpha)_{t-s}\,\mathrm{d}s - \int_0^\infty \varphi_\alpha(s)(\dot{\rho}_i^\alpha - \omega_{ij}\rho_j^\alpha)_{t-s}\,\mathrm{d}s$$
$$-2T\mu\lambda_\alpha\rho_i^\alpha(t) + R_{\alpha\beta}\Phi_i^\beta(t)\text{: } (s \geqslant 0) \qquad (1.6)$$

where β_α is the kernel of external friction and φ_α is the kernel of internal friction.

The nucleus of external friction $\beta_\alpha(s)$ was selected by Pokrovsky and Volkov [28, 29] in the form of the sum of the purely viscous and viscoelastic, or aftereffect, terms:

$$\beta_\alpha(s) \equiv \beta(s) = 2\xi\delta(s) + \frac{\xi B}{\tau}\mathrm{e}^{-s/\tau} \qquad (1.7)$$

For undiluted systems, the hydrodynamic forces described by the first term on the right-hand side of eqn (1.7) make a contribution only at high frequencies of motion, when the displacements of the subchains are not large. Here the motion of a subchain is determined only by its closest 'monomer' surroundings. At low frequencies, the hydrodynamic forces are chiefly determined by the second term which considers entanglements and relaxation of subchain–medium interaction; B shows the increase in the coefficient of friction ξ of a particle due to entanglements.

The second term from the right in eqn (1.6) describes the forces of internal viscosity. With a view to the physical meaning of the force of internal friction for undiluted systems, the nucleus of internal friction was chosen in the form

$$\varphi_\alpha(s) \equiv \varphi(s) = \begin{cases} 0, & \alpha = 0 \\ \dfrac{\xi E}{\tau} \mathrm{e}^{-s/\tau}, & \alpha = 1, 2, \ldots, N \end{cases} \tag{1.8}$$

where E is the coefficient of the growth in internal viscosity due to entanglements. Since the zero mode corresponds to the conditions of the absence of motion of a chain as a whole [13] when it experiences no strains, the zero eigenvalue is $\varphi_0 = 0$.

The parameters B and E were assumed to be constant by Pokrovsky and Volkov [28, 29], though in the general case they depend on the number of the mode.

Integration of eqn (1.6) over all the times s preceding the given moment t considers the entire previous history of a macromolecule in the system.

The last right-hand term in Eq. (1.6) characterizes the random Brownian force. The latter, as in the problem with viscous surroundings, is distributed by a Gaussian law with a zero mean.

Hence, in the approximation being considered, the dynamics of a macromolecule in an undiluted system with entanglements depend on the parameters B, E, τ^*, and $\chi = \tau/2B\tau^*$ determined by the chemical nature of a polymer. For polymers with a linear structure, the dependence of these parameters on the chain length and concentration is universal [29]. The chemical nature of the polymers is considered by the theory in terms of the coefficient of friction ξ.

Equation (1.6) of the dynamics of a macromolecule, together with postulates (1.7) and (1.8) on the nature of interaction of a polymer chain with its surroundings introduced into it, form the basis for calculating the physical quantities we are interested in. They are determined by the simultaneous and nonsimultaneous second moments of the normal coordinates and their derivatives $\langle \rho_i^\alpha \rho_j^\beta \rangle$, $\langle \rho_i^\alpha \dot{\rho}_j^\beta \rangle$, and $\langle \dot{\rho}_i^\alpha \dot{\rho}_j^\beta \rangle$.

For the interaction of a macromolecule with its surroundings by laws (1.7) and (1.8), the following expressions were obtained by Pokrovsky and Volkov [28, 29] for the equilibrium moment of the normal coordinates and the function $\mu_\alpha(s)$ at the moment s:

$$M_\alpha^0(s) = \frac{H_\alpha}{2\mu\lambda_\alpha}\left(M_+ \mathrm{e}^{-s/2\tau_\alpha^+} - M_- \mathrm{e}^{-s/2\tau_\alpha^-}\right) \tag{1.9}$$

$$\mu_\alpha(s) = H_\alpha\left(\mu_+ \mathrm{e}^{-s/2\tau_\alpha^+} - \mu_- \mathrm{e}^{-s/2\tau_\alpha^-}\right) - E(s) \tag{1.10}$$

$$\mu_\pm = 1 + B - \frac{\tau}{2\tau_\alpha^\pm}, \quad M_\pm = E + \mu_\pm, \quad H_\alpha = \frac{\tau_\alpha^R}{\tau_\alpha^+ - \tau_\alpha^-} \tag{1.11}$$

where

$$2\tau_\alpha^\pm = \tau_\alpha \pm \sqrt{(\tau_\alpha^2 - 2\tau\tau_\alpha^R)}, \quad 2\tau_\alpha = \tau + 2\tau_\alpha^R(1 + B + E) \tag{1.12}$$

are two sets of the relaxation times of a macromolecule appearing when it is introduced into a viscoelastic medium.

Inspection of eqns (1.12) reveals that the relaxation spectrum also includes the Rouse relaxation times because the limiting process to a viscous medium without a memory and entanglements, i.e. B, E, and $\tau \mapsto 0$ in formulae (1.12), leads to the Rouse spectrum τ_α^R. In the other limiting case, namely in an undiluted system with numerous entanglements (B, $E \gg 1$), superslow relaxation times $\tau_\alpha^+ > \tau$ appear.

Consequently, generalization of the equation for the dynamics of a macromolecule placed in a viscous Newtonian medium (see eqn (1.3)) undertaken for viscoelastic surroundings in the form of eqns (1.6)–(1.8) also leads to the relevant generalization of the relaxation spectrum (*cf.* expressions (1.4) and (1.12)). Instead of the single Rouse set τ_α^R, two sets of relaxation times closely related to τ_α^R appear in the second case.

It is important to stress that eqn (1.6) with determination of the random force can be taken as the first approximation (linearity with respect to the coordinates and velocities, linearity with respect to the velocity gradients) for nonequilibrium phenomena in undiluted polymers. Here effects are excluded that are associated with nonlinear addends such as the reptations indicated by de Gennes [24], which appear when considering the differences in the mobility of a particle along and across a chain. The latter can be described with the aid of addends of the second or higher order in the consecutive theory of motion of a macromolecule in an undiluted system (see below).

1.2.3 Viscoelasticity of Undiluted Polymers

1.2.2.1 The Stress Tensor

The rheological properties of a dilute suspension of test macromolecules—Brownian particles bound into a chain—are determined by the tensor of additional stresses $\sigma_{ik}(t)$ [33] appearing when the test chains are introduced into a viscoelastic liquid. Its general expression for equations of dynamics (1.3)–(1.6) has the form [26, 27]:

$$\sigma_{ik}(t) = -nT\delta_{ik} + nT\sum_{\alpha=1}^{N}\int_0^\infty \Big\{2\mu\lambda_\alpha\langle\rho_i^\alpha(t)\rho_k^\alpha(t)\rangle - \frac{m}{T}\langle\dot\rho_i^\alpha(t)\dot\rho_k^\alpha(t)\rangle$$
$$+\frac{1}{T}\int_0^\infty \varphi(s)[\langle\rho_i^\alpha(t)\dot\rho_k^\alpha(t-s)\rangle - \omega_{kl}(t-s)\langle\rho_i^\alpha(t-s)\rho_i^\alpha(t)\rangle]\Big\}ds$$

By inserting the moments of the normal coordinates (eqn (1.9)) into this expression, we arrive at an expression for σ_{ik} with an accuracy up to the terms of the first order with respect to the velocity gradient of the medium γ_{ik}:

$$\sigma_{ik}(t) = -nT\delta_{ik} + 2nT\sum_{\alpha=1}^{N}\int_0^\infty \Big\{\Big[2\mu\lambda_\alpha\mu_\alpha(s)M_\alpha^0(s) - \frac{m}{T}\dot\mu_\alpha(s)\dot M_\alpha^0(s)\Big]\gamma_{ik}(t-s)$$
$$+\frac{1}{2T}\int_0^\infty \varphi(s)[\mu_\alpha(s+U)\dot M_\alpha^0(U)$$
$$+\dot\mu_\alpha(U)M_\alpha^0(s+U)]\gamma_{ik}(t-s-U)\,dU\Big\}ds \tag{1.13}$$

where s is the current time of orientation and U is the current time of deformation.

Expression (1.13) for the stress tensor holds for any of the systems being considered containing macromolecules, viz. for a solution, melt, or blend of polymers.

The microinertial effects are not significant in slow motion, therefore the mass of the subchains—Brownian particles—can be disregarded in the following, and we can assume in eqn (1.13) that $m \to 0$.

1.2.2.2 Dynamic Modulus and Superslow Relaxation Times

Polymer systems have a noticeable elasticity, which has been associated with the existence in a system of a network of links—entanglements formed by long macromolecules [11]. By analogy with the theory of elasticity of linear polymers [4], the expression for the modulus of elasticity was written as

$$G_e = nT\frac{M}{M_e} \tag{1.14}$$

where M_e is the length of a macromolecule between adjacent entanglements.

The viscoelastic behavior of a polymer system is best shown when a periodic shear strain is applied to it (periodic regimes) [1]. In this case $\gamma_{ik} \propto \exp(-i\omega t)$, where ω is the circular frequency of deformation.

The storage modulus G' and loss modulus G'' are quantities measured experimentally as functions of ω and characterizing the dynamic viscoelastic properties of a polymer system. They are the real and imaginary parts of the complex dynamic modulus [1]:

$$G^*(\omega) = G'(\omega) - iG''(\omega) \tag{1.15}$$

The latter is related to the complex dynamic viscosity by the expression [1]:

$$G^*(\omega) = -i\omega\eta^*[\omega] = -i\omega\{\eta'[\omega] - i\eta''[\omega]\} \tag{1.16}$$

where the brackets signify the one-sided Fourier transform of the memory function $\eta(s)$. Expression (1.13) for the stress tensor has the form:

$$\sigma_{ik}(t, \omega) = -nT\delta_{ik} + 2\eta[\omega]\gamma_{ik}[t, \omega] \tag{1.17}$$

From eqns (1.13), (1.15), (1.16), and (1.9)–(1.12), we can calculate the values of G' and G'' vs ω [7, 55]:

$$G'(\omega) = nT \sum_{\alpha=1}^{N} \left[\frac{P^+(\omega\tau_\alpha^+)^2}{1+(\omega\tau_\alpha^+)^2} + \frac{P^{++}(\omega\tau_\alpha^{++})^2}{1+(\omega\tau_\alpha^{++})^2} \right]$$

$$G''(\omega) = nT \sum_{\alpha=1}^{N} \left[\frac{P^+(\omega\tau_\alpha^+)}{1+(\omega\tau_\alpha^+)^2} + \frac{P^{++}(\omega\tau_\alpha^{++})}{1+(\omega\tau_\alpha^{++})^2} \right] \tag{1.18}$$

Here ω is the frequency of deformation of a system, n is the density of the number of macromolecules, T is the temperature, $\alpha = 0, 1, \ldots,$ P^+ and P^{++} are weights of relaxation times in the relaxation spectrum, and N is the mode number.

It can be seen from eqn (1.18) that two relaxation processes are manifest in the low-frequency region with relaxation times of

$$\tau_\alpha^+ = \tau/2 + \tau_\alpha^R(B+E), \quad \tau_\alpha^{++} = 2\tau\tau_\alpha^+/(\tau + 2\tau_\alpha^+) \tag{1.19}$$

and a spectrum with its own weight is present in each of them:

$$P^+ = (\tau_\alpha^R/\tau_\alpha^+)^2 B^2, \; P^{++} = (\tau_\alpha^R/\tau_\alpha^+)^2 BE(\tau_\alpha^+/\tau_\alpha^{++}) \tag{1.20}$$

where τ is the postulated time of relaxation of the medium surrounding the arbitrarily singled out macromolecule and formed by identical entangled macromolecules; and B, as mentioned above, is a parameter of theory showing the increase in the coefficient of friction of a subchain ξ at the expense of entanglements. The latter was determined [52, 53] as:

$$B \propto C^{4.8} M^{2.4} \tag{1.21}$$

where C is the concentration when dealing with concentrated solutions, E is a measure of the increase in the 'internal' viscosity at the expense of entanglements, and τ_α^{R} are the Rouse relaxation times of a macromolecule [1] that, as it were, is in a low-molecular viscous liquid formed from the same monomers as the polymer.

Expanding the expression for the complex dynamic modulus into a series relative to a low frequency, we have

$$G^*(\omega) \equiv G'(\omega) - iG''(\omega) = A_G\omega^2 - i\eta\omega$$

From eqns (1.18)–(1.20) the coefficients of elasticity and viscosity of the system are written as

$$A_G = nT(B\tau^*)^2 \sum_{\alpha=1}^{N} \left[\frac{1}{\alpha^4} + \frac{2\chi\psi(\chi\alpha^2+\psi+1)}{\alpha^2(2\chi\alpha^2+\psi+1)}\right] = A_G^{+} + A_G^{++}$$

$$\eta = nTB\tau^* \sum_{\alpha=1}^{N} \left[\frac{1}{\alpha^2(\chi\alpha^2+\psi+1)} + \frac{\psi}{\alpha^2(\chi\alpha^2+\psi+1)}\right] = \eta^{+} + \eta^{++} \tag{1.22}$$

For high frequencies, the expressions for the modulus on the plateau have the form

$$G'_e = nT \sum_{\alpha=1}^{N} \left[\frac{1}{(\chi\alpha^2+\psi+1)^2} + \frac{\psi}{2\chi\alpha^2(\chi\alpha^2+\psi+1)}\right] = G'^{+}_e + G'^{++}_e \tag{1.23}$$

where $\tau^* = \tau_1^{R}$ are the greatest Rouse relaxation times. The values of the parameters $\psi = E/B$ and $\chi = \tau/2B\tau^*$ for melts and concentrated solutions of polymers with a large number of entanglements per macromolecule N_e are in the region [51–54]

$$\psi \gg 1, \quad \chi \ll 1 \tag{1.24}$$

For long macromolecules ($N \gg 1$) formulae (1.22) and (1.23) become

$$A_G = nT(B\tau^*)^2\left(\frac{\pi^4}{90} + \frac{\pi^3}{3}\chi\right), \quad \eta = nTB\tau^*\frac{\pi^2}{6}\left(\frac{1}{\psi}+1\right) \tag{1.25}$$

$$G'_e = nT\left(\frac{\pi}{4}\frac{1}{\chi^{1/2}\psi^{3/2}} + \frac{\pi^2}{12}\frac{1}{\chi}\right) \tag{1.26}$$

As previously, the first addend in eqns (1.25) corresponds to the relaxation component with relaxation time τ_α^+, and the second with time τ_α^{++}. For the latter in the indicated region, eqns (1.19) lead to the relations

$$\tau_\alpha^+ \geqslant \tau, \quad \tau_\alpha^{++} \approx \tau \propto C^{2,8} M^{3,4} \tag{1.27}$$

The second relation points to the self-consistency of the theory—the calculated macroscopic relaxation time of the system coincides with the postulated time τ. On the other hand, the first of relations (1.27) indicates that slower relaxation processes with relaxation times τ_α^+ are manifested in a system.

In the region of the parameter values (1.24), the moduli on the plateau are determined as follows [7, 52]:

$$G'_e = \frac{\pi^2}{12}(nT\chi^{-1}) \propto C^3 M^0 \tag{1.28}$$

The quantity χ in expressions (1.22)–(1.28) is a small parameter of theory and has the meaning of the reciprocal of the number of entanglements:

$$\chi = \frac{\pi^2}{4}\frac{M_e}{M}$$

The general form of the plots of G' and G'' vs ω calculated for the region (1.24) is shown in Fig. 1.1. This figure reproduces behavior in the region of low frequencies or, in other words, in the region of the slowest relaxation processes considered heretofore.

Analysis of the available experimental results [1, 6, 7] (see also Chapter 2) shows that only the $(++)$ component (Fig. 1.1) was registered in the terminal zone, as a rule. It was then used to determine the relevant initial values of the coefficients A_G^{++} and η^{++}. Calculations show that another slower relaxation process with relaxation times τ_α^+ is observed in the terminal zone at low frequencies. It is manifested on the plot of G' vs ω in the form of a 'step' below which a linear section with a slope equal to 2 is observed on the plot of G' vs ω. The relevant step on the plot of G'' vs ω both in theory [28] and in the experimental results of Raju *et al.* [6] is either less pronounced or absent.

All these features can be seen well in Fig. 1.2 presenting the results of theoretical calculations of G' and G'' by the above expressions for polybutadienes. (For a comparison with experimental results see Chapter 2).

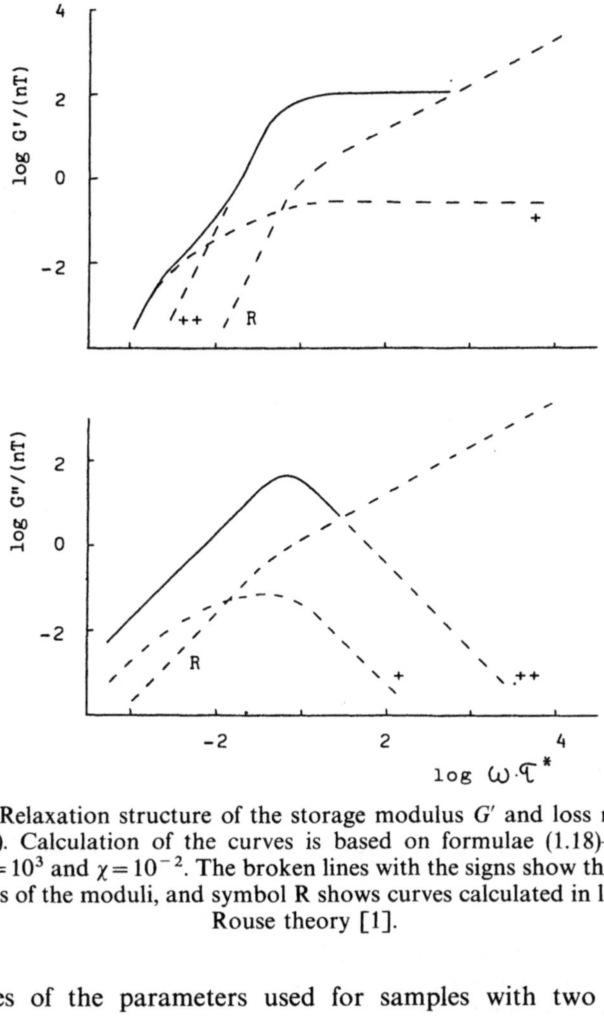

FIG. 1.1. Relaxation structure of the storage modulus G' and loss modulus G'' (solid lines). Calculation of the curves is based on formulae (1.18)–(1.20) with $B=10^2$, $E=10^3$ and $\chi=10^{-2}$. The broken lines with the signs show the relaxation components of the moduli, and symbol R shows curves calculated in line with the Rouse theory [1].

The values of the parameters used for samples with two molecular weights are given in Table 1.1.

The parameter χ according to formula (1.28) determines the height of the rubbery plateau on the plot of G' vs ω. The values of χ in Table 1.1 are well consistent with the values of $\chi_1=1{\cdot}96\times10^{-2}$ and $\chi_2=1{\cdot}4\times10^{-2}$ appraised by formula (1.29) for $M_e=1900$ [7]. The parameter B_2 can be selected arbitrarily and can be quite large, and B_1 will then be determined by it according to eqn (1.21):

$$B_2/B_1 \propto (M_2/M_1)^{2\cdot4} = (1{\cdot}42)^{2\cdot4}$$

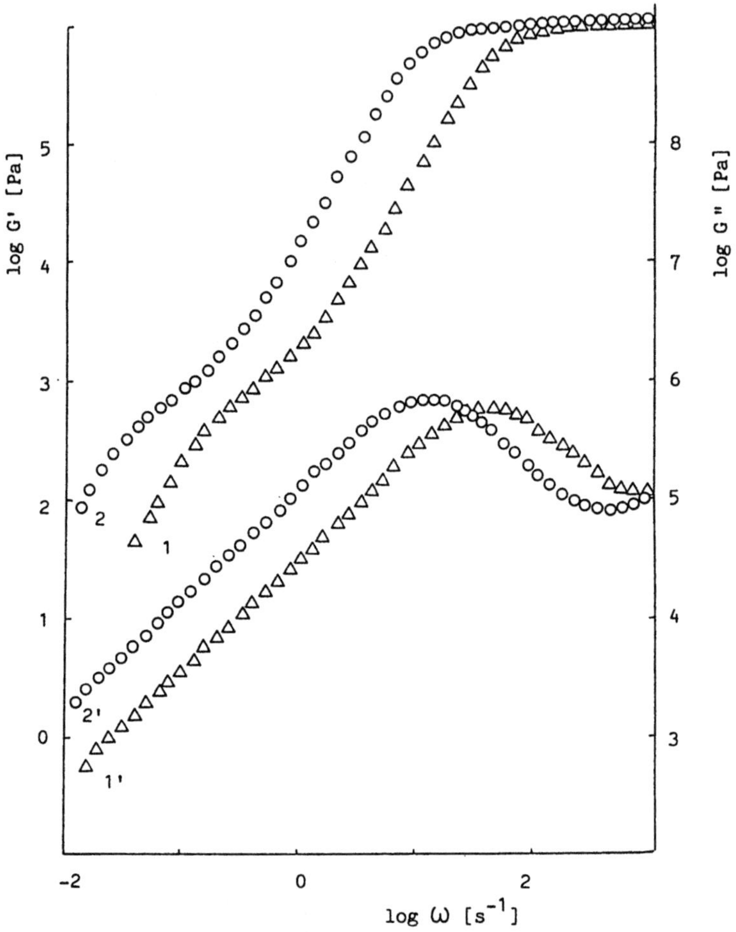

FIG. 1.2. Results of theoretical calculation of G' and G'' vs ω for polybutadienes. The theoretical parameters were determined by Table 1.1 with a view to the experimental results obtained in Ref. 7. Curves 1, 1′ and 2, 2′ are for molecular weights of $1{\cdot}0\times10^5$ and $1{\cdot}4\times10^5$, respectively; the temperature was 333 K.

The parameter $\psi = E/B$ determines the height of the low-frequency projection connected with the superslow relaxation times. It is selected from the conditions of the best coincidence of the theoretical and experimental steps [7], and then together with the parameters B is used to find the values of E_1 and E_2.

Yanovsky *et al.* [7] were the first to establish the dependence of E on

TABLE 1.1
Theoretical and Experimental Characteristics of PB at 333 K

$M \times 10^{-5}$	$B \times 10^{2}$	$E \times 10^{-3}$	$\chi \times 10^{-2}$	$\tau^* \times 10^{3}$ (s)	$n \times 10^{-18}$ (cm^{-3})
1·4	10	5	0·9	6·3	3·8
1·0	4·31	2·15	1·2	2·5	5·4

M and C. They showed that with a change in the molecular weight, the height of the step in the region of superslow times on the plot of G' vs ω remains practically constant. With a view to expression (1.21) for B and ψ, it was found that $E \propto M^{2\cdot 4}$. Hence with account of the known relation $\tau^* \propto M^2$ and expression (1.19), we obtain $\tau_\alpha^+ \propto M^{4\cdot 4}$.

The values of τ^* were determined according to the horizontal displacement of the theoretical curves up to their best coincidence (simultaneously for G' and G'') with the experimental results in the terminal zone [7]. The values of n for determining the vertical displacement were found by the formula $n = \rho N_A/M$, where the Avogadro constant $N_A = 6{\cdot}025 \times 10^{23}$, and ρ is the density of the polybutadiene (PB) samples at 333 K, equal to 0·9 g cm^{-3}.

In summarizing, we can conclude that the relaxation processes observed experimentally in linear amorphous polymers, including superslow ones, are naturally and without special assumptions described by theory [52–55, 57]. Actually, the superslow relaxation processes had been predicted by Pokrovsky and Volkov [28], but at that time they were not interpreted adequately.

The concentration dependence of the superslow relaxation times τ_α^+ can be established experimentally. This enables one to determine all the parameters of the above theory, including the dependence of E on C. It is best to run such experiments using PB, with a low value of M_e [7, 11].

The value of the modulus on the plateau.

$$G'_e = G_e^{++} = \frac{1}{6} nT \frac{R^2}{\xi^2} \tag{1.29}$$

can be used to appraise the characteristic scale ξ in a system naturally obtained in theory [53], and not introduced into it from outside as in the 'tubular' reptation models [2, 5]. The latter formula, coinciding with the known one [11] but obtained from phenomenological considerations, reminds one of the first model interpretations of a concentrated solution (melt) as a system with a fluctuation network having cross-links (entanglements). However, as can be seen from the above reasoning and from several publications [7, 28, 52–58], the use of the concept of a network of entanglements or reptations to explain the features of the viscoelastic behavior of a concentrated linear polymer system is not mandatory because the approach employing a characteristic length ξ is quite physical, in our opinion. This quantity was previously considered to be the distance between neighboring cross-links of the entanglement network.

1.2.2.3 Viscoelasticity of Polymer Blends

Study of a blend of two polymers of which one is present in a small amount provides a unique possibility of obtaining direct information on the dynamics of an isolated selected macromolecule in a viscoelastic liquid formed by the macromolecules of the matrix [52, 65].

A system formed by a linear polymer with a molecular weight M_1 and a small additive of the same polymer with a greater molecular weight M_2 is quite a convenient model. If the content of the high-molecular-weight additive is so small that its macromolecules interact only with molecules of the matrix, the system formed by the linear polymer with the molecular weight M_1 is the medium in which the molecules move.

The change in the stress when a small number of macromolecules of another species are introduced is determined by the dynamics of the singled-out noninteracting macromolecules among the macromolecules of the other length. Hence, the case being considered is of interest from the standpoint of the theory of viscoelasticity of linear polymers.

The regions of low frequencies at which the dynamic modulus can be represented in the form of expansion (1.16) and also the region of low concentrations of the high-molecular-weight additive are of considerable interest. The intrinsic values of the viscosity and elasticity coefficients are determined, respectively, as follows:

$$[\eta] = \lim_{C\to 0} \frac{\eta_2 - \eta_1}{C\eta_1}, \quad [A_G] = \lim_{C\to 0} \frac{A_{G2} - A_{G1}}{CA_{G1}} \tag{1.30}$$

(where the subscript 1 relates to the quantities characterizing the matrix, and C is the concentration of the additive. Of interest is the dependence of the indicated intrinsic quantities on the length (or molecular weights) of the macromolecules of the matrix and additive.

The viscoelastic behavior of dilute blends of polymers has been studied experimentally by a number of authors [64–67]. Yanovsky *et al.* [64–66] determined the values of $[\eta]$ and $[A_G]$ for blends of polybutadienes of various lengths and narrow molecular weight distributions (see also Chapter 3). The results were approximated by the relations

$$[\eta] \propto M_1^{-0.8} M_2^{0.5}, \quad [A_G] \propto M_1^{-(1.8-2.2)} M_2^{(1.3-3.0)} \tag{1.31}$$

Calculations of the intrinsic quantities by the theory of Pokrovsky and Kokorin [52] yield the following relations:

$$[\eta] \propto M_1^{-1} M_2, \quad [\mathrm{A_G}] \propto M_1^{-1} M_2 \tag{1.32}$$

These results are not affected by how the quantity B depends on the length (molecular weight) of a macromolecule.

It should be noted that consideration of the competing mechanism of thermal motion—reptations of the macromolecules—leads [68] to relations of the intrinsic viscosity of a blend:

$$[\eta] \propto M_1^{-3} M_2^3 \tag{1.33}$$

A comparison of expressions (1.32) and (1.33) obtained theoretically with the experimental results shows that the values of the experimental indices are closer to those in expressions (1.32) than to those in (1.33). This confirms that in their influence on the motion of a singled-out macromolecule the surrounding macromolecules are equivalent to a viscoelastic medium and shows that the model of reptation motion is not completely suitable for describing relaxation processes. As already noted, this is due to the reptation effect being connected with addends of an order higher than the first in the equation of motion of a macromolecule, whereas the first-order addends are the main ones when treating relaxation phenomena. An attempt to describe viscoelasticity with exclusion of the main linear terms distorts the picture. This explains the lack of success when describing the viscoelasticity of polymers in the reptation model.

The above results hold for $M \gg M_1$. This is due to the fact that in our reasoning we omitted the effect of replacement of matrix macromolecules by additive ones.

1.2.3 Enclosing

We find it expedient to terminate the discussion of the very simple corollaries following from the theory being considered with the examples in Section 1.2.2. For a more detailed analysis, one can turn to the original publications cited above.

The basic ideas of the above theory of relaxation processes in undiluted polymer systems, which can be defined as a self-consistent first-order theory, were initially formulated by Pokrovsky and Volkov [28], and later less completely by Renca [69]. A number of authors have continued to study the problem [53–58]. To date, no appreciable contradictions of the theory with the large set of experimental results obtained have been discovered [6, 7, 64–67, 70, etc.], showing that the theoretically described slow relaxation processes correspond to the real ones observed in an undiluted polymer system.

The theoretical results may be applied to slow (low-frequency) motion, do not depend on the structural details of a macromolecule chain, and in this sense are universal; the chemical nature of a polymer is included in terms of the values of the theoretical parameters. In the region of high frequencies, the relaxation processes and viscoelastic behavior are determined by the details of chain structure and details of interaction of a macromolecule with its neighbors. In the latter case, one must proceed from a macromolecule model that is more detailed than the model of subchains and consider the correlation of the orientations of adjacent sections of chains [3].

1.3 IMITATION MODELING BY THE METHOD OF BROWNIAN DYNAMICS OF THE VISCOELASTIC BEHAVIOR OF POLYMERS

When speaking of the merits and shortcomings of the theoretical approaches used to describe the viscoelastic behavior of polymer melts (including the single molecule approach), we must remember that in all the above cases we dealt with macromolecules far from the surface. However, the latter circumstance is exceedingly important, in particular towards describing the viscoelastic behavior of filled polymers (a heterophase system). Near a surface one must consider the interaction of separate parts of a macromolecule with the surface. Here we should expect changes in the mobility of a macromolecule and the relaxation time. The equation of dynamics of a macromolecule is modified and becomes nonlinear. Analytical appraisal of the results in the simpler variant that has proved successful [52–58] becomes impossible. Numerical methods of study come to the forefront.

With a view to the specific nature of the model proposed by Pokrovsky and Volkov [28] that schematizes a macromolecule by a set of Gaussian subchains, including $(N+1)$ Brownian particles bound consecutively by elastic forces, it is logical to employ this model for numerical calculations by the method of Brownian dynamics [3]. For this purpose, the equation of dynamics of a macromolecule must be written in a form that is convenient for the modification associated with the presence of surface forces, and also convenient for performing numerical calculations, because in this case one has to work with a nonlinear system of equations.

It should be noted that a correctly compiled system of equations for the dynamics of a macromolecule [28] is nonlinear, and such systems have been simplified and linearized in the literature [7, 52, 53, 55, 57] only to enable them to be considered analytically. Study of a model in its correct nonlinear formulation by the method of Brownian dynamics gives a deeper understanding of the mechanism of formation of the viscoelastic properties and substantiates the simplifications made earlier [7, 28] in analytical study of the dynamics of a macromolecule.

We can thus state that the use of the method of Brownian dynamics is an essential and important stage in studying the mechanism of the viscoelasticity of polymers in boundary layers.

1.3.1 Modified Equation of the Dynamics of a Macromolecule

After modification of the system of equations and performing some calculations, the equation of dynamics of a macromolecule (1.6) can be written as a system of linear stochastic equations:

$$\frac{\mathrm{d}r_i^\alpha}{\mathrm{d}t}=u_i^\alpha$$

$$m\frac{\mathrm{d}u_i^\alpha}{\mathrm{d}t}=-\xi(u_i^\alpha-\delta_{i1}qr_2^\alpha)+F_i^\alpha+G_i^\alpha-2T\mu A_{\alpha\gamma}r_i^\gamma-\delta_{i2}p^\alpha(r_2^\alpha)+\Phi_i^\alpha(t)$$

$$\tau\frac{\mathrm{d}F_i^\alpha}{\mathrm{d}t}+F_i^\alpha=-\xi[B\delta_{ij}-H(e_i^\alpha e_j^\alpha-\tfrac{1}{3}\delta_{ij})](u_j^\alpha-\delta_{1j}qr_2^\alpha) \tag{1.34}$$

$$\tau\frac{\mathrm{d}G_i^\alpha}{\mathrm{d}t}+G_i^\alpha=-\xi\{E[u_i^\alpha-\tfrac{1}{2}q(\delta_{i1}r_2^\alpha-\delta_{i2}r_1^\alpha)]+G_{\alpha\gamma}(u_j^\alpha-u_j^\gamma)e_j^{\alpha\gamma}e_i^{\alpha\gamma}\}$$

where r_i^α is the coordinate of the α particle ($\alpha=0,\ldots,N$; $i=1,2,3$); u_i^α is the velocity of the α particle; m is the mass of a particle; ξ is the coefficient of friction; $q=\mathrm{d}v_1/\mathrm{d}t_2$ is the velocity gradient; $2T\mu$ is the elasticity coefficient; $e_i^\alpha=(r_i^{\alpha+1}-r_i^{\alpha-1})/(|r^{\alpha+1}-r^{\alpha-1}|)$; $e_i^{\alpha\gamma}=(r_i^\alpha-r_i^\gamma)/(|r^\alpha-r^\gamma|)$; $p^\alpha(r_2^\alpha)$ is the force of attraction to (repulsion from) the surface; τ is the relaxation time; $B, H, E, C_{\alpha\beta}$ are numerical parameters; and Φ_i^α is a random force discussed below. The matrix

$$[A_{\alpha\gamma}] = \begin{Bmatrix} 1 & -1 & 0\ldots0 \\ -1 & 2 & -1\ldots0 \\ 0 & -1 & 2\ldots0 \\ \cdots & \cdots & \cdots \\ 0 & 0 & 0\quad 1 \end{Bmatrix}$$

has the dimension $(N+1)(N+1)$. Excluding the forces F_i^α and G_i^α from system of equations (1.34), we have

$$\mathrm{d}u_i^\alpha/\mathrm{d}t = a_i^\alpha$$

$$\mathrm{d}r_i^\alpha/\mathrm{d}t = u_i^\alpha$$

$$\begin{aligned} m\tau\frac{\mathrm{d}a_i^\alpha}{\mathrm{d}t} &+ ma_i^\alpha + \tau\xi a_i^\alpha + \xi u_i^\alpha - \delta_{i1}q\xi(\tau u_2^\alpha + r_2^\alpha) + 2T\mu\tau A_{\alpha\gamma}u_i^\gamma + 2T\mu A_{\alpha\gamma}r_i^\gamma \\ &+ \tau\delta_{i2}\frac{\mathrm{d}p^\alpha}{\mathrm{d}r_2^\alpha}u_2^\alpha + \delta_{i2}p^\alpha + \xi B(u_i^\alpha - \delta_{i1}qr_2^\alpha) - \xi H(e_i^\alpha e_j^\alpha - \tfrac{1}{3}\delta_{ij}) \\ &\times(u_j^\alpha - \delta_{j1}qr_2^\alpha) + \xi Eu_i^\alpha + \xi C_{\alpha\gamma}(u_j^\alpha - u_j^\gamma)e_j^{\alpha\gamma}e_j^{\alpha\gamma} - \tfrac{1}{2}\xi Eq(\delta_{i1}r_2^\alpha + \delta_{i2}r_1^\alpha) = \xi_i^\alpha(t) \end{aligned} \qquad (1.35)$$

where the new random force ξ_i^γ has been introduced as

$$\xi_i^\alpha = \tau\frac{\mathrm{d}\Phi_i^\alpha}{\mathrm{d}t} + \Phi_i^\alpha \qquad (1.36)$$

For numerical realization, the system of equations (1.35) will be more convenient in the dimensionless form after determining the variables

$$\begin{aligned} r_i^\alpha &= \sqrt{\left[\frac{2(B+E)T\tau}{\xi}\right]}R_i^\alpha \\ u_i^\alpha &= \frac{1}{\tau}\sqrt{\left[\frac{2(B+E)T\tau}{\xi}\right]}U_i^\alpha \\ a_i^\alpha &= \frac{1}{\tau}2\sqrt{\left[\frac{2(B+E)T\tau}{\xi}\right]}A_i^\alpha \end{aligned} \qquad (1.37)$$

The equations of dynamics (in the absence of velocity gradients) acquire the form

$$\mathrm{d}R_i^\alpha/\mathrm{d}S = U_i^\alpha \qquad \mathrm{d}U_i^\alpha/\mathrm{d}S = A_i^\alpha$$

$$M\frac{\mathrm{d}A_i^\alpha}{\mathrm{d}S} + MA_i^\alpha + A_i^\alpha + U_i^\alpha + ZA_{\alpha\gamma}U_i^\gamma + ZA_{\alpha\gamma}R_{i2}^\gamma + \delta_{i2}\frac{\mathrm{d}p^\alpha}{\mathrm{d}R^\alpha}U_2^\alpha \tag{1.38}$$
$$+\delta_{i2}P^\alpha + BU_i^\alpha - H(e_i^\alpha e_j^\alpha - \tfrac{1}{3}\delta_{ij})U_j^\alpha + EU_i^\alpha + C_{\alpha\gamma}(U_j^\alpha - U_j^\gamma)e_j^{\alpha\gamma}e_i^{\alpha\gamma} = f_i^\alpha(s)$$

where

$$M = m/\xi\tau \qquad Z = 2T\mu\tau/\xi \qquad P^\alpha = \tau p^\alpha/\xi$$

$$f_i^\alpha(s) = \frac{\tau}{\xi}\sqrt{\left[\frac{\xi}{(B+E)2T\tau}\right]}\xi_i^\alpha(t)$$

The quantities M and Z are the ratios of three characteristic relaxation times of the system between which, for the case of large N considered, the relations $m/\xi \ll \xi/2T\mu \ll \tau$ hold, so that we have

$$M \ll Z^{-1} \ll 1,\ Z \gg 1 \tag{1.39}$$

When comparing the results calculated for various values of the number of particles N, one must compare the values of the quantities for an identical value of the parameter M and various values of Z related by the expression

$$Z_2 = \frac{N_2^2}{N_1^2}Z_1 \tag{1.40}$$

1.3.2 Normal Coordinates, Random Forces, Calculated Quantities

It is convenient to analyze the dynamics of a macromolecule in normal coordinates determined as $P_i^\alpha = \Theta_{\gamma\alpha}R_i^\gamma$.

The orthogonal and normalized transformation matrix is determined by the requirement $\Theta_{\mu\lambda}A_{\lambda\gamma}\Theta_{\gamma\beta} = \lambda_\mu\delta_{\mu\beta}$ and has the form:

$$\Theta_{\gamma\alpha} = \left(\frac{2-\delta_{\gamma 0}}{N+1}\right)^{1/2}\cos\left(\frac{\alpha(1+2\gamma)\pi}{2(N+1)}\right) \tag{1.41}$$

where

$$\lambda_\alpha = \left(\frac{\pi\alpha}{N}\right)^2 \qquad \alpha = 0, 1, 2, \ldots \ll N \tag{1.42}$$

The zero normal coordinate P^0 is proportional to the coordinate of the center of mass R_i:

$$P_i^0 = \frac{1}{\sqrt{(N+1)}} \sum_{\gamma=0}^{N} R_i^\alpha = [\sqrt{(N+1)}]R_i \tag{1.43}$$

All the variables determined above are random quantities whose properties are determined by the system of equations (1.38).

Usually the mean characteristics of a random force are determined from the requirement that known results must be obtained for equilibrium. For instance, for the square of the velocity of any particle we must have $\Sigma\langle(u_i^\alpha)^2\rangle = 3T/m$ regardless of its number.

This requirement determines the correlation for the linear case when $P^\alpha = 0$, $H = 0$, $C_{\alpha\gamma} = 0$ and $M \to 0$, $B \gg 1$:

$$\langle \xi_i^\alpha(t)\xi_j^\gamma(t')\rangle = 2T\xi(B+E)\delta_{ij}\delta_{\alpha\gamma}\delta(t-t')$$

Here and below the angle brackets signify averaging over the realizations:

$$\langle A\rangle = \frac{1}{n}\sum_{i=1}^{n} A_i \tag{1.44}$$

In formulae (1.37), introducing dimensionless variables, the above relations have the form

$$\left\langle \sum_{i=1}^{3} (U_i^\alpha)^2 \right\rangle = \frac{3}{2}\frac{1}{M(B+E)} \tag{1.45}$$

$$\langle f_i^\alpha(s) f_j^\alpha(s')\rangle = \delta_{ij}\delta_{\alpha\gamma}\delta(s-s') \tag{1.46}$$

The mean values of the force are presumed to equal zero, while the distribution is Gaussian, i.e. for any component

$$p(f_i) = \frac{e^{-f_i^2/2}}{\sqrt{(2\pi)}} \tag{1.47}$$

When nonlinear terms are included ($p^\alpha = 0$, $H = 0$, $C_{\alpha\gamma} = 0$), the distribution law may be more involved and must be established empirically.

The behavior of a macromolecule is appraised primarily by its mobility as a whole and by the relaxation times.

The mobility is customarily [3] characterized by the diffusivity, which is defined as the coefficient of proportionality between the mean square value of the displacement and the time $\langle \Delta q^2\rangle = 6DT$.

This is why the dependence of the mean square displacement on the time and other parameters of a problem is important. Of special interest here is seeing how the nonlinear terms affect the mobility of a macromolecule, which cannot be determined by analytical procedures.

The relaxation times as fundamental quantities characterizing the dynamics of a macromolecule are manifested when studying various phenomena such as viscoelasticity, optical anisotropy, light and neutron scattering. However, there is no need to consider these phenomena to determine the relaxation time. The same fundamental relaxation times can be evaluated as characteristics of the time dependence of the correlation functions for each of the nonzero modes (formula (1.49)). Here the modification of the relaxation time values when including the described nonlinearities is of interest. Hence, the mean square displacement of the center of mass of a macromolecule is

$$L(s)=\left\langle \sum_{i=1}^{3} [R_i(s)-R_i(0)]^2 \right\rangle$$

This quantity, with a view to eqn (1.43), can be written in terms of the zero normal coordinate:

$$L(s)=\frac{1}{N+1}\sum_{i=1}^{3} \langle [P_i^0(s)-P_i^0(0)]^2 \rangle \tag{1.48}$$

The nonzero normal coordinates are limited. To characterize the dynamics, the following quantities are calculated as a function of s:

$$\sum_{i=1}^{3} \langle P_i^\alpha(0) P_i^\alpha(s) \rangle$$

It is also convenient to determine the relation

$$K^\alpha(s)=\frac{\sum_{i=1}^{3} \langle P_i^\alpha(0) P_i^\alpha(s) \rangle}{\sum_{i=1}^{3} \langle P_i^\alpha(0) P_i^\alpha(0) \rangle} \tag{1.49}$$

and to find the relaxation time $\theta_\alpha=\tau_\alpha/\tau$ by the expression

$$K^\alpha(s)=\exp\left(-\frac{s}{2\theta_\alpha}\right) \tag{1.50}$$

Consequently, the calculations yield plots of L and K^α vs the time s ($\alpha=1, 2, \ldots, N$), and also the values of the relaxation times θ_α.

1.3.3 Algorithm of Imitation Process and Its Realization

In numerical realization, we register the positions, velocities, and accelerations of the Brownian particles at moments of time at the interval $\Delta s = h$ and replace the differential equations (1.38) with difference ones:

$$R_i^\alpha(s+h) = R_i^\alpha(s) + hU_i^\alpha$$

$$U_i^\alpha(s+h) = U_i^\alpha(s) + hA_i^\alpha$$

$$\begin{aligned} A_i^\alpha(s+h) = A_i^\alpha(s) + h\frac{1}{M}[&-MA_i^\alpha - U_i^\alpha - A_i^\alpha - ZA_{\alpha\gamma}R_i^\gamma - ZA_{\alpha\gamma}U_i^\gamma \\ &-\delta_{i2}\frac{dP^\alpha}{dR_2^\alpha}U_2^\alpha - \delta_{i2}P^\alpha - BU_i^\alpha + H\sum_{i=1}^{3}(e_i^\alpha e_j^\alpha - \tfrac{1}{3}\delta_{ij})U_j^\alpha - EU_i^\alpha \\ &-C_{\alpha\gamma}\sum_{i=1}^{3}(U_j^\alpha - U_j^\gamma)e_j^{\alpha\gamma}e_i^{\alpha\gamma}] + \frac{1}{M}g_i^\alpha(s,h) \end{aligned} \tag{1.51}$$

where the random forces are designated as

$$g_i^\alpha(s,h) = \int_s^{s+h} f_i^\alpha(h)\,dh$$

with the dispersion

$$\int_s^{s+h}\int_s^{s+h} \langle f_i^\alpha(h) f_i^\alpha(\alpha)\rangle\, dh\, d\alpha = h \tag{1.52}$$

To realize the algorithm, a program package has been developed consisting of a main program and subprograms. Appendix 1 shows the structure of the program package for solving the problem.

The program of parameter adjustment makes it possible to alter the values of the basic variables controlling a process. These parameters include the number of particles, the step of differentiation with respect to time, the maximum time, the number of arguments of the correlation function, and the number of slides of the state of a molecule.

The values of the parameters are stored in the file 'Parameter MLK', and are then read by the 'calculation' program. The structure of the latter is shown in Appendix 1. In the course of work, the program solves the system of differential equations and calculates the process characteristics.

The result is the formation of data for the graphical program, and also a file of the results containing the calculation parameters, process characteristics, and appraisals of the relaxation time for each mode.

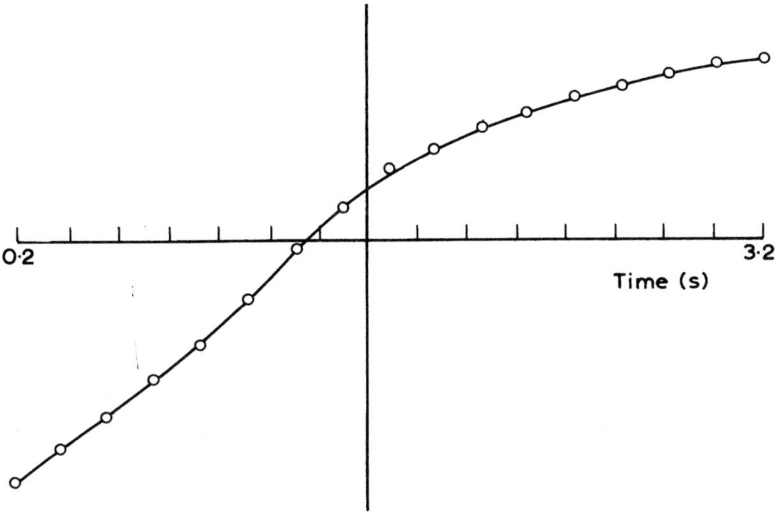

FIG. 1.3. The mean value of the square of the displacement of a macromolecule's center of mass vs time.

The graphical program draws the state of a macromolecule in space according to the first and second coordinates in equal intervals of time. The number of slides is set by the input parameter (Appendix 1).

1.3.4 Test Calculation

The parameters and results of the control calculation are presented in Fig. 1.3 which shows the mean of the square of the displacement of a macromolecule's center of mass vs time (for 11 particles in moments of time 0·2). It is significant that the square of the displacement of the center of mass is not a linear function of time, as for a macromolecule in a viscous liquid, but extends out to a plateau, which corresponds to earlier results procured analytically [55].

The consistency of the results with the analytical ones [55] shows that the program satisfactorily reporoduces the motion of a macromolecule.

1.4 STRUCTURAL APPROACH TO MODELING THE BEHAVIOR OF FILLED POLYMERS

Filled polymer materials are dispersed in homogeneous bodies whose structure and properties depend not only on the properties of the matrix and filler, but also on various kinds of interaction between the components, on their mutual arrangement and volume ratio (the geometric factor), on the parameters of their processing and production (the technological factor), etc. These very important factors, whose change causes alteration of the properties of a polymer material, must be considered both when predicting the viscoelastic, stress–strain properties, and service life of already known materials and when developing new polymer composites. In addition to the theoretical approaches (phenomenological, single molecular, etc.) that describe quite reliably the properties of a polymer marix, i.e. a single-phase system, the involved nature of the problem has led to the appearance of empirical and semi-empirical 'structural and rheological' models aimed at predicting the behavior of heterophase polymer materials to a certain extent. The construction of various models and the need to study them require a search for new analytical and numerical methods for their analysis.

Mathematical modeling is one way of solving technical problems encountered in the development and prediction of the properties and processing technology of composite materials, and of describing their properties under diverse service conditions. Lately computers are coming into great favor for conducting analytical studies on models, and the relevant programs are being developed. Privalko *et al.* [71] describe programs for analyzing a system of modeling the properties of composites whose logical basis is the percolation theory. The finite-element method has become very popular, and there is a great diversity of programs for numerical calculations by the finite-element method to describe the behaviour of individual systems [72–80]. A detailed analysis of the models and numerical methods for calculating the behavior of fluid single-phase viscoelastic systems that are relatively simple from the structural viewpoint has been given by Mitsoulis [81].

Matters are considerably more involved in describing the properties of objects with a complicated structure, i.e. filled polymers and composites [71]. The problem consists primarily in selecting a physically substantiated model.

1.4.1 Phenomenological Models

The first estimates of the elastic properties of such complicated systems from the model viewpoint were undertaken on the basis of the well-known phenomenological models advanced by Voigt [82] and Reuss [83].

In a very simple variant, the behavior of a two-component polymer material is modeled by using two Voigt elements connected in parallel and reflecting the properties of each of the components. The properties of the system as a whole are described by the rule of mixtures (the additivity principle). Many authors have employed different variants of this procedure. The modulus of elasticity of such a composition E_c is calculated as:

$$E_c = \varphi_m E_m + \varphi_f E_f \tag{1.53}$$

where (here and below) φ is the volume fraction of a component, and the subscripts m, f, and c signify the polymer matrix, filler, and composite, respectively.

Similarly, using elements of the Reuss model, we have:

$$\frac{1}{E_c} = \frac{\varphi_m}{E_m} + \frac{\varphi_f}{E_f} \quad \text{or} \quad \frac{1}{E_c} = \frac{M_m}{E_m^\alpha} + \frac{M_f}{E_f^\alpha} \tag{1.54}$$

where M is the mass fraction of a component, and α is a constant ranging from 3·6 to 3·4.

Since experiments have shown that these very simple models and expressions do not generally describe the actually observed properties of two-component compositions satisfactorily within a broad interval of changes in the content of the components, numerous attempts have been made to improve these models by employing parallel (||) and consecutive (⊥) connection of elements A and B as shown in Fig. 1.4 [84]. For the

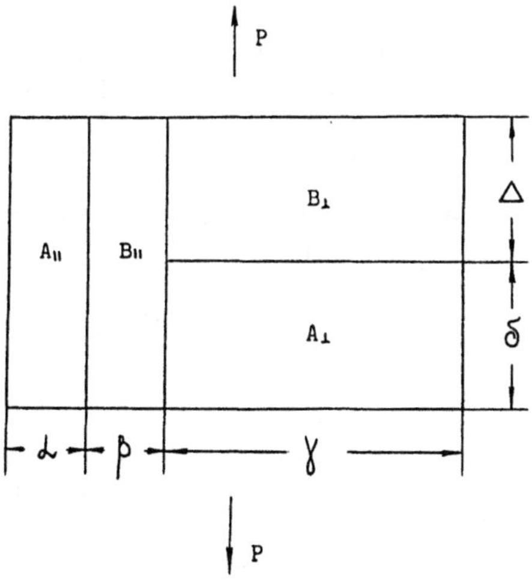

FIG. 1.4. Model of two-phase composition of elements A and B. P is the force.

model in Fig. 1.4, the values of E_c are calculated as follows:

$$E_c = E_A\varphi_{A\|} + E_B\varphi_{B\|} + \frac{E_A E_B(\varphi_{A\perp} + \varphi_{B\perp})}{E_A\varphi_{B\perp} + E_B\varphi_{A\perp}} \tag{1.55}$$

where

$$\varphi_A = \varphi_{A\|} + \varphi_{A\perp};\ \varphi_B = \varphi_{B\|} + \varphi_{B\perp};\ \varphi_A + \varphi_B = 1$$
$$\varphi_{A\|} = \alpha;\ \varphi_{A\perp} = \delta\gamma;\ \varphi_{B\|} = \beta;\ \varphi_{B\perp} = \Delta\gamma$$

The model of the geometric mean of the above cases relates to similar attempts [85, 86]:

$$\log E_c = x_m \log E_m + x_f \log E_f \tag{1.56}$$

or

$$E_c = E_m^{x_m} E_f^{x_f} \tag{1.57}$$

where x is the mole fraction.

Calculations performed for filled polymers have shown that the values of E_c obtained by eqns (1.53)–(1.57) are consistent satisfactorily with the experimental results, but generally only for materials with a low filler content (up to 20–30%). When there were considerable discrepancies between the theoretical and experimental values, attempts were made to bring them into agreement by introducing empirical correction factors taking certain structural and physical features of a filled system into account.

For example, to consider the adhesion of the filler to the polymer matrix, a coefficient K was introduced [87], and eqn (1.53) became:

$$E_c = E_m\varphi_m + E_f\varphi_f K \tag{1.58}$$

To describe the effect of reinforcing a polymer with spherical particles, Kerner [88] proposed the equation

$$E_c = \left\{1 + \frac{\varphi_f}{\varphi_m}\left[\frac{15(1-\mu_m)}{8-10\mu_m}\right]\right\}E_m \tag{1.59}$$

where μ_m is Poisson's ratio of the matrix.

Christensen [89], proceeding from energy ideas, obtained an equation for the shear modulus (G) that differs somewhat from Kerner's equation:

$$\frac{G_c}{G_m} = 1 - \frac{15(1-\mu_m)(1-G_f/G_m)\varphi_f}{7-5\mu_m+2(4-5\mu_m)(G_f/G_m)} \tag{1.60}$$

With absolutely rigid inclusions and an incompressibe matrix ($\mu_m = 0{\cdot}5$), eqn. (1.60) transforms as follows:

$$G_c/G_m = 1 + 2{\cdot}5\varphi_f \tag{1.61}$$

Expression (1.61) is the well-known Einstein relation for a suspension with a viscous medium and spherical particles.

Kerner's equation [88], which became quite popular in practice, was repeatedly modified by various authors who introduced coefficients considering either the degree of adhesion between the particles of the filler and matrix, or the geometry of the filler particles and the closeness of their packing [90–93]. For instance, for good adhesion between the filler and matrix particles, the equation was modified to the form:

$$E_c = E_m \frac{G_f\varphi_f/[(7-5\mu_m)G_m + (8-10\mu_m)G_f] + \varphi_m/[15(1-\mu_m)]}{G_m\varphi_m/[(7-5\mu_m)G_m + (8-10\mu_m)G_f] + \varphi_m/[15(1-\mu_m)]} \tag{1.62}$$

After considering the above examples, we can terminate our treatment of the known models describing the elastic properties of two-component systems because the principle of their construction is clear, as is their low predictability.

Models predicting not only elastic, but also viscoelastic, properties are the next step in this direction. Here we must note first of all the model advanced by Takayanagi *et al.* [92] by which the complex dynamic modulus E_c^* of a system in which filler particles are distributed in the matrix can be calculated as follows:

$$\frac{1}{E_c^*} = \frac{\Phi}{(1-K)E_m^* + KE_f^*}\,\frac{1-\Phi}{E_m^*} \tag{1.63}$$

For spherical particles, K and Φ are determined as

$$K = 2 - 3\varphi_f/5; \quad \Phi = 5\varphi_f/(2 + 3\varphi_f)$$

Simple transformations yield the following expressions for the storage and loss moduli:

$$E_c' = \frac{P}{P^2 + Q^2} \qquad E_c'' = \frac{Q}{P^2 + Q^2} \tag{1.64}$$

where P and Q are complicated functions of K and Φ.

Apart from the above expressions, many other empirical relations have been employed in practice with various degrees of success. In particular, Kenunen *et al.* [93] also used Kerner's model [88] to analyze the components of the complex dynamic modulus of two-component systems or mixtures. Here the determining relations had the form:

$$E'_{\mathrm{c}}=\frac{AE'_{\mathrm{m}1}-BE'_{\mathrm{m}2}}{C} \qquad E''_{\mathrm{c}}=\frac{AE''_{\mathrm{m}1}+BE''_{\mathrm{m}2}}{C} \tag{1.65}$$

where the subscript 'm' relates to the matrix, and A, B, and C are coefficients depending on (1) the volume fraction of the matrix, (ii) the maximum filler content when its particles do not yet interact, and (iii) empirical constants. The results of calculations by this model [93] have shown that expressions (1.65) are applicable when the shape of the particles is close to spherical, the components are immiscible and their moduli differ quite greatly, and deformation is not attended by stratification of the components or their sliding along the boundary. It is also significant that the model presumes a low filler content.

Analysis of models of the above type and similar ones and verification of their adequacy (see, for example, Refs 71, 86 and 94) has shown that at present it is difficult to recommend universal models that are correct for calculating the properties of a composition within a broad range of the change in the filler concentration and the conditions of deformation. No appreciable changes have been introduced by a number of other models proposed recently, in particular those in Ref 95–97 for describing viscoelastic, elastic, dissipative, and other properties of filled polymers. They have also found only a limited region of application.

Not being able to consider in greater detail the causes of the failure of the above models to describe the properties of filled polymers observed experimentally, we must note that the lack of a strictly physical substantiation of the proposed models was one of the main causes.

1.4.2 Three-Element Structural Model

From this standpoint, a three-element model can be considered more systematic and improved in this sense for filled polymers. It considers not only the properties of the filler and matrix, but also the intermediate layer appearing on the boundary of the filler particles (interphase layer), which has specific properties that differ from those of the matrix [87, 98, 99] (Fig. 1.5).

We consider the particles to be spherical and their distribution in the matrix homogeneous, and assume that the volume fraction of the filler is low (to exclude the mutual influence of the particles). We can thus

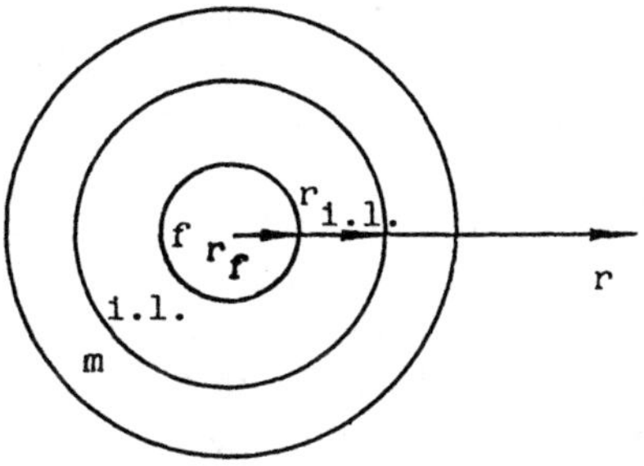

FIG. 1.5. Three-element model of a reinforced polymer.

write the expression for the modulus of elasticity of a composite E_c in the form:

$$\frac{1-2\mu_c}{E_c}=\frac{(1+\mu_f)^2(1-2\mu_m)^2}{(1+\mu_m)^2(1-2\mu_f)E_f}\varphi_f+\frac{1-2\mu_m}{E_m}\varphi_m$$

$$+\frac{3\varphi_f(1-2\mu_m)^2}{(1+\mu_m)^2 r_f^3}\int_{r_f}^{r_i}\frac{[1+\mu_i(r)]^2}{(1-2\mu_i)(r)E_i(r)}r^2\,dr \qquad (1.66)$$

The filler and matrix are assumed to be elastic, homogeneous, and isotropic bodies, whereas the modulus of elasticity and Poisson's ratio, i.e. $E_i(r)$ and $\mu_i(r)$ in the interphase layer, are functions varying along the position vector. With this in view, for periodic deformation eqn (1.66) is transformed into the following expressions for the elastic (storage) E'_c and dissipative (loss) E''_c components of the complex modulus:

$$E'_c=\frac{(1-2\mu_c)z}{z^2+u^2} \qquad E''_c=\frac{(1-2\mu_c)u}{z^2+u^2} \qquad (1.67)$$

where z and u depend in a complicated fashion on $E_i(r)$ and $\mu_i(r)$. The specific form of the functions $E_i(r)$ and $\mu_i(r)$ differs depending on the law of distribution of the properties in the interphase layer (linear, parabolic, hyperbolic).

In terminating this short review of the models employed for specific practical calculations of the properties of filled polymers or two-component systems, we note that apart from mechanical or mathematical approaches, thermodynamic models are also discussed in the literature. The latter presume that the properties of any heterogeneous systems depend on the energy of local and collective interaction at an interface with the formation of an interphase layer. It is considered here that the interphase layers are dissolved in the polymer. Such an approach treats the filled composition as an equilibrium multicomponent system. This problem was discussed in greater detail by Korsakov *et al.* [100].

Investigation of the numerous models describing the structure of a filled polymer and its properties, and also a review of the available computer programs, reveal that the additivity principle cannot generally be employed as the initial physical prerequisite when modeling such systems. Indeed, the major role of surface forces at the polymer–filler boundary of contact in the phenomenon of adhesion is well known. This must also tell on the structure of the polymer layer bounding on the filler. The mechanism of the formation of such layers at the interface of a polymer with another phase (the interphase layer) was considered, in particular, in Refs 98 and 101. And, although the concept of boundary layers has been criticized highly [102, 103], the influence of the surface undoubtedly affects the properties of a polymer in the boundary region (see also Section 1.3).

Direct experimental results obtained by birefringence and photoelasticity techniques [104, 105], as well as information on the successful use of a three-element model with a varying composition over the thickness of the boundary layer, point to the physical substantiation of the existence of an interphase layer. The properties of this layer should differ from those of the polymer matrix. The three-element model in which the boundary layer properties vary by a parabolic law [99] seems to be the logically best one. Its experimental verification showed that it describes satisfactorily the viscoelastic parameters of some composites with a filler content up to 25%. The three-element approach has been employed in developing a three-element geometric model of a filled polymer. The principle of constructing such a model and employing it for specific calculations is significant and is described briefly in the following section.

1.5 A GEOMETRIC APPROACH TO MODELING THE BEHAVIOR OF FILLED POLYMERS

1.5.1 Three-Element Geometric Model: One-Dimensional Variant

Since filled polymer systems at high filler concentrations are media with a limited fluidity (see Chapter 5), while their behavior in deformation at definite rates (frequencies) is close to elastic, to a rough approximation and as a first stage it is quite justified to construct a composite model with elements differing in elasticity. The model can subsequently be made more involved with a view to the specific features of the structure of a system and the existence, for instance, of interphase layers.

A rod with a cross-sectional area S is an element of such a model (Fig. 1.6). It consists of three parts (sections OA, AB, BC), each of which has its own properties described by the moduli of elasticity E_i and Poisson's ratios μ_i. The left-hand end of the composite rod is fixed, while the force P_0 producing the displacement U_0 acts on its right-hand end.

The stressed state in rod OA can be expressed by the following equations and boundary conditions [106] $(0<x<l_1)$:

$$\sigma = \frac{E_1}{1-\mu_1^2}\frac{dU}{dx}, \quad \frac{d\sigma}{dx}=0$$

$$U(0)=0, \quad \sigma(l_1-0)=\sigma(l_1+0)-\frac{P_0'}{S}$$

where P_0' and σ are the force and stress in rod OA.

The boundary value problems for rods AV $(x_1<x<x_2=l_1+l_2)$ and BC $(x_2\leqslant x\leqslant x_3=l_1+l_2+l_3)$ are similar:

$$\sigma = \frac{E_2}{1-\mu_2^2}\frac{dU}{dx}, \quad \frac{d\sigma}{dx}=0, \quad U(l_1-0)=U(l_1+0), \quad \sigma(x_2-0)=\sigma(x_2+0)$$

and

$$\sigma = \frac{E_3}{1-\mu_3^2}\frac{dU}{dx}, \quad \frac{d\sigma}{dx}=0, \quad U(l_2-0)=U(l_2+0), \quad u(x_3)=U(0)=U_0$$

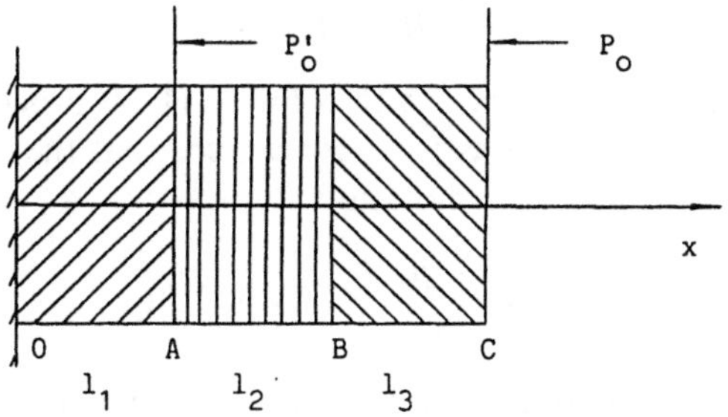

FIG. 1.6. One-dimensional three-element geometric model of a filled polymer.

Solution of the boundary value problem for rod BC leads to $\sigma=\sigma_0=\text{const.}$ or $P_0=S\sigma_0=\text{const.}$:

$$U(x)=\sigma_0\frac{1-\mu_3^2}{E_3}(x-x_3)+U_0 \quad (x_2\leqslant x\leqslant x_3)$$

The boundary value problems for rods AB and OA are solved similarly. As a whole after simple calculations, we obtain for the entire rod:

$$P_0=\frac{U_0+\dfrac{1-\mu_1^2}{E_1}\dfrac{l_1}{S}P_0'}{\dfrac{1-\mu_1^2}{E_1}\dfrac{l_1}{S}+\dfrac{1-\mu_2^2}{E_2}\dfrac{l_2}{S}+\dfrac{1-\mu_3^2}{E_3}\dfrac{l_3}{S}} \tag{1.68}$$

The above model approach employs procedures of the one-dimensional theory of elasticity [106]. Equations such as (1.68) obtained with its aid can also be obtained when employing the principle of electromechanical analogy.

Indeed, if we consider an electric circuit consisting of three resistances $R_i(i=1,2,3)$ and two current sources with emf's of ε_0 and ε_0', we can obtain the following equations by applying Ohm's and Kirchhoff's laws for the currents (I_i) and voltages (U_i^e) to the given circuit (the indexation of the parameters is shown in Fig. 1.7):

$$\begin{gathered} I_1+I_0'=I_2, \qquad I_2=I_3, \qquad U_1^e=I_1R_1 \\ U_2^e=I_2R_2, \quad U_3^e=I_3R_3, \quad U_1^e=\varepsilon_0 \\ U_1^e+U_2^e+U_3^e=\varepsilon \end{gathered} \tag{1.69}$$

If we introduce the notation $I_2=I_0$, the solution of system (1.69) acquires the form:

$$I_0=\frac{\varepsilon+I_0'R_1}{R_1+R_2+R_3} \tag{1.70}$$

By comparing this expression with solution (1.68) proceeding from the one-dimensional theory of elasticity, we can see that they are completely analogous if we introduce the notation

$$R_i=\frac{1-\mu_i^2}{E_i}\frac{l_i}{S} \quad (i=1,2,3) \tag{1.71}$$

Hence, with such analogy, the current I_i is equivalent to the force P inside a rod, the voltage U_i^e to the displacement U, while the role of the

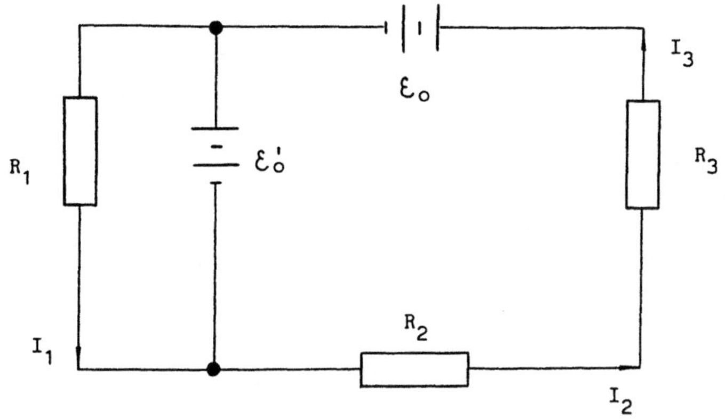

FIG. 1.7. Electrical analog of a one-dimensional three-element geometric model of a filled polymer.

resistance is played by the set of parameters (1.71). With a view to this analogy, we shall conditionally call the quantity R_i (1.71) the 'elastic resistance'.

It should be noted that such analogies are encountered quite often in the theory of electricity, elasticity, heat conductivity, and other theories [71]. Such problems can be reduced to the calculation of graphs by employing the 'generalized Kirchhoff laws' [107]. In our case, the graph of the problem has the form shown in Fig. 1.8. The calculations of this graph lead to formulae (1.68) and (1.70).

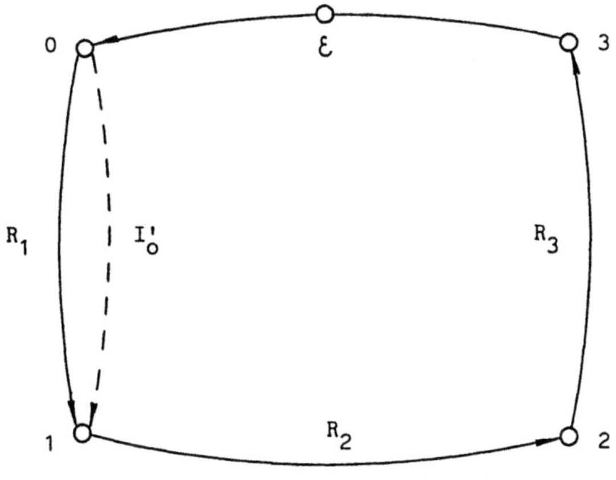

FIG. 1.8. Graph of an electrical analog of a one-dimensional three-element geometric model of a filled polymer.

1.5.2 Two-Dimensional Variant

The above results can be used for developing a two-dimensional mathematical model. We can consider as an example a triangular element (Fig. 1.9) whose legs are approximated by three composite rods (similar to those shown in Fig. 1.6). Of interest are the calculations of the stressed state of such a triangle by an approximate method. Its essence consists in that each rod is characterized by a 'resistance' R_i determined by formulae (1.71) or their reciprocals, conditionally termed 'rigidities' for convenience of interpretation:

$$G_i = \frac{1}{R_i} = \frac{E_i}{1-\mu_i^2} \frac{S_i}{l_i} \tag{1.72}$$

All the quantities in formulae (1.72) are generally known except the cross-sectional areas of the rods S_i. The latter are selected from the condition of equivalence of the stressed state of the triangular region and the 'rod' model (see below).

It is convenient to employ the method of finite elements (MFE) to calculate the stressed state of triangular element $M_1 M_2 M_3$ (Fig. 1.9).

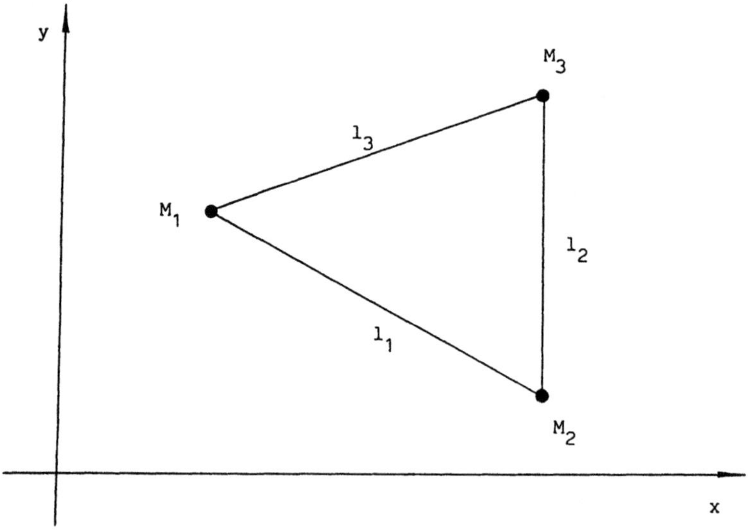

FIG. 1.9. Simplex element of a two-dimensional geometric model of a filled polymer.

Here triangle $M_1M_2M_3$ is a simplex element (A is its area). The displacements $U(x, y)$ and $V(x, y)$ are taken as polynomial functions. For example [108]:

$$U(x, y)=N_1U_1+N_2U_2+N_3U_3 \quad (1.73)$$

where $U_i=U(x_i, y_i)$, $i=1, 2, 3$, and N_i are shape functions determined as

$$N_i=\frac{1}{2A}(a_i+b_ix+c_iy) \quad (1.74)$$

Designating the lengths of the triangle legs by $M_1M_2=l_1$, $M_2M_3=l_2$, $M_1M_3=l_3$, and their projections on the coordinate axes by l_{ix} and l_{iy}, the constants in eqn (1.74) can be written as:

$$\begin{aligned} b_1&=-l_{2x}, \quad b_2=l_{3x}, \quad b_3=l_{1x} \\ c_1&=-l_{2x}, \quad c_2=-l_{3x}, \quad c_3=l_{1x} \end{aligned} \quad (1.75)$$

(The coefficients for a_i will not be required in the following, therefore we do not give their expressions.)

Expressions (1.73) for the displacement components U and V can be written as

$$\begin{aligned} U&=N_1U_1+N_2U_2+N_3U_3 \\ V&=N_1V_1+N_2V_2+N_3V_3 \end{aligned} \quad (1.76)$$

and presented in the matrix form.

By the method of finite elements, the matrix of the gradients **B** depends on the relations between the relative elongations ε and the displacements U and V, which have the forms [106]:

$$\varepsilon_x=\frac{\partial U}{\partial x}, \quad \varepsilon_y=\frac{\partial V}{\partial y}, \quad \gamma_{xy}=\frac{\partial U}{\partial y}+\frac{\partial V}{\partial x}$$

Inserting eqns (1.76) into these expressions, we get

$$\begin{aligned} \varepsilon_x&=U_1\frac{\partial N_1}{\partial x}+U_2\frac{\partial N_2}{\partial x}+U_3\frac{\partial N_3}{\partial x} \\ \varepsilon_y&=V_1\frac{\partial N_1}{\partial y}+V_2\frac{\partial N_2}{\partial y}+V_3\frac{\partial N_3}{\partial y} \\ \gamma_{xy}&=U_1\frac{\partial N_1}{\partial y}+V_1\frac{\partial N_1}{\partial x}+U_2\frac{\partial N_2}{\partial y}+V_2\frac{\partial N_2}{\partial x}+U_3\frac{\partial N_3}{\partial y}+V_3\frac{\partial N_3}{\partial x} \end{aligned} \quad (1.77)$$

Relations (1.77) can also easily be presented in the matrix form. The partial derivatives of a function of the form of N_i with respect to the coordinates x and y are calculated by eqn (1.74).

According to Segerlind [108], the matrix of rigidity **G** is determined by the volume integral

$$\mathbf{G}=\iiint\limits_V \mathbf{B}^*\mathbf{D}\mathbf{B}\,\mathrm{d}V \tag{1.78}$$

where **B*** signifies the transposed [108] matrix of the gradients, and **D** the matrix of the elastic characteristics.

Owing to linearity of the polynomial functions (1.73), the integrand in eqn (1.78) is a constant quantity. Therefore

$$\mathbf{G}=\mathbf{B}^*\mathbf{D}\mathbf{B}V$$

where V is the volume of a triangular element equal to the product of its area and unit thickness, i.e. $V=A\times 1$. Hence:

$$\mathbf{G}=A\mathbf{B}^*\mathbf{D}\mathbf{B}$$

The marix of the rigidity **G** is generally very involved and multiplication of its component matrices is relegated to a computer.

The equivalence of the stressed states for a triangular element and a rod model consists in the approximate coincidence of their characteristics. Denisov [109] established for a heat conductivity process that the elements of the 'heat conductivity' matrix (in our case it is the rigidity matrix) obtained by the method of finite elements are the heat conductivities between the relevant joints of a simplex element taken with the opposite sign. This is also confirmed by analysis of the very simple problems of the theory of elasticity: the elements of the rigidity matrix $-a_{12}$, $-a_{13}$, and $-a_{23}$ are the rigidities between the relevant joints of a simplex element, i.e.

$$G_1=-a_{12},\ G_2=-a_{23},\ G_3=-a_{13} \tag{1.79}$$

After the required calculations, we can find the values of a_{12}, a_{13}, and a_{23}. Proceeding from these results and expressions (1.79) for G_i, i.e. for the rigidity between the joints of a triangular element, we can write:

$$G_1=-\frac{E}{1-\mu^2}\frac{1}{4A}\left(b_1b_2+\frac{1-\mu}{2}c_1c_2\right)$$

$$G_2 = -\frac{E}{1-\mu^2}\frac{1}{4A}\left(b_2 b_3 + \frac{1-\mu}{2}c_2 c_3\right) \tag{1.80}$$

$$G_3 = -\frac{E}{1-\mu^2}\frac{1}{4A}\left(b_1 b_3 + \frac{1-\mu}{2}c_1 c_3\right)$$

Comparing eqns (1.80) and (1.72), we can determine the expressions for the 'effective' rod cross-sections as follows:

$$S_1 = -\frac{l_1}{4A}\left(b_1 b_2 + \frac{1-\mu}{2}c_1 c_2\right)$$

$$S_2 = -\frac{l_2}{4A}\left(b_2 b_3 + \frac{1-\mu}{2}c_2 c_3\right) \tag{1.81}$$

$$S_3 = -\frac{l_3}{4A}\left(b_1 b_3 + \frac{1-\mu}{2}c_1 c_3\right)$$

Substituting from eqns (1.75) for the constants b_i and c_i (i = 1, 2, 3), we can write expressions (1.81) in the form

$$S_1 = \frac{l_1}{4A}\left(l_{2y} l_{3y} - \frac{1-\mu}{2} l_{2x} l_{3x}\right)$$

$$S_2 = \frac{l_2}{4A}\left(-l_{1y} l_{3y} + \frac{1-\mu}{2} l_{1x} l_{3x}\right) \tag{1.82}$$

$$S_3 = \frac{l_3}{4A}\left(l_{1y} l_{2y} + \frac{1-\mu}{2} l_{1x} l_{2x}\right)$$

The geometric meaning of the effective cross-sections can be understood if we superpose axis Ox of the coordinate system on leg M_1M_2 of the simplex element being considered (Fig. 1.10).

Inspection of the figure reveals that $l_{2y} = l_{3y} = h$, $l_{1x} = l_1$ and $l_{1y} = 0$. Using these relations and eqns (1.75), we can write formulae (1.80) as follows:

$$G_1 = \frac{1}{4A}(\lambda_1 h^2 - \lambda_2 l_{2x} l_{3x})$$

$$G_2 = \frac{1}{4A}\lambda_2 l_1 l_{3x} \tag{1.83}$$

$$G_3 = \frac{1}{4A}\lambda_2 l_1 l_{2x}$$

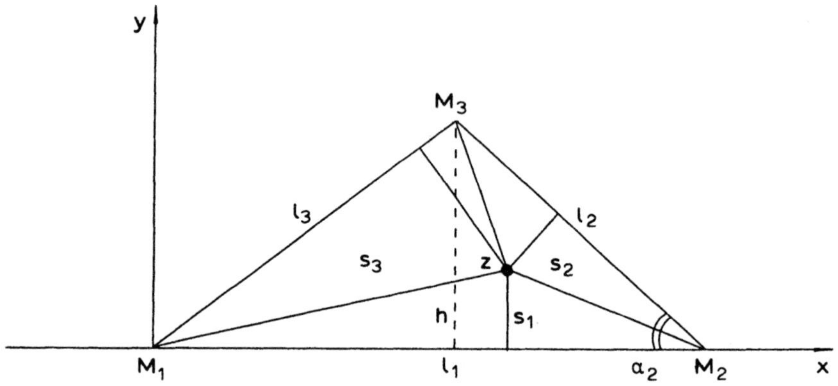

FIG. 1.10. Effective cross-sections of a simplex element.

where the following notation has been introduced:

$$\lambda_1 = \frac{E}{1-\mu^2}, \quad \lambda_2 = \frac{1}{2}\frac{E}{1-\mu} \tag{1.84}$$

Analysis of expressions (1.83) shows that the problem of the stressed state of an isotropic body reduces to the problem of nonisotropic heat conductivity [109]. Indeed, isotropic heat conductivity occurs provided that $\lambda_1 = \lambda_2$, which is not observed in our case. If $\lambda_1 = \lambda_2 = \lambda$, point Z in Fig. 1.10 should coincide with the center of the circle circumscribing the triangle, while the rigidities of the legs are determined by the formulae

$$G_i = \lambda \frac{S_i}{l_i} \quad (i = 1, 2, 3) \tag{1.85}$$

Here the effective cross-sectional areas of the legs S_i are the lengths of perpendiculars extending from the center of the circumscribed circle to the relevant legs, i.e. s_i in Fig. 1.10.

For the case when $\lambda_1 \neq \lambda_2$, we shall introduce a new effective coefficient λ_e:

$$G_i = \lambda_e \frac{s_i}{l_i} \quad (i = 1, 2, 3) \tag{1.86}$$

The effective cross-sectional areas of the legs of triangle $M_1M_2M_3$ (Fig. 1.10) here are the lengths of perpendiculars extending from point z to the relevant legs.

The area A of triangle $M_1M_2M_3$ consists of the sum of the areas of triangles M_1zM_2, M_2zM_3 and M_1zM_3. Hence

$$A=\tfrac{1}{2}l_1s_1+\tfrac{1}{2}l_2s_2+\tfrac{1}{2}l_3s_3 \tag{1.87}$$

From expressions (1.86) and (1.83), we find

$$\begin{aligned} s_1&=\frac{l_1}{4A}\lambda_e(\lambda_1h^2-\lambda_2l_{2x}l_{3x})\\ s_2&=\frac{l_2}{4A}\lambda_e(\lambda_2l_1l_{3x})\\ s_3&=\frac{l_3}{4A}\lambda_e(\lambda_2l_1l_{2x}) \end{aligned} \tag{1.88}$$

Since (Fig. 1.9)

$$l_{2x}+l_{3x}=l_1;\ l_{2x}^2+h^2=l_2^2 \tag{1.89}$$

we can write the expressions for the effective cross-sectional areas in the form

$$\begin{aligned} S_1&=\frac{l_1}{4A}\lambda_e[(\lambda_1-\lambda_2)h^2+\lambda_2(l_2^2-l_{2x}l_1)]\\ S_2&=\frac{\lambda_2}{4A}\lambda_el_1l_2(l_1-l_{2x})\\ S_3&=\frac{\lambda_2}{4A}\lambda_el_1l_{2x}l_3 \end{aligned} \tag{1.90}$$

Inserting eqns (1.90) into eqn (1.87) and using the formulae for the cosine of $\angle M_1M_2M_3=\alpha_2$ of the triangle being considered, simple geometric transformations yield:

$$h^2l_1^2(\lambda_1+\lambda_2)=8A^2\lambda_e$$

Having in view that

$$h^2l_1^2(\lambda_1+\lambda_2)=8\lambda_e\tfrac{1}{4}h^2l_1^2$$

and using the notation introduced in eqns (1.84), we find:

$$\lambda_e=\frac{\lambda_1+\lambda_2}{2}=\frac{E}{4}\left(\frac{3-\mu}{1-\mu^2}\right)$$

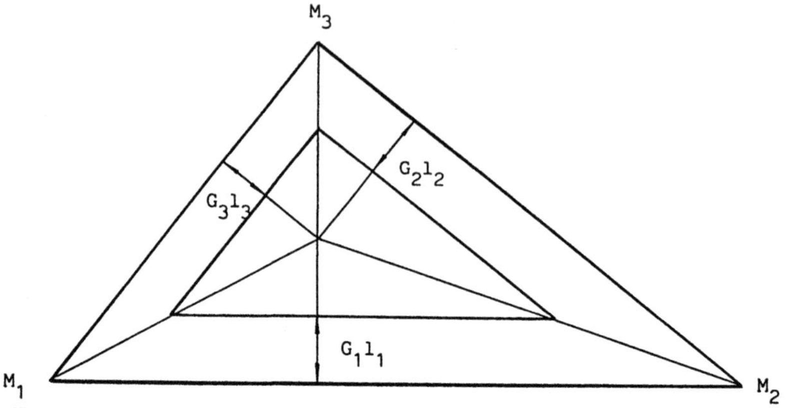

FIG. 1.11. Determination of the center of rigidity of a simplex element.

Hence, the introduction of the effective elastic constant into the problem of the stressed state of an isotropic body enables one to reduce it to the case equivalent to consideration of the process of nonisothermal heat conductivity. The 'center of rigidity' of the triangle, viz. point z, can be calculated [109] by the geometric method shown in Fig. 1.11. Indeed, if we draw straight lines parallel to the legs of the triangle at a distance of $G_i l_i$ therefrom $(i=1, 2, 3)$ and then draw rays through the points of intersection of these lines and the relevant apices of the triangle, the point of intersection of the rays will be point z.

The use of a geometric model for specific calculations of the properties of a number of polymers, including epoxide ones, yielded good results. The discrepancy between the theoretical and experimental data did not exceed 3%.

With this in view, the author with his co-workers proposed a mathematical procedure for describing the properties of composites with the use of a two- and three-dimensional geometric model and computer programs including the method of finite elements.

1.5.3 Program Based on the Method of Finite Elements

The developed program, GIP/SAP, is a system of completely compatible programs for forming, loading and calculating models by the method of finite elements, and for analyzing the results with printed output (Appendix 2). The GIP/SAP system is built up of the following programs:

GIMP—for preprocessing the model and generating the image in the machine language of the program for the method of finite elements;

FESA—for calculating the stressed–strained state of a model;
PP2D—for postprocessing (visualization of the calculation results);
AUTOCA—for the preparation and output of the graphical documents which the program PP2D is compatible with.

The scheme of solution of the problems and their analysis consists of the following steps;

(1) formation and editing of the model and the key points setting the geometry of the model;
(2) formation of segments and subregions with connection of the elements to the subassemblies or generation of subregions;
(3) setting of the loads, material, type and cross-section of the elements and superposition of the boundary conditions;
(4) assembly of a model from individual subregions;
(5) renumbering of the subassemblies of the working data base and generation of an image in the input language of the program for the method of finite elements;
(6) calculation of the stressed and strained state of the MFE model.

The FESA program performs postprocessing to realize the results in various forms of representation.

The PP2D program analyzes the results and establishes the need to alter the geometry and material of the model, repeat calculation stages, and peform further analysis until a stressed–strained section is obtained that corresponds to the strength requirements.

Let us consider results of specific calculations as an example of using the GIP/SAP program for predicting the stressed state of filled polymers on the basis of numerical experiments using the method of finite elements. Figure 1.12 presents variants for a polymer filled with spherical particles. A cross-section with a filler particle surrounded by the polymer matrix has been selected as the demonstration volume. The model reproduces the stressed–strained state in simple shear and compression. The calculation results are presented in the form of a picture of the deformed section and the values of the displacement of the loaded boundary in the direction of the load. The reaction is also reflected in the distribution of the color intensity by structure elements.

Figure 1.12 shows elements of the cross-sectional structure of a filled polymer containing (a) ‘one spherical particle, (b) the same particle with a singled-out interphase layer, (c) a cross-section with two identical

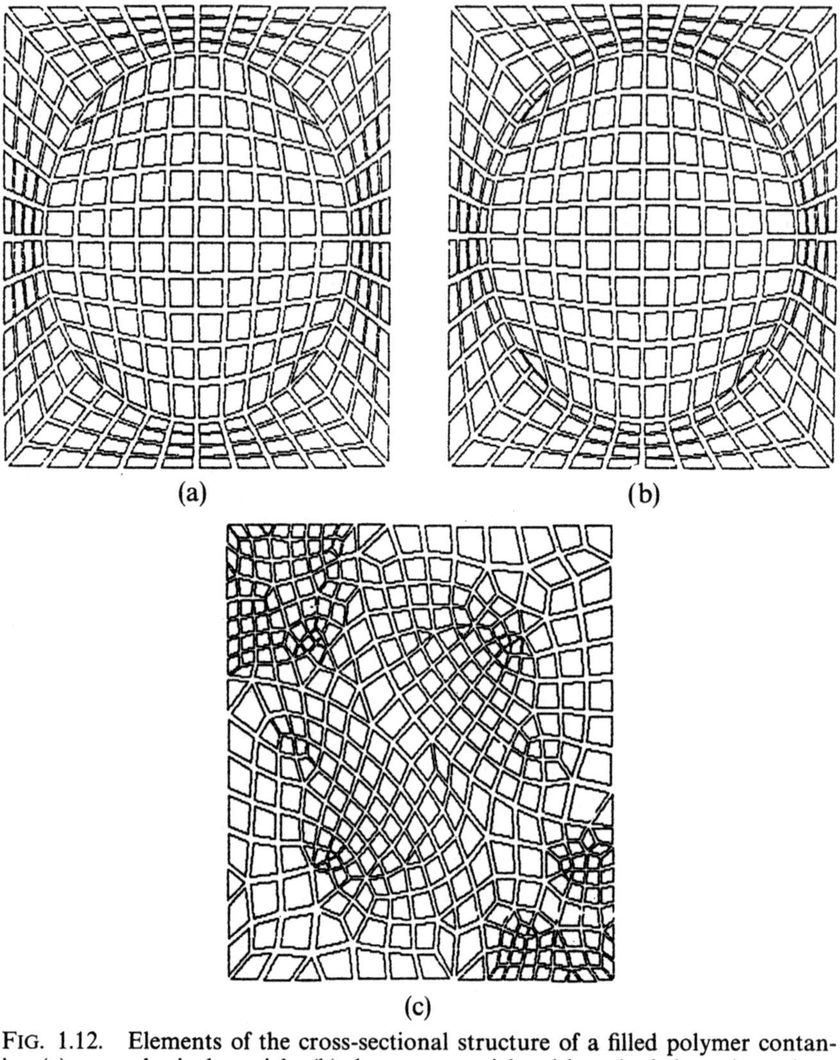

FIG. 1.12. Elements of the cross-sectional structure of a filled polymer contan-ing (a) one spherical particle, (b) the same particle with a singled-out interphase layer, and (c) two identical particles.

particles. Figures 1.13–1.18 present the results of the effect of (A) compressive and (B) shear forces on the structure (Fig. 1.12(a)–(c)). The directions of force action are shown by arrows.

Matrix–filler variants with various properties of the components were considered. In Figs 1.13–1.15, 1.17 and 1.18, the polymer matrix M1 had the

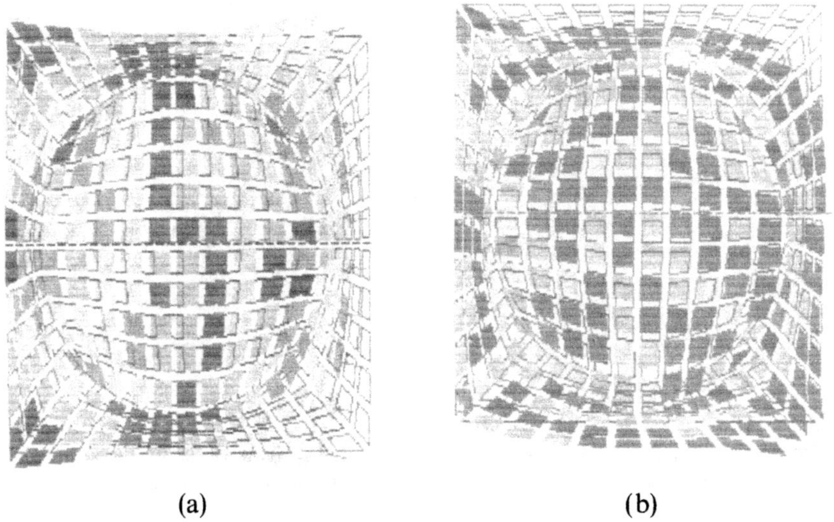

FIG. 1.13. Effect of (A) compressive and (B) shear forces on structure of a filled polymer with one particle (for designation of parameters see text).

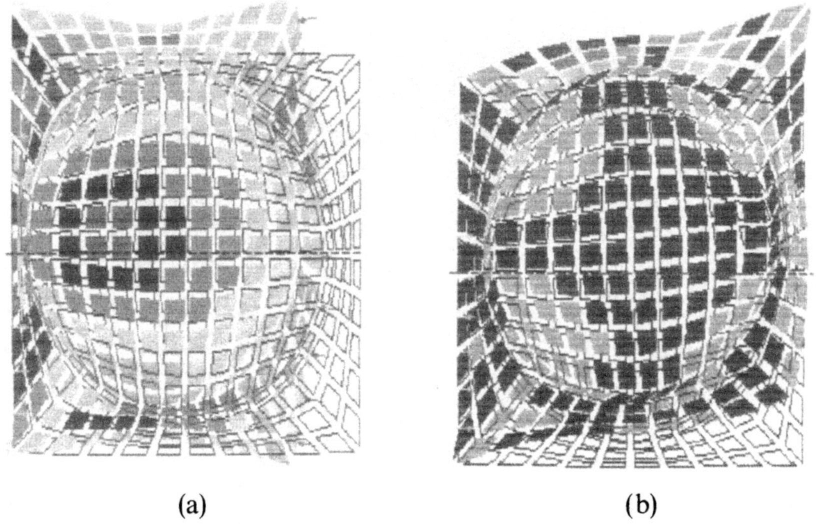

FIG. 1.14. Effect of (A) compressive and (B) shear forces on structure of a filled polymer with one particle and interphase layer (for designation of parameters see text).

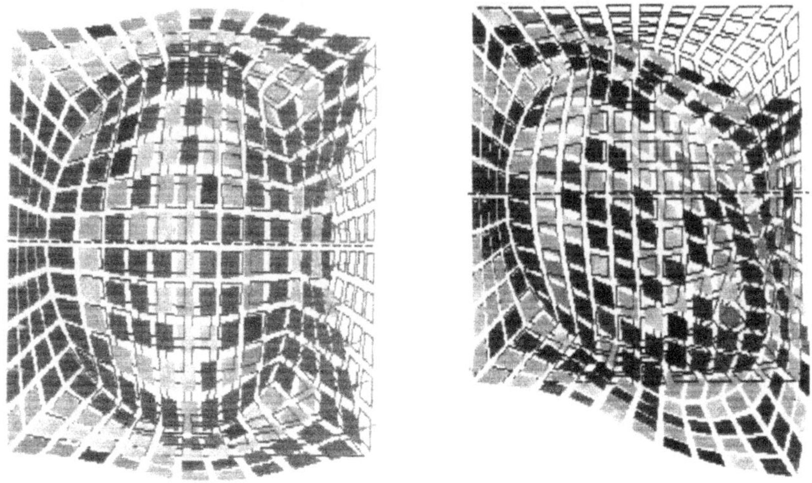

(a) (b)

FIG. 1.15. Effect of (A) compressive and (B) shear forces on structure of a filled polymer with one particle (for designation of parameters see text).

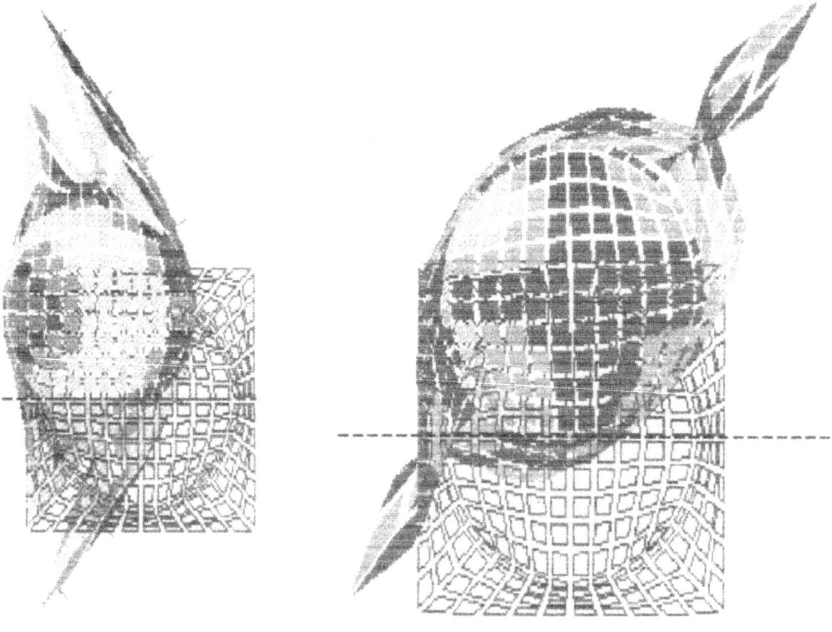

(a) (b)

FIG. 1.16. Effect of (A) compressive and (B) shear forces on structure of a filled polymer with one particle (for designation of parameters see text).

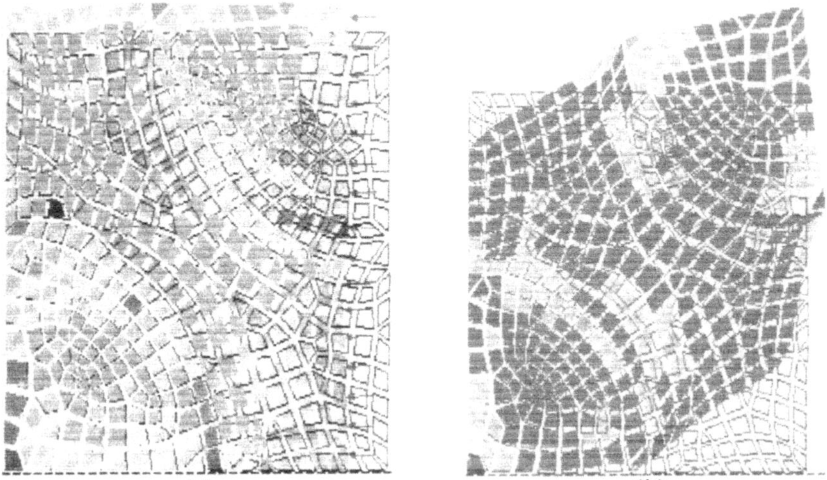

(a) (b)

FIG. 1.17. Effect of (A) compressive and (B) shear forces on structure of a filled polymer with two particles (for designation of parameters see text).

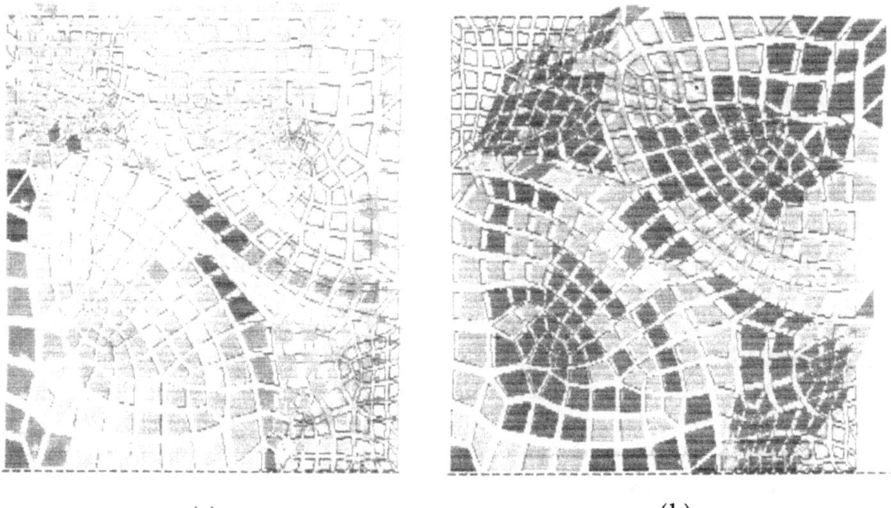

(a) (b)

FIG. 1.18. Effect of (A) compressive and (B) shear forces on structure of a filled polymer with two particles (for designation of parameters see text).

following characteristics: density $\rho_{M1} = 1{\cdot}22\,g\,cm^{-3}$, modulus of elasticity $E_{M1} = 4{\cdot}4\,GPa$, ultimate elongation $\varepsilon_{M1}^{ult} = 6{\cdot}0\%$, and Poisson's ratio $\nu_{M1} = 0{\cdot}37$. In Fig. 1.16, the matrix M2 had $E_{M2} = 0{\cdot}044\,GPa$, $\varepsilon_{M2}^{ult} = 600\%$, and $\nu_{M2} = 0{\cdot}46$.

The filler in Figs 1.13, 1.14 and 1.16–1.18 was formed of particles of high-modulus silicate glass (F1) with the following characteristics: $\rho_{F1} = 2{\cdot}58\,g\,cm^{-3}$, $E_{F1} = 0{\cdot}93\,GPa$, $\varepsilon_{F1}^{ult} = 4{\cdot}8\%$, and $\nu_{F1} = 0{\cdot}25$. In Fig. 1.15 the filler F2 had the characteristics $E_{F2} = 4{\cdot}4\,GPa$, $\varepsilon_{F2}^{ult} = 6{\cdot}0\%$, and $\nu_{F2} = 0{\cdot}37$.

In Fig. 1.14, the interphase layer I.L. had the following characteristics: $E_{I.L.} = 0{\cdot}44\,GPa$, $\varepsilon_{I.L.}^{ult} = 80\%$, and $\nu_{I.L.} = 0{\cdot}41$.

Figures 1.13, 1.15, and 1.16 model the structure of Fig. 1.12(a), Fig. 1.14 the structure of Fig. 1.12(b), and Figs 1.17 and 1.18 that of Fig. 1.12(c). Moreover, in Figs 1.17 and 1.18 the distance between the filler particles was varied from one commensurable with the particle size (Fig. 1.17) to their contact (Fig. 1.18).

In all the examples, the stress scale is shown in the upper part of the figures.

These examples show that the pictures of stress distribution in the elements differ substantially depending on the properties of the matrix and filler and on the structure of the selected demonstration volume.

1.6 CONCLUSION

The above analysis of molecular theories and model approaches and also the discussion of the relevant conclusions and corollaries cannot be exhaustive for a number of objective and practical reasons. The former include primarily the novelty of the approaches and, consequently, the need for a comprehensive and broad experimental verification, which requires time. Although the examples given here, with respect to the experimental appraisals, do witness their good prognostic ability, the range of examples and their effects could undoubtedly be expanded substantially. But this is impossible owing to the restricted volume of the present book. For justification, I can only refer to Appendices 1 and 2 presenting prototypes of programs for performing numerical calculations by the methods of Brownian dynamics and finite elements (geometrically, two- and three-dimensional models). If desired, this will allow the reader to undertake independent steps for performing specific calculations when other methods, in his or her opinion, are not effective enough.

REFERENCES

1. FERRY, J.D. *Viscoelastic Properties of Polymers*, 3rd ed., Wiley, New York, 1980.
2. DE GENNES, P.G. *Scaling Concepts in Polymer Physics*, Cornell Univ. Press, Ithaca, NY, 1979.
3. GOTLIB, YU.YA., DORINSKY, A.A. and SVETLOV, YU.E. *Fizicheskaya Kinetika Makromolekul* (Physical Kinetics of Macromolecules), Khimiya, Moscow, 1986.
4. LODGE, A.S. *Elastic Liquids*, Academic Press, New York, 1964.
5. DOI, M. and EDWARDS, S.F. *The Theory of Polymer Dynamics*, Clarendon Press, Oxford, 1986.
6. RAJU, V.R., RACHAPUDY, H. and GRAESSLEY, W.W. *J. Polym. Sci., Polym. Phys. Ed.*, **17**(7) (1979) 1223.
7. YANOVSKY, YU.G., POKROVSKY, V.N., KOKORIN, YU.K., *et al. Polym. Sci. USSR*, **30**(5) (1988) 1037.
8. LODGE, T.P. and SCHRAG, J.L. *Macromolecules*, **15**(5) (1982) 1376.
9. BIRD, B.B., ARMSTRONG, R.C. and HASSAGER, C. *Dynamics of Polymeric Liquids*, Wiley, New York, 1977.
10. ASTARITA, G. and MARRUCCI, G. *Principles of Non-Newtonian Fluid Mechanics*, McGraw-Hill, New York, 1974.
11. GRAESSLEY, W.W. *Adv. Polym. Sci.*, **16**(1) (1974) 1.
12. ROUSE, P.E. *J. Chem. Phys.*, **21**(7) (1953) 1272.
13. ZIMM, B.H. *J. Chem. Phys.*, **24** (1956) 269.
14. CERF, R. *J. Phys. et Radium*, **19**(1) (1958) 122.
15. PETERLIN, A. *J. Polym. Sci.*, **5**(A2) (1967) 179.
16. BUSSE, W.F. *J. Phys. Chem*, **36** (1932) 2862.
17. BALLARD, D.G.H., WIGNALL, G.D. and SCHELTON, J. *Eur. Polym. J.* **9** (1973) 965.
18. COTTON, J.P., DECKER, D., BENOIT, H., *et al. Macromolecules*, **7**(6) (1974) 863.
19. FLORY, P.J. *J. Chem. Phys.*, **17**(3) (1949) 303.
20. EDWARDS, S.F. *Proc. Phys. Soc.*, **91** (1967) 513.
21. PRAGER, S. *J. Chem. Phys.*, **46**(4) (1976) 1475.
22. EDWARDS, S.F. and GRANT, J.W.V. *J. Phys. A*, **6**(8) (1973) 1169.
23. EDWARDS, S.F. and GRANT, J.W.V. *J. Phys. A*, **6**(8) (1973) 1186.
24. DE GENNES, P.G. *J. Chem. Phys.*, **55**(3) (1971) 572.
25. DOI, M. and EDWARDS, S.F. *J. Chem Soc., Faraday Trans.*, **74**(10) (1978) 1789.
26. DOI, M. and EDWARDS, S.F. *J. Chem Soc., Faraday Trans.*, **74**(10) (1978) 1802.
27. DOI, M. and EDWARDS, S.F. *J. Chem Soc., Faraday Trans.*, **74**(10) (1978) 1818.
28. POKROVSKY, V.N. and VOLKOV, V.S. *Vysokomol. Soed.*, **20**(2) (1978) 255.
29. POKROVSKY, V.N. and VOLKOV, V.S. *Vysokomol. Soed.*, **20**(12) (1978) 2700.
30. BUECHE, F. *J. Chem. Phys.*, **20** (1952) 1959.
31. BUECHE, F. *J. Chem. Phys.*, **40**(2) (1952) 484.

32. POKROVSKY, V.N. and YANOVSKY, YU.G. *Rheol. Acta*, **12**(4) (1973) 280.
33. POKROVSKY, V.N. *Statisticheskaya Mekhanica Razbavlennykh Suspenzii* (Statistical Mechanics of Dilute Suspensions), Nauka, Moscow, 1978.
34. TANNER, J. *Macromolecules*, **4**(6) (1971) 748.
35. KLEIN, I. *Macromolcules*, **11**(5) (1978) 852.
36. KLEIN, I. *Proc. Roy. Soc., London*, **365A** (1979) 53.
37. KLEIN, I. *Macromolecules*, **14**(4) (1981) 460.
38. LEGER, L., HERVET, H. and RONDELES, F. *Macromolecules*, **14**(6) (1981) 1732.
39. SMITH, B.A., SAMULSKI, E.T., VU, L.P., *et al. Phys. Rev. Lett.*, **52**(1) (1984) 45.
40. NEMOTO, N., *et al. Macromolecules*, **18**(2) (1985) 308.
41. FLEISCHER, G. *Polym. Bull.*, **9**(2) (1983) 152.
42. VAN NEERWALL, E.D. *Rubber Chem. and Techno.*, **58**(3) (1985) 527.
43. TIRRELL, M. *Rubber Chem. and Technol.*, **57**(3) (1985) 523.
44. DESCHAMPS, H. and LEGER, L. *Macromolecules*, **19**(11) (1986) 2760.
45. KRANBUEHL, D.E. *Macromolecules*, **18**(8) (1985) 1638.
46. MARMONIER, M.F. *Phys. Rev. Lett.*, **55**(10) (1985) 1078.
47. DES CLOISEAUX, J. *J. Phys. Lett.*, **45**(1) (1984) 17.
48. VOLKOV, V.S. and VINOGRADOV, G.V. *Rheol. Acta*, **23**(3) (1984) 231.
49. VOLKOV, V.S. and VINOGRADOV, G.V. *Vysokomol. Soed.*, **29A**(12) (1987) 2602.
50. GREBNER, B.L. and POKROVSKY, V.N. *Vysokomol. Soed.*, **29B**(9) (1987) 704.
51. VOLKOV, V.S. and VINOGRADOV, G.V. *J. Non-Newtonian Fluid Mech.*, **25**(3) (1987) 261.
52. POKROVSKY, V.N. and KOKORIN, YU.K. *Vysokomol. Soed.*, **26B**(8) (1984) 573.
53. POKROVSKY, V.N. and KOKORIN, YU.K. *Vysokomol. Soed.*, **27B**(10) (1985) 793.
54. VOLKOV, V.S., POKROVSKY, V.N. and VINOGRADOV, G.V. *Vysokomol. Soed.*, **28A**(1) (1986) 117.
55. POKROVSKY, V.N. and KOKORIN, YU.K. *Vysokomol. Soed.*, **29A**(10) (1987) 2173.
56. PUSHNOGRAI, G. V. and POKROVSKY, V. N. *Vysokomol. Soed.*, **30A**(11) (1988) 2497.
57. KOKORIN, YU.K. and POKROVSKY, V.N. *Vysokomol. Soed.*, **32A**(12) (1990) 2409.
58. POKROVSKY, V.N. and PUSHNOGRAI, G.V. *Mekhanika Zhidkosti i Gaza*, **1** (1991) 71.
59. FLORY, P.J. *Principles of Polymer Chemistry*, Cornell Univ. Press, Ithaca, NY, 1953.
60. VERDIER, P.H. *J. Chem. Phys.*, **45**(6) (1966) 2122.
61. KARGIN, V.A. and SLONIMSKY, G.L. *DAN SSSR*, **62**(2) (1948) 239.
62. AL-NOSIMI, G.F., MARTINES-MEKLER, G.C. and WILSON, C.A., *J. Phys. Lett.*, **39**(1978) 373.
63. FREED, K.F. and METIN, H., *J. Chem. Phys.*, **68** (1978) 4604.
64. YANOVSKY, YU.G., VINOGRADOV, G.V., VOLKOV, V.S. *et al.* In: *Proceedings of International Congress on Rheology*, Plenum Press, New York, 1980, p. 483.

65. YANOVSKY, YU.G., VINOGRADOV, G.V. and IVANOVA, L.I., *New Developments in Polymer Rheology*, Inst. Petrochem. Sym., Moscow, 1980.
66. VINOGRADOV, G.V., YANOVSKY, YU.G. and IVANOVA, L.I., *Intern. J. Polym. Mater.*, **9** (1982) 257.
67. MONTFORD, J.P., MARIN, G., and MONGE, PH., *Macromolecules*, **17** (1984) 1551.
68. DAOUND, M., and DE GENNES, P.G. *J. Polymer Sci., Polymer Phys. Ed.*, **17** (1979) 1971.
69. RONCA, G., *J. Polym. Phys.*, **79** (1983) 1031.
70. RICHTER, D., FARAGO, B.K., FETTEAS, L.J. *et al. Phys. Rev. Letters*, **64** (1990) 1389.
71. PRIVALKO, V.P., NOVIKOV, V.V. and YANOVSKY, YU.G. *Osnovy Teplofiziki i Reofiziki Polimernykh sistem* (Fundamentals of Thermophysics and Rheophysics of Polymer systems), Naukova Dumka, Kiev, 1990.
72. BUSH, M.B. and TANNER, R.I. *Int. J. Num. Meth. Fluids*, **3** (1983) 71.
73. MITSOULIS, E. and VLACHOPOULOS, *J. Adv. Polym. Technol.*, **4** (1984) 107.
74. BUSH, M.B., MILTHORE, J.F. and TANNER, R.I. *J. Non-Newtonian Fluid Mech.*, **22** (1987) 129.
75. KISTLER, S.F. and SCRIVENM, L.E. *Int. J. Num. Meth. Fluids*, **4** (1984) 207.
76. MARCHAL, J.M. and CROCHET, M.J. *J. Non-Newtonian Fluid Mech.*, **17** (1985) 157.
77. MALKUS, D.S. *Finite Element Methods for Viscoelastic Flow, Viscoelasticity and Rheology*, Academic Press, New York, 1985.
78. BERIS, A.N., ARMSTRONG, R.C. and BROWN, R.A. *J. Non-Newtonian Fluid Mech.*, **22** (1987) 129.
79. MARCHAL, J.M. and CROCHET, M.J. *J. Non-Newtonian Fluid Mech.*, **26** (1987) 77.
80. PAPANASTASION, A.C., SCRIVEN, L.E. and MACOSKO, C.W. *J. Non-Newtonian Fluid Mech.*, **22** (1987) 271.
81. MITSOULIS, E. Numerical simulation of viscoelastic fluids. In: *Encyclopedia of Fluid Mechanics*, Ed. N.P. Cheremisinoff, Gulf Publishing, Houston, TX, 1990, Chapter 21.
82. VOIGT W. *Lehrbuch der Kristalphysik*, Abhandl. Akad. Wiss, Gottingen, 1899.
83. REUSS, R. *Z. Angew, Math. und Mech.*, **9**(1) (1929) 49.
84. NIELSEN, L.E. *J. Appl. Polym. Sci.*, **21**(6) (1977) 1579.
85. CAREY, J.F. *Polym. Eng. Sci.*, **25**(16) (1985) 1017.
86. MANSON, J.A. and SPERLING, L.H. *Polymer Blends and Composites*, Plenum Press, New York, 1976.
87. SPATHIS, E.P. and THEOCARIS, P.S. *Int. J. Adhesion and Adhesives*, **1**(4) (1981) 195.
88. KERNER, E.H. *Proc. Phys. Soc.*, **69B** (1956) 808.
89. CHRISTENSEN, R. *Mechanics of Composite Materials*, Wiley, New York, 1979.
90. LEWIS, T.B. and NIELSEN, L.E. *J. Appl. Polym. Sci.*, **14B** (1970) 1449.
91. NIELSEN, L.E. *Ann. Phys.*, **41**(11) (1970) 4626.
92. TAKAYANAGI, M., UEMURA, S. and MINAMI, S. *J. Polym. Sci.*, **5**(1) (1967) 113.

93. KENUNEN, I.V., VOLODIN, V.P., LISHANSKY, I.C., *et al. Mekhanika Komposit. Mater.*, **4** (1986) 746.
94. PROZELHOF, R.C., THRONE, J.H. and RUETSCH, R.R. *Polym. Eng. Sci.*, **16**(9) (1976) 615.
95. JAMES, F.C. *Polym. Eng. Sci.*, **25**(16) (1985) 1017.
96. IAHAI, O. and COHEN, H.J. *Int. J. Mech. Sci.*, **9**(3) (1987) 539.
97. TONG SUN, DAJNN CHEN and HANXIN ZHON. *J. China Text. Univ.*, **5**(1) (1988) 1.
98. LIPATOV, YU.S. *Fiziko-khimiya Polimerov* (Physicochemistry of Polymers), Khimiya, Moscow, 1975.
99. SIDERIDIS, E. *Comp. Sci. Technol.*, **37**(2) (1986) 305.
100. KORSAKOV, YU.K., YARTSEV, I.K., BARANOVSKY, V.M., *et al. Plastmassy*, **12** (1980) 19.
101. LIPATOV, YU.S. *Sintez i Modifikatsiya Polimerov* (Synthesis and Modification of Polymers), Nauka, Moscow, 1976.
102. BIRNSTEIN, T.M., SKVORTSOVA, A.M. and SARIBAN, A.A. *Vysokomol. Soed.*, **17A**(11) (1975) 2558.
103. SKVORTSOVA, A.M., GORBUNOVA, A.A., ZHULINA, E.B., *et al. Vysokomol. Soed.*, **20A**(4) (1978) 816.
104. DADIVANIAN, A.K. and AGRANOVA, S.A. *Vysokomol. Soed.*, **22A**(7) (1980) 1499.
105. SARKISIAN, V.A., ASRATIAN, M.G., MHATARIAN, A.A., *et al. Vysokomol. Soed.*, **27** (1985) 1331.
106. RABOTNOV, YU.N. *Mekhanika Deformiruemogo Tverdogo Tela* (Mechanics of Solid Deformation), Nauka, Moscow, 1988.
107. BRAMMELER, A., ALLAN, R. and HEMEN, YA. *Slabozapolnenye Matritsy* (Slightly Filled Matrices), Energiya, Moscow, 1975.
108. SEGERLIND, L. *Applied Finite Element Analysis*, Academic Press, New York, 1976.
109. DENISOV, E.E. *Geometrischeskii Smysl Anisotropnoi Teploprovodnosti* (Geometric Interpretation of Anisotropic Heat Conduction), Msyl', Minsk, 1987.

Chapter 2

Rheological Properties, Relaxation Behavior, and Rupture of Polymers at Temperatures above their Glass Transition Temperature

2.1 INTRODUCTION

At temperatures above the melting point and glass transition temperature, the physical state of a polymer system depends on the ratio of the rates of external action and of relaxation processes occurring therein caused by thermal motion. This problem is of paramount importance for understanding the features of the viscoelastic properties of polymer systems depending on their nature, rate of deformation, and temperature. It is important both for understanding the features of the structure and for determining the workableness of polymer systems.

When dealing with the properties of polymer systems in the fluid state, the matter of the boundaries of this state is of primary importance. The upper temperature boundary is determined by thermo-oxidizing destruction, while the lower one is due to crystallization and glass transition.

A very important feature of high-molecular-weight compounds is their ability to be in a rubbery state. The latter is between the fluid and glassy states. The features of the passage from the fluid to the forced rubbery state under the influence of a change in the rate of the external action was discussed in detail in Refs 1 and 2. An example was used of objects with molecular weights (MW) exceeding certain critical values [1–4]. The latter are of two kinds. On the one hand, within a narrow interval of molecular weights (M_c), a sharp change in the nature of the viscosity dependence on the MW is observed. The exhibition by polymers of large recoverable strains is also associated with this [5]. On the other hand, the conditions of the appearance of a rubbery plateau are associated with the achievement of a critical molecular weight M_e [1, 2] that characterizes the weight of the macromolecular chain (dynamic segment) between the links of the spatial network. These critical molecular weights are interrelated [3]. They are convenient for normalizing the molecular weights of polymers, i.e. determining the position of a sample in the polymer homologous series relative to the boundary molecular weights.

A special place is occupied by high-molecular-weight samples whose molecular weight M considerably exceeds the critical value M_e ($M \geqslant 10M_e$). Here every macromolecule is linked to its neighbors at a large number of points—the three-dimensional network is quite homogeneous, and its links can withstand high stresses without failure because the loads are statistically distributed uniformly over a large number of links. On the other hand, breaking of the links may acquire an avalanche nature, which leads to a loss of continuity of the polymers.

The lowest values of M_e are typical for linear polyethylene and 1,4-polybutadiene (PB) [1, 2]. Already at quite low molecular weights, these polymers have a comparatively dense network of entanglements and, consequently, all their properties are determined by the presence of this network.

A transition from one physical state to another is observed most sharply in polymers with a narrow molecular-weight distribution (MWD). Such virtually monodispersed polymers are the most convenient for studying how their properties depend on the molecular weight. When they are subjected to shear causing high stresses, Newtonian flow can be observed similar to the flow of simple liquids. However, when critical stresses and deformation rates are reached, the process terminates in a spurt effect, when the increase in the deformation rates causes a polymer to pass over from the fluid to the rubbery state. Since in the latter state the deformability of polymers is always restricted and the conditions of polymer adhesion to the solid walls confining them alter, a spurt occurs, in which an involved set of processes takes place—breaking of a polymer and its separation from the walls. Depending on the test conditions, this results either in an exceedingly rapid growth in the volumetric flow rate or in a sharp drop in the resistance to deformation (Fig. 2.1(a)).

The entanglement network in high-molecular-weight and monodispersed linear flexible-chain polymers has an especially high density and homogeneity. Therefore, in deformation the individual entanglements are subjected to low effects. At low stresses, this corresponds to the first limiting conditions of polymer deformation when it can be concluded that the polymer entanglement network remains practically unchanged.

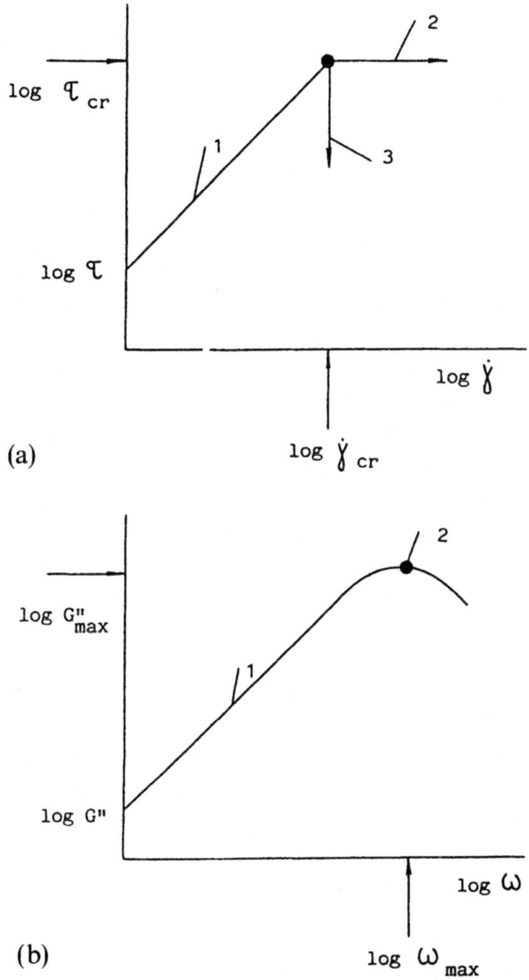

FIG. 2.1. Correlation of the conditions of transition of linear high-molecular-weight polymers of narrow MWD from the fluid to the high-elastic structure based on dynamic low-amplitude data and the result of experiments on capillary and rotational viscometers. (a) Shear stress vs shear rate: 1—steady-state Newtonian flow; 2—spurt, a jumpwise increase in the flow rate; 3—spurt, a sharp decrease of torque to loss of adhesion of the polymer to the walls of the duct and inner cylindrical rotor. (b) Loss modulus vs frequency: 1—terminal zone—the region of the proportionality between loss modulus and frequency; 2—maximum corresponding to the transition to the high-elastic state; τ_{sp} is independent of molecular weight and temperature; ω_{max} and $\dot{\gamma}_{cr}$ are inversely proportional to the initial viscosity.

In the general case (not only for high-molecular-weight monodispersed polymers), these conditions correspond to extrapolation of the parameters characterizing the rheological properties of polymers to the initial values of the shear stresses and rates.

When high stresses are applied to a polymer material, dynamic equilibrium may shift in the direction of network destruction. In the limiting case, the polymer is destroyed or separates from the solid wall. This corresponds to the second limiting condition of deformation. It should be noted that the beginning of destruction of a polymer and its separation from a wall are similar in nature. The limiting cond'tions of polymer deformation are characterized by a network completel; retaining its initial structure and determined by the initial values of the rheological parameters. They can therefore be described by simple relations of the theory of linear viscoelasticity [6, 7].

Proceeding from theoretical considerations [4, 8, 9] and also, in particular, the results of studying polybutadienes and polystyrenes [2, 10–12], we can see that the passing of polymers from one physical state into another can be monitored most strictly and simply by employing harmonic deformation conditions with small amplitudes (dynamic measurements). It is important to find the general laws of the passing from the fluid to the rubbery state for homologous series of polymers, and also criteria for comparing polymers of various molecular weights and series with a view to their glass transition temperatures. Of no less importance is the determination of the criteria for comparing the rheological characteristics determined under diverse deformation conditions, e.g. dynamic ones with low amplitudes, and tests involving the flow of polymer systems, when large strains may develop (experiments in capillary viscometers, Fig. 2.1(b), or in tension).

2.2 VISCOELASTIC BEHAVIOR IN LOW-AMPLITUDE SHEAR

Studies of the rheological behavior of polymer systems on the basis of dynamic experiments (periodic low-amplitude deformation) have come into great favor [7]. Two experimental approaches are employed here, viz. (i) with a varying frequency under isothermal conditions, and (ii) at a constant frequency with a varying temperature. In the first case, the reaction of a system to mechanical effects, i.e. the frequencies (rates) of deformation, is determined. For samples with a high molecular weight under isothermal conditions, the most complete information can be obtained only with wide variation of the frequency of an experiment, namely more than 10 decimal orders of magnitude, depending on the MW and nature of the polymer. This involves considerable experimental difficulties. In the second case, simpler in practice, information is obtained mainly on the temperature and relaxation transitions in a system by altering the temperature at a fixed frequency [7, 12, 13].

Temperature–frequency reduction, which is an indirect way of determining the relaxation properties of a material, is widely described in the literature [7]. But we must determine whether its use is correct within a broad range of frequencies and temperatures of an experiment covering various physical states of a system. It should be noted that a procedure for determining the transitions of an amorphous polymer from one physical state into another on the basis of the temperature dependence of the modules was first given in Ref. 14. Here the following typical states were observed: glassy, leather-like, rubbery, transition to a fluid, and a fluid state. A number of authors [15, 16] discuss the concept of the temperature of a T_{11} transition for flowing polymers. It characterizes the completion of the transition from the rubbery state, in which fluidity is already manifest, to a purely fluid state.

Major achievements were made in the field of dynamic studies as a result of studying polymers of a narrow fractional composition that are model ones to a certain extent [12, 17]. Interesting results were obtained, in particular, in Ref. 18 for polymer-homologous series of linear polymers with a narrow MWD. They include polybutadienes (PB), polystyrenes (PS), polyisoprenes (PI), and poly(methylphenyl siloxanes) (PMPS). Of special interest are the results of experimental study of PB with a narrow MWD. The magnitudes of the molecular weights of PB at which the transition from polymers to oligomers occurs, i.e. when the system begins to behave like a viscoelastic one, are about one-seventh of those for PS [19]. On the other hand, the low glass transition temperature of polybutadienes allows studies to be conducted within a broad range of temperatures whose upper boundary (depending on the PB microstructure) may be 150–200°C above the glass transition temperature T_g.

The dynamic viscoelastic behavior of polymers has been and continues to be widely discussed in the literature [7, 17]. However, the study of systems with a narrow MWD sometimes allows us to observe new features of the relaxation nature of high-molecular-weight systems that have received inadequate attention. This includes slow (superslow) relaxation processes in the region of the fluid state, the features of temperature–frequency behavior in the region of the rubbery state, generalization of the dynamic characteristics for polymers of various chemical structures, and problems of correlation of the results of low-amplitude deformation with those of experiments with tension or continuous shear, i.e. in the region of large strains (up to failure of a polymer). In view of the above, we shall analyze the results of dynamic studies of polymers with a narrow MWD in greater detail.

2.2.1 Influence of Frequency: Generalized Plot of Dynamic Characteristics

Typical curves obtained for isothermal conditions with a varying frequency for 1,4-PB with a narrow MWD are shown in Fig. 2.2. At low frequencies, the slope of the curves of G' vs ω tend to unity, and of G'' vs ω to 2, which is typical of the region of the fluid state of a polymer (the terminal zone). The presence of a plateau and peaks is a typical manifestation of the viscoelastic behavior of polymers with a narrow MWD. For homologous series of polymers with a narrow MWD at $M/M_e>10$, the height of the plateau and the size of the peak were found to be practically independent of the molecular weight (Table 2.1).

The rheological behavior of a polymer system in the terminal zone is determined by parameters such as the initial viscosity η_0, the initial high elasticity coefficient A_G^0, and the initial high elasticity modulus G_e^0. They are found from the expressions:

$$\eta_0=\lim_{\omega\to 0}(G''/\omega),\quad A_G^0=\lim_{\omega\to 0}(G'/\omega^2),\quad G_e^0=\eta_0^2/2A_G^0 \tag{2.1}$$

As a function of the molecular weight, $\eta_0=K_1M^\alpha$ and $A_G^0=K_2M^{2\alpha}$, where K_1 and K_2 are constants. (The values of α for some polymers are also given in Table 2.1.)

According to Ref. 18, the values of the frequency corresponding to the peak on the curve of G'' vs ω, i.e. to the quantity ω_{max}, depending on the molecular weight, are inversely proportional to the initial viscosity η_0.

Figure 2.3 (a, b) shows a typical picture of the change in the frequency plots of the loss modulus with the temperature for the same polymer (1,4-PB). Shifting of the quantity ω_{max} depending on the temperature is similar to the change in the initial viscosity of the system η_0 with temperature (Fig. 2.3(c)).

Very useful information can be procured when plotting the magnitudes of the complex dynamic modulus $[G^*]=\sqrt{(G')^2+(G'')^2}$ vs the frequency for different temperatures (Fig. 2.3). The dashed line on Fig. 2.3(b) corresponding to the peaks of G'' vs ω on Fig. 2.3(a) separates the fluid

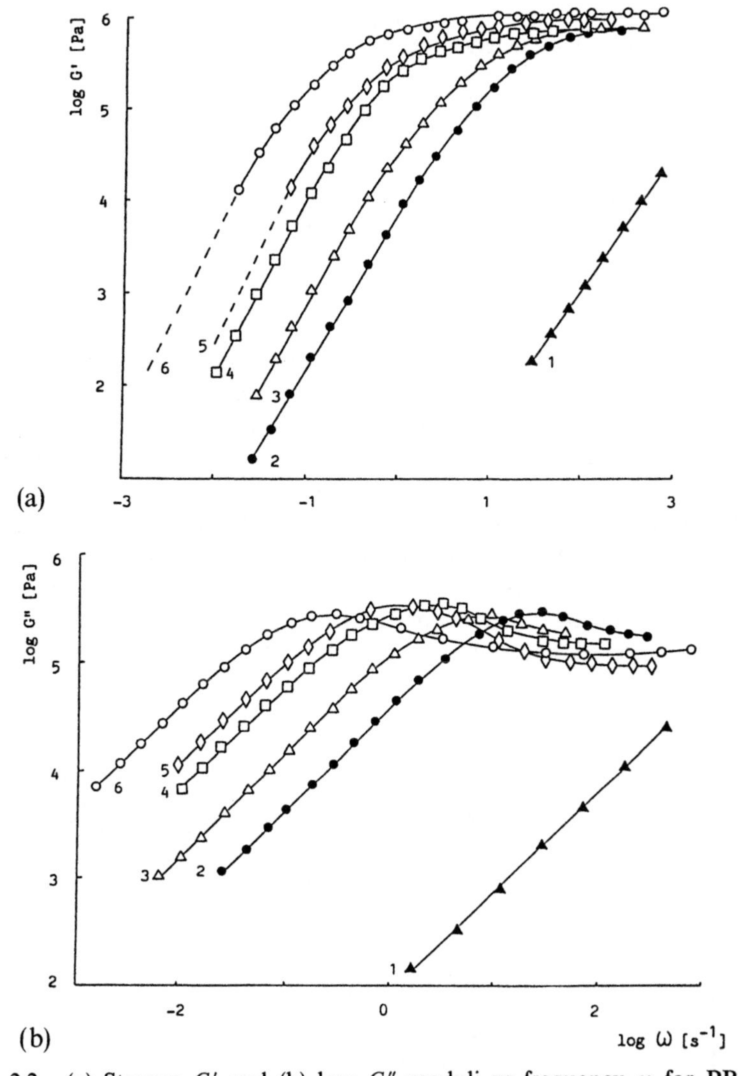

FIG. 2.2. (a) Storage G' and (b) loss G'' moduli vs frequency ω for PB with $M_v \times 10^{-5} = 0{\cdot}2$ (curve 1), 0·8 (2), 1·3 (3), 1·9 (4), 2·4 (5) and 3·2 (6).

state of the system (bottom part of Fig. 2.3(b)), the transition to the rubbery state, and the rubbery state (top part). The expansion of the transition state region along the frequency scale with lowering of the temperature indicates a higher viscoelastic anomaly in high-molecular-

TABLE 2.1
Viscoelasticity Parameters (Rubbery State) of Monodisperse Polymers

Polymer	$G'_{pl} \times 10^{-5}$ (*Pa*)	$G''_{max} \times 10^{-5}$	G''_{max}/G'_{pl}	*a*
Polybutadienes (PB)	6·3–10	3·0	0·35	3·5
Polyisoprenes (PI)	3·0–4·0	1·4	0·4	3·3
Polystyrenes (PS)	1·8–2·0	0·7	0·4	3·7
Polymethylphenyl siloxanes (PMPS)	1·0–1·1	0·35	0·35	3·3

weight polymers in deformation. This conclusion has been confirmed experimentally many times for flow in capillaries (see Section 2.3).

The need to systematize the results of dynamic measurements obtained within a broad frequency range in order to compare them in the relevant states resulted in a search for comparison criteria.

It is logical to use first of all the parameter M_e as such a criterion. This quantity can be found proceeding from the elementary theory of the rubbery state of crosslinked elastomers, assuming that there is an analogy in the manifestation of viscoelastic properties between the spatial network in noncrosslinked polymers and the network formed by the lateral

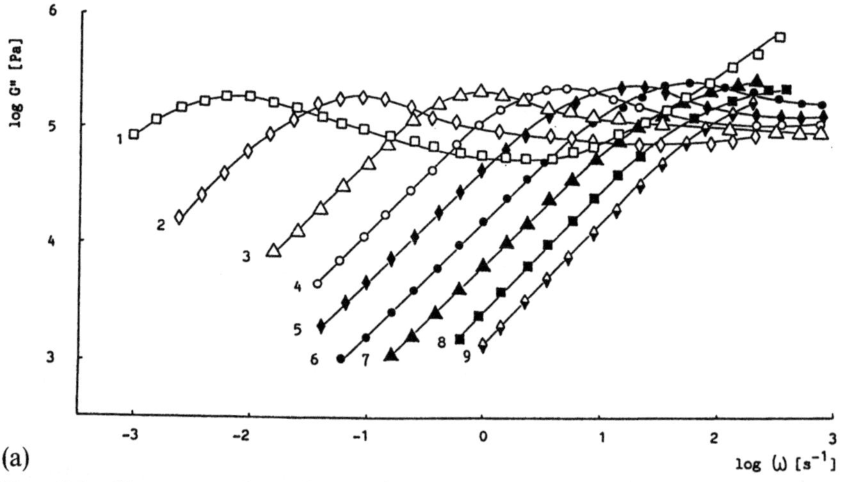

(a)

FIG. 2.3. Frequency dependence of (a) loss modulus G'' and (b) magnitude of complex modulus $|G^*|$ for PB at 210 K (curve 1), 233 K (2), 253 K (3), 273 K (4), 293 K (5), 313 K (6), 333 K (7), 353 K (8) and 373 K (9); (c) temperature dependence of unit viscosity $\eta_0(\diamond)$ and values of ω_{max} (○).

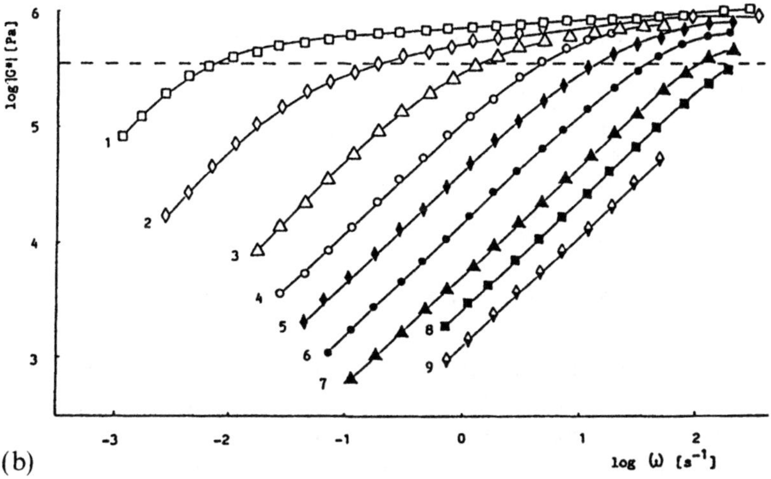

(b)

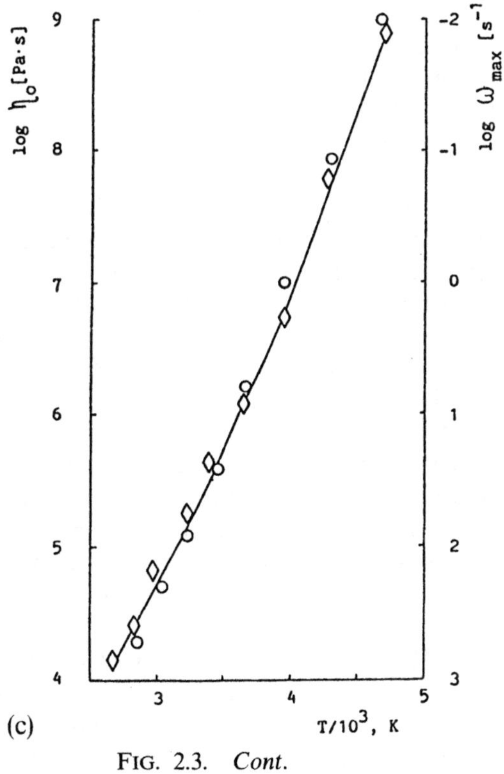

(c)

FIG. 2.3. *Cont.*

TABLE 2.2
Some Molecular Parameters of Polymers of Different Homologous Series

Polymer	$M_v \times 10^{-5}$	M_0	$M_e \times 10^{-3}$	M/M_e	$Z=M_e/M_0$	$(M/Z)\times 10^{-3}$	T_{com}
PB	1·02	54	3·45	29	64	1·6	280
PS	3·51	104	20·0	17	190	1·8	473
PMPS	2·70	136	23·0	12	170	1·6	346
PI	1·85	68	7·0	26	103	1·8	300
PMMA	2·10	100	13·0	16	130	1·6	405
PVA	2·30	100	14·5	16	145	1·6	405
PDMS	5·36	74	22·5	24	300	1·8	250

chemical bonds. The quantity M_e is determined quite simply from the expression $M_e = K\rho RT/G'_{pl}$, where ρ is the density, R is the molar gas constant, T is the absolute temperature, G'_{pl} is the storage modulus on the rubbery plateau, and K is a coefficient usually taken as equal to unity. (The values of M_e for selected polymers are given in Table 2.2.)

Let us see whether the quantity M_e or another one related to it can be used for a quantitative comparison of the viscoelastic behavior of polymers belonging to various homologous series as was done, for example, in Ref. 17. We must also consider the remoteness of this state from the glass transition temperature.

Coincidence of the plots of both G' and G'' vs ω for various polymer homologous series can be achieved with the following normalization conditions: the storage and loss moduli are normalized according to their values on the plateau and at the peak points, respectively; the molecular weights are normalized by the value of M_e/M_0, where M_0 is the molecular weight of a monomer unit; and the comparison temperature T_{com} is selected to be 100°C above T_g. The results of such normalization are shown in Fig. 2.4 and Table 2.2.

Hence, not the magnitudes of the molecular weights but their normalized values are important in characterizing the properties of polymers. Moreover, the properties of polymers must be compared at the relevant temperatures. Here polymers differing in their molecular weights by a factor of 10 or more and in their temperatures by 200°C can be in the corresponding states (PB on the one hand, PS on the other).

2.2.2 Superslow Relaxation Transitions and Their Interpretation

Many features of the rheological behavior of polymer melts and their concentrated solutions are associated with the relaxation processes in a system (see Chapter 1). The description and interpretation of

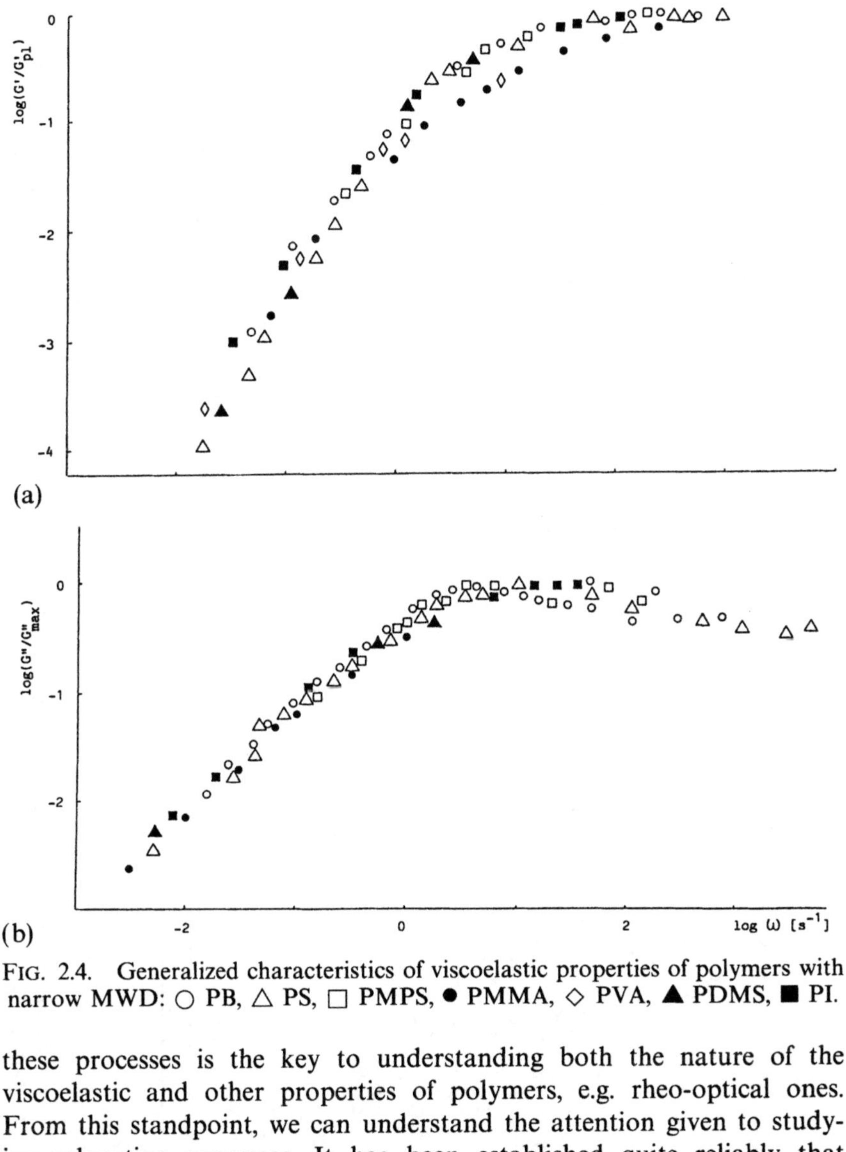

FIG. 2.4. Generalized characteristics of viscoelastic properties of polymers with narrow MWD: ○ PB, △ PS, □ PMPS, ● PMMA, ◇ PVA, ▲ PDMS, ■ PI.

these processes is the key to understanding both the nature of the viscoelastic and other properties of polymers, e.g. rheo-optical ones. From this standpoint, we can understand the attention given to studying relaxation processes. It has been established quite reliably that polymers with a high MW ($M \geqslant 10M_e$) are characterized by a group of rapid processes and also by a group of slow processes with considerably different relaxation times. Information [8, 20, 21] has shown that in addition to the above relaxation processes, there may be a still slower

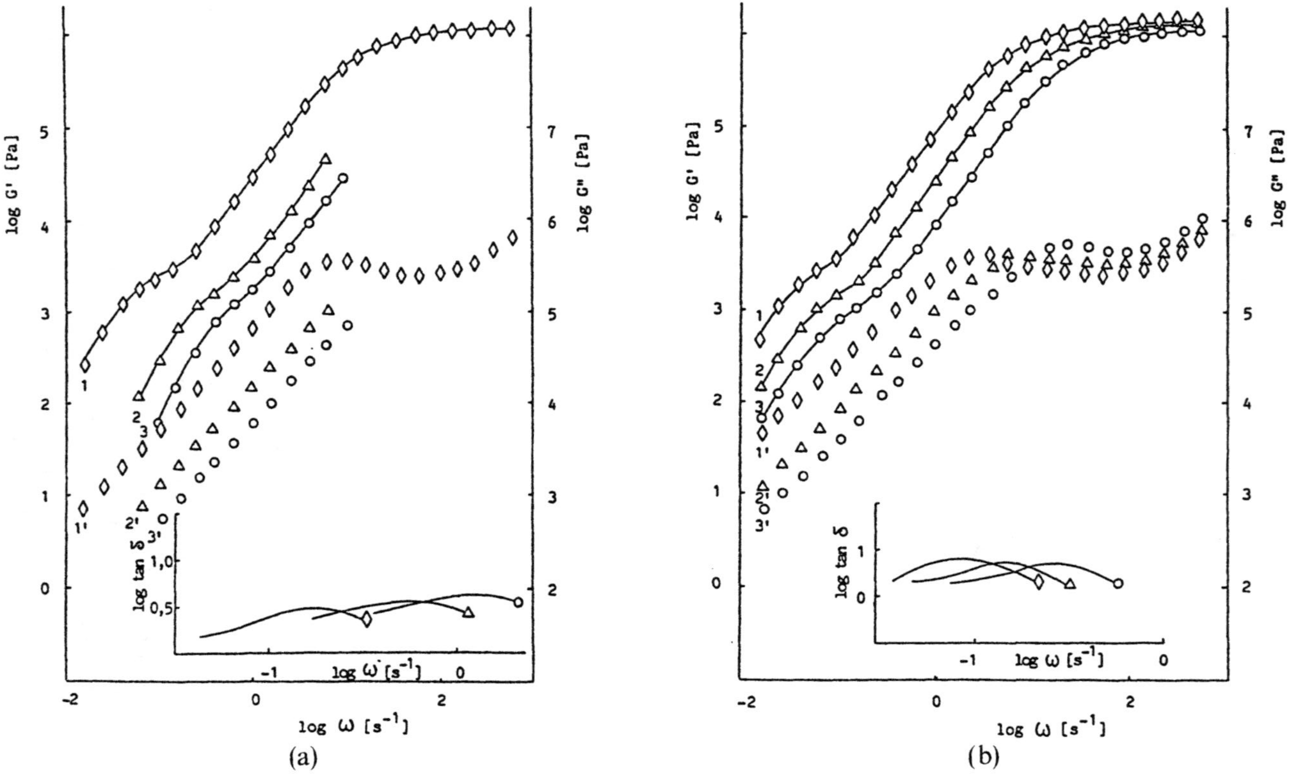

FIG. 2.5. Frequency dependence of storage G' (curves 1–3) and loss G'' (1–3) moduli, and $\tan\delta$, for 1,4-PB with (a) $M = 1{\cdot}0 \times 10^5$ and (b) $M = 1{\cdot}4 \times 10^5$, at 298 K (curves 1 and 1′), 333 K (2 and 2′) and 363 K (3 and 3′).

(superslow) one whose mechanism was discussed in the preceding chapter.

In this connection, of interest are new experimental data obtained in Ref. 22, where an attempt was made to directly appraise such superslow processes.

It can be seen from Fig. 2.5 containing plots of G' and G'' vs ω for 1,4-PB with a narrow MWD ($\bar{M}_w/\bar{W}_n = 1{\cdot}05$) that a low-temperature relaxation transition is observed in the terminal zone. At all temperatures it is especially clear on the plot of G' vs. ω. On the curves of G'' vs ω, this transition is hardly noticeable, and it virtually vanishes in logarithmic coordinates. But judging from the plots of $\tan\delta$ vs ω shown in the bottom part of the figure, the indicated relaxation process is registered quite reliably. It can also be seen from the above data that with elevation of the temperature the low-frequency relaxation process (on the plot of G' vs ω) is less noticeable, but on the curves of $\tan\delta$ vs. ω its position and change with elevation of the temperature are determined quite reliably as previously.

A comparison of the dynamic characteristics of objects with various molecular weights reveals that a reduction in the molecular weight of a polymer displaces the observed transition to the region of high frequencies (Fig. 2.6).

The existence of a superslow relaxation process is also undoubtedly evident when calculating the values of the initial constants of a system such as η_0, $A^0_{G'}$ and the equilibrium (reversible) shear compliance $J^0_e = 1/G^0_e$ determined in the terminal zone by the linear theory of viscoelasticity and its generalizations.

The results of calculating the quantities η_0, $A^0_{G'}$ and J^0_e in the terminal zone above and below (along the frequency scale) the superslow relaxation process show that, depending on the range, the values of A^0_G and J^0_e differ greatly (Fig. 2.7). The position of the peak on the plot of $\tan\delta$ vs ω corresponding to the relaxation process being considered has been shown to change with the molecular mass in proportion to $M^{4{\cdot}5}$. According to Refs 1, 2 and 18 the peak on the curve of G'' vs ω characterizing the transition of the system from the rubbery to the fluid state varies for PB in proportion to $M^{3{\cdot}5}$, while the initial viscosity of the system and the initial high elasticity coefficient increase according to the laws $\eta_0 \propto M^{3{\cdot}5}$ and $A^0_G \propto M^{7{\cdot}0}$ with a growth in MW. With consideration of the superslow relaxation process, the initial parameters of the system change in a somewhat different way with MW (see Chapter 1). The temperature dependence of the indicated quantities for regions above and below the

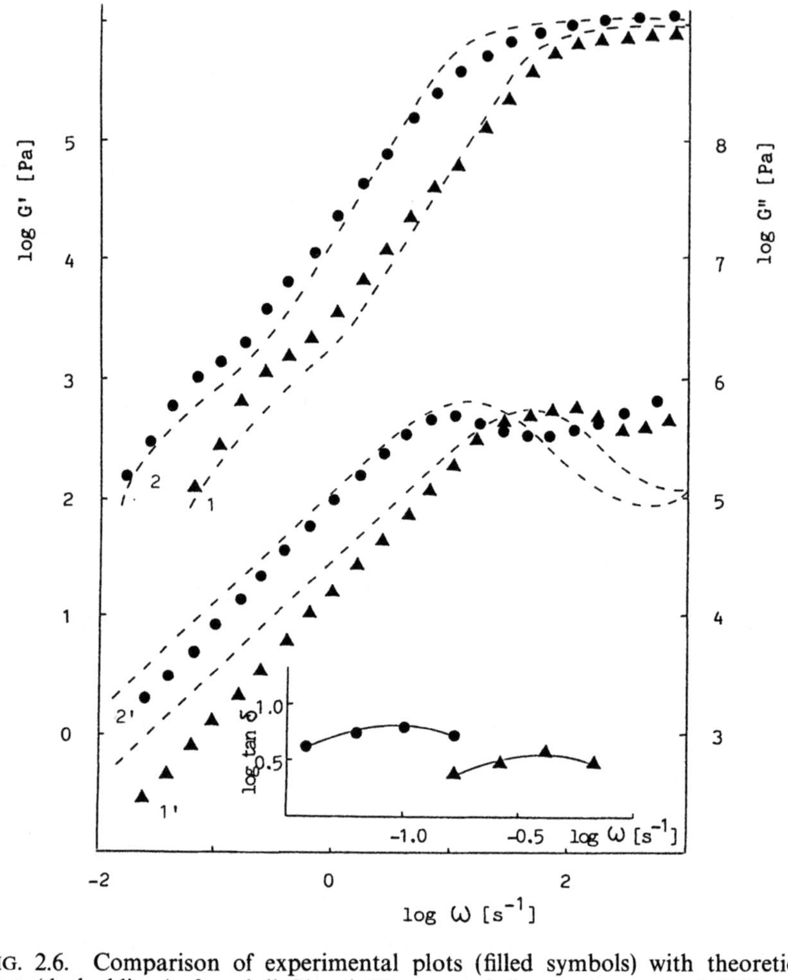

FIG. 2.6. Comparison of experimental plots (filled symbols) with theoretical plots (dashed lines) of moduli G' and G'' vs ω, and also tan δ vs ω, for 1,4-PB with $M = 1{\cdot}0 \times 10^5$ (curves 1 and 1′) and $1{\cdot}4 \times 10^5$ (2 and 2′) at 333 K.

superslow relaxation transition is also different, viz. the activation energy of viscous flow has been determined as 27·6 kJ mol^{-1} [2, 17] and 37 kJ mol^{-1} [22].

Analysis of the rheological behavior in the terminal zone and a description of the superslow relaxation process are also of interest for other reasons. The traditional techniques of relaxation spectrometry [23] under conditions of scanning over the temperature reveal a whole set of

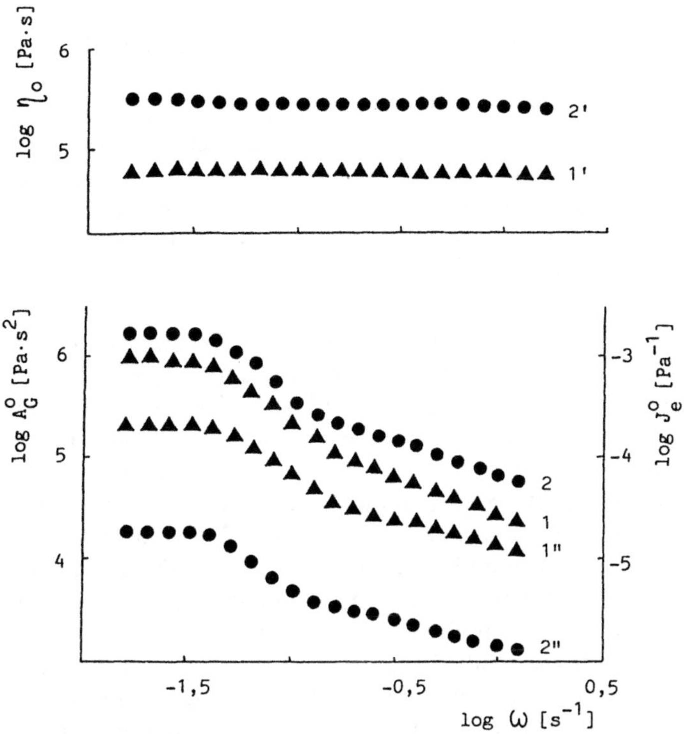

FIG. 2.7. Values of initial elasticity coefficient A_G^0 (curves 1 and 2), η_0 (1′ and 2′) and equilibrium reversible shear compliance J_e^0 (1″ and 2″) at various frequencies in the terminal zone for 1,4-PB with $M \times 10^5 = 1{\cdot}0$ (1, 1′ and 1″) and 1·4 (2, 2′ and 2″) at 298 K.

relaxation transitions in a polymer system that reflect the mobility of a macromolecule or parts thereof. It is not surprising that this method is also sensitive to the molecular ordering in a polymer, i.e. to a certain extent to the nature of the supermolecular formations. It is therefore natural to assume that under conditions of scanning with respect to the frequency in the region of the fluid state, relaxation processes may be manifested that are associated with large-scale mobility, i.e. with the mobility of associations of macromolecules and supermolecular structures with density fluctuations of various natures [23].

In each case, the manifestation of these relaxation processes can be discovered only under the definite conditions of running an experiment. Moreover, it is logical to presume the existence of several superslow relaxation processes in the region of low frequencies of various physical natures. They are meanwhile not predicted theoretically from model considerations. The assumption concerning the relation between superslow relaxation processes and the structure of a melt or solution of a viscoelastic medium opens up prospects for determining the structural features of an object by rheological techniques. This is why the experimental study of the viscoelastic characteristics in the terminal zone in the region of very low frequencies is important not only for theory, but also for practical applications.

2.2.3 Influence of Temperature: Temperature–Frequency Reduction

Consideration of a set of results obtained under conditions of dynamic experiments at various frequencies and temperatures has revealed that in various physical states a high-molecular-weight polymer does not react regularly to a change in these parameters. This is reflected in a qualitative transformation of the nature of the dynamic relations. We can give as an example here the growth in the anomaly of viscoelasticity at lowered temperatures, and the degeneration of the rubbery plateau on the temperature dependences of the moduli of storage. Such effects are not predicted by the existing theoretical concepts.

This problem was studied experimentally in detail in Ref. 12. It will be expedient to consider some practically important conclusions here.

The results of dynamic studies of 1,2-PB (Fig. 2.8) obtained for a broad range of frequencies show that the nature of the curves of G' and G'' vs ω depends appreciably on the temperature of the experiment. At elevated temperatures, the rubbery plateau degenerates (curves 7, 8), and with an increase in the frequency, the system passes virtually directly from the fluid state to the glassy one. (The region of frequencies in which superslow relaxation processes are manifested was not reached in Ref. 12. This is explained, in particular, by the different glass transition temperatures T_g of 1,4-PB and 1,2-PB, i.e. 177 and 238 K, respectively.) The temperature coefficients for the region of the fluid (a_T^f) state and of the transition to the glassy state (a_T^g) differ sharply (Fig. 2.9). This indicates the existence of various temperature dependences of the relaxation times corresponding to the long-time and short-time parts of a system's relaxation spectrum, respectively.

The latter is an illustration of the broad, though restricted, possibilities of employing the principle of temperature–frequency reduction. Indeed, within the confines of physical states such as the fluid one or the transition to a glassy one, the temperature dependence of the viscoelastic

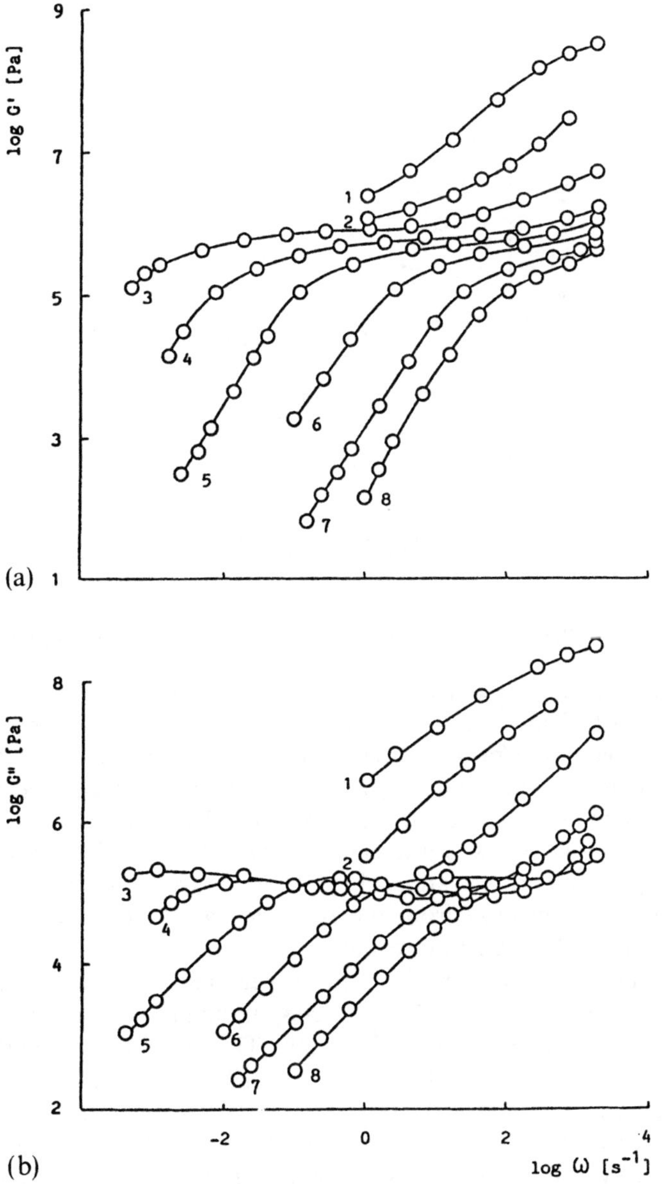

FIG. 2.8. Dependence of (a) log G' and (b) log G'' on log ω at 240 K (curve 1), 246 K (2), 258 K (3), 273 K (4), 295 K (5), 323 K (6), 363 K (7) and 393 K (8) for 1,2-PB with $M = 1{\cdot}35 \times 10^5$.

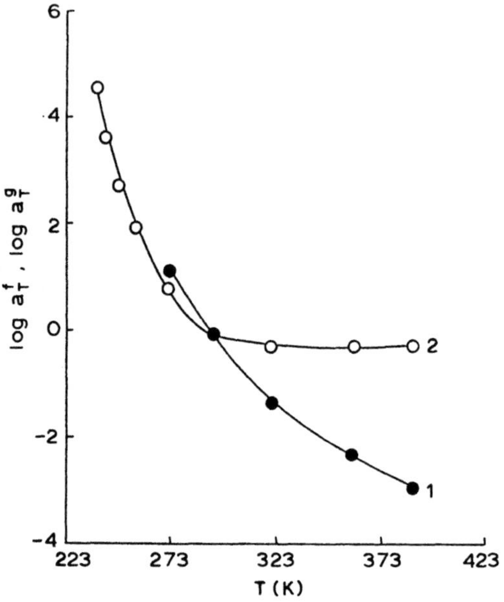

FIG. 2.9. Temperature dependence of log a_T^f (curve 1) and log a_T^g (curve 2).

dynamic characteristics is unambigously described by the relevant temperature coefficient a_T^f or a_T^g; consequently the method of temperature–frequency reduction may be used here. At the same time, this method cannot be used in the region of the rubbery state in this case because it is not clear how we can consider the observed change in the shape of the curves of G' and G'' vs ω (the effects of narrowing of the region of the rubbery plateau with elevation of the temperature, of an increase in the region of the transition from the rubbery state to the fluid one with diminishing of the temperature, etc.).

The values of the loss moduli G''_{max} at peak and G''_{min} at the minimum and the frequencies ω_{max} and ω_{min} corresponding to them change noticeably with the temperature (Fig. 2.10). The use of the temperature coefficients a_T^f and a_T^g in Ref. 12 made it possible to plot reduced dynamic relations, viz. G'_{red} and G''_{red} vs ω, for a very wide range of reduced frequencies (about 12 decimal orders of magnitude) covering various states of a system, from fluid to glassy (Fig. 2.11). It should be noted that direct determinations of the dynamic characteristics were also conducted for the frequency region log $\omega = -3{\cdot}4$ to $+3{\cdot}2$. Their results for the region of the fluid state are well consistent with the reduced characteristics.

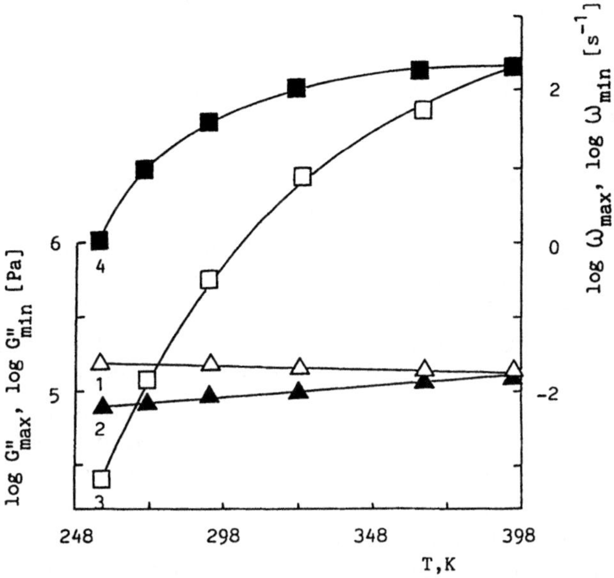

FIG. 2.10. Temperature dependences of $\log G''_{max}$ (curve 1) and $\log G''_{min}$ (2), $\log \omega_{max}$ (3) and $\log \omega_{min}$ (4).

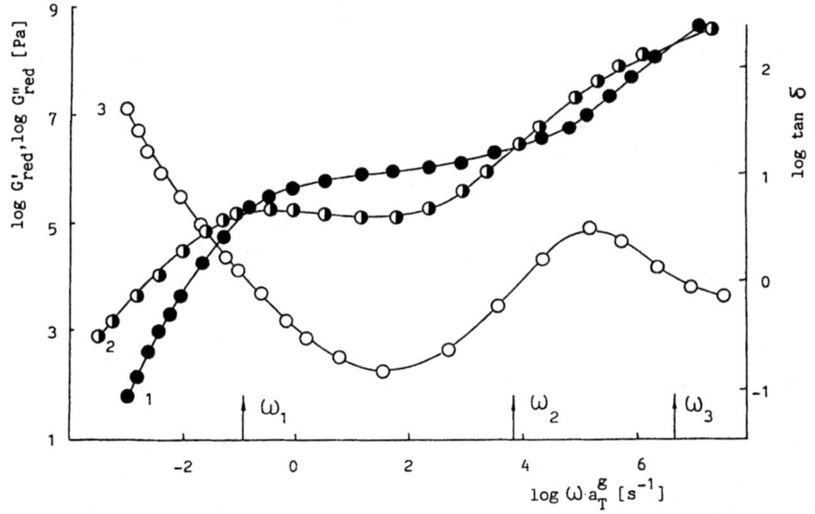

FIG. 2.11. Dependence of $\log G'_{red}$ (curve 1), $\log G''_{red}$ (2), and $\log \tan \delta$ (3) on logarithm of reduced frequency ωa_T^g for 1,2-PB with $M = 1{\cdot}35 \times 10^5$ at 295 K.

The plots of G'_{red}, G''_{red}, and tan δ vs ω_{red} show that values of tan δ from the maximum ones to unity correspond to the fluid state, while the point of intersection of the curves of G' and G'' vs ω at low frequencies, which corresponds to $\omega_1 \approx \omega_{max}$, can be its boundary. The region of the rubbery state terminates at the second point of intersection of the curves of G' and G'' vs ω, i.e. in the range of medium frequency values. Here tan δ, after passing through a minimum, again equals unity at the frequency ω_2. The plot of G'' vs ω also passes through a minimum. It should be noted that the points of the minima of G'' and tan δ vs ω, figuratively speaking, divide the region of the rubbery state practically in half. Beyond the limits of this state (the right-hand part of Fig. 2.11), the values of G' and G'' increase sharply, that of G' always more rapidly than G'' (this is the region of the transition to the glassy state). The values of tan δ also grow here, passing through a minimum, and then for the third time tan $\delta = 1$. The frequency corresponding to the peak on the curve of tan δ vs ω, ω_{max}, divides the transition region to the glassy state in half. Above the value ω_3 is the region of the glassy state.

Similar results for polybutadienes of other molecular weights showed [18] that $G''_{min} \propto M^{0.4}$ and $\omega_{min} \propto M^{-0.6}$, whereas the values of ω_2, ω_3, and ω_{max} remain practically constant with a change in the molecular weight of a polymer.

2.2.4 Calculation of Initial Viscoelastic Parameters

The effectiveness of employing any method for determining the viscoelastic behavior of a system when using dynamic measurements can be seen quite well when comparing the plots of G' and G'' vs ω or G' and G'' vs T (Fig. 2.12). Figure 2.12(c) shows how the physical states are determined according to Ref. 14. Scanning over the temperatures can be seen to be just as effective as scanning over the frequencies. A typical peak—the T_{11}-transition—can be seen on the plots of both G'' vs T and G'' vs ω. In both cases, the T_{11}-transition is registered quite clearly by the discontinuity on the curves of tan δ vs ω and tan δ vs T. In this connection, attention must also be given to the relations a_T^f and a_T^g vs T (Fig. 2.9). The point of intersection of these curves corresponds to the temperature of T_{11}-transition.

It is known from the literature and can be seen from Fig. 2.12(a) that the attainment of the region of the fluid state on the curves of G' and G'' vs ω is associated primarily with the need to realize relatively low values of the deformation frequencies (under $10^{-2}s^{-1}$) and is determined by the molecular weight of a polymer, its MWD, the temperature of the

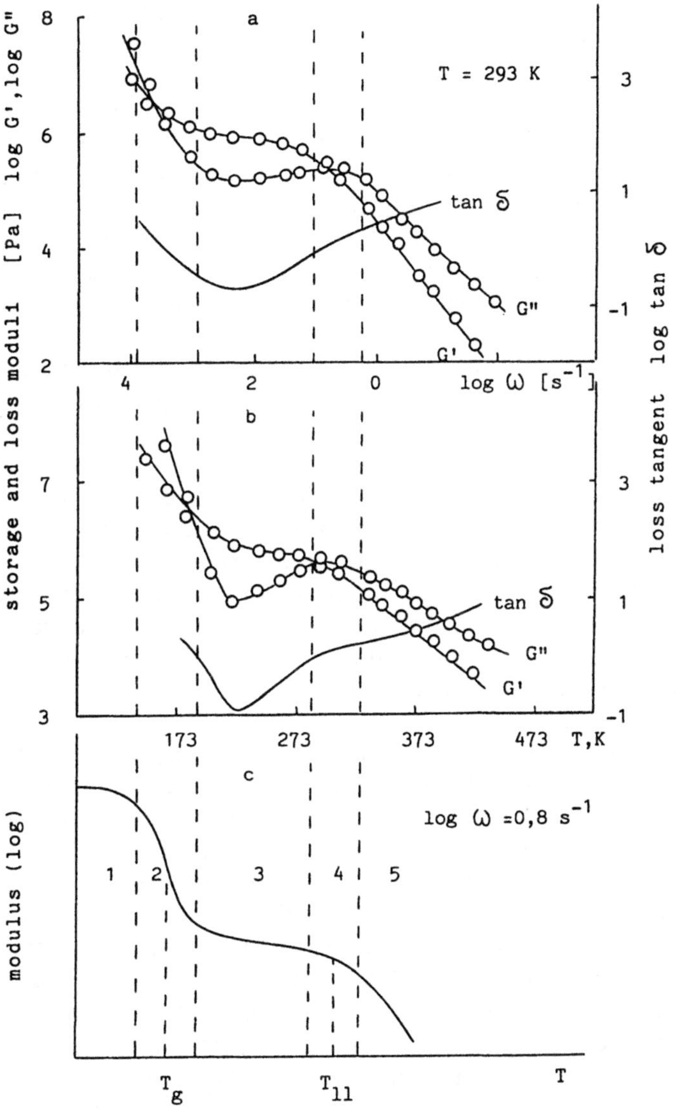

FIG. 2.12. (a) Frequency and (b) temperature dependences of storage G' and loss G'' moduli and $\tan\delta$ for 1,2-PB; (c) scheme to determine the physical states of a polymer system on the basis of determination of its moduli, according to Ref. 14 (1—glass, 2—leathery, 3—rubbery, 4—transition to fluid, 5—viscous flow).

experiment, and the nature of the polymer. It is difficult to run experiments in this region of frequencies, while extrapolation of the results to low values is not always correct and leads to substantial errors. In this connection, the determination of the initial viscoelastic constants of a system is often problematic, and sometimes even impossible, but this is exceedingly important from the experimental viewpoint. The quantities η_0 and A_G^0 determine not only the viscoelastic behavior of a polymer system in the region of the fluid and rubbery states, but also the conditions of normalization of the rheological properties of polymers of different homologous series. They are needed when plotting the generalized temperature-invariant relations [17].

The experimental difficulties indicated above can be alleviated to a certain extent by raising the temperature of an experiment. This enables one to employ the principle of temperature–frequency reduction, which is an indirect way of expanding the experimental possibilities of the dynamic method. But as shown above, the temperature reduction functions only within definite limits. Moreover, variation of the experimental temperature is often limited by the thermal stability of a system.

Possibilities of determining the initial viscoelastic constants of polymers also appear from an analysis of the temperature dependences of the dynamic characteristics obtained at a constant and relatively high (from the experimental standpoint) frequency, of order of magnitude 0.1 to $1\ s^{-1}$ (see Figs 2.12 and 2.13).

A comparison for 1,4–PB of the temperature dependences of dynamic characteristics such as the storage modulus G', the loss modulus G'', and $\tan\delta$, obtained for several values of the circular frequencies, reveals that, depending on the frequency used in an experiment, there are changes in the length of the rubbery plateau, the distance between the peak and minimum on the curves of G'' vs T, and also the slopes of the curves in the terminal zone. The curve of $\tan\delta$ vs T has a typical minimum in the region of the rubbery state, while in the transition to the fluid state the curves of $\tan\delta$ vs T have typical discontinuities that can be identified with the reaching of the T_{11}-transition temperature. It is significant that depending on the frequency of an experiment, the rate of growth of the plot of $\tan\delta$ vs T at elevated temperatures varies. This is apparently associated with the magnitude of the anomaly in the viscoelastic properties of a system in the fluid state.

Using expressions (2.1) and data such as those presented in Figs 2.12 and 2.13 (curve 3), the values of η_0 and A_G^0 can readily be calculated for various temperatures in the fluid state region. For the case being

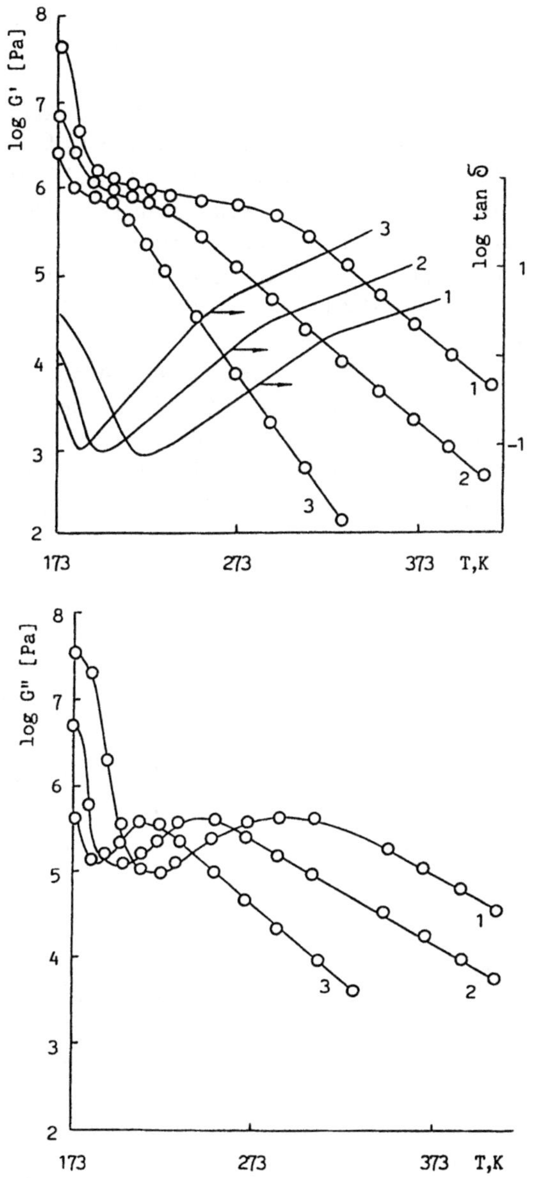

FIG. 2.13. Temperature and frequency dependences of storage G' and loss G'' moduli and tan δ for 1,4-PB at circular frequencies $\log \omega$ (s^{-1}) of $-0{\cdot}8$ (curves 1), 0·0 (curves 2) and $+1{\cdot}0$ (curves 3).

considered, the results of calculations are presented in Fig. 2.14. The values of η_0 and A_G^0 determined from the frequency dependences of the dynamic characteristics in the region of the fluid state, where $G' \propto \omega^2$ and $G'' \propto \omega$ for temperatures of 293, 323, and 383 K according to [24], are also shown on the curves of η_0 and A_G^0 vs 1/T (the dark symbols). It can be clearly seen that the values of η_0 and A_G^0 obtained in these two ways are plotted on one curve and are well consistent. The choice of the required frequency ω for realizing the T_{11}-transition for an experiment scanning over the temperature is determined to an appreciable extent by the molecular weight and MWD, the nature of the polymer, and its glass transition temperature.

Hence, we can see that the temperature dependences of the dynamic viscoelastic characteristics allow us not only to determine the most important initial viscoelastic parameters of a system that are generally found from the frequency dependences of the dynamic characteristics, but also to establish the temperature dependences of these parameters.

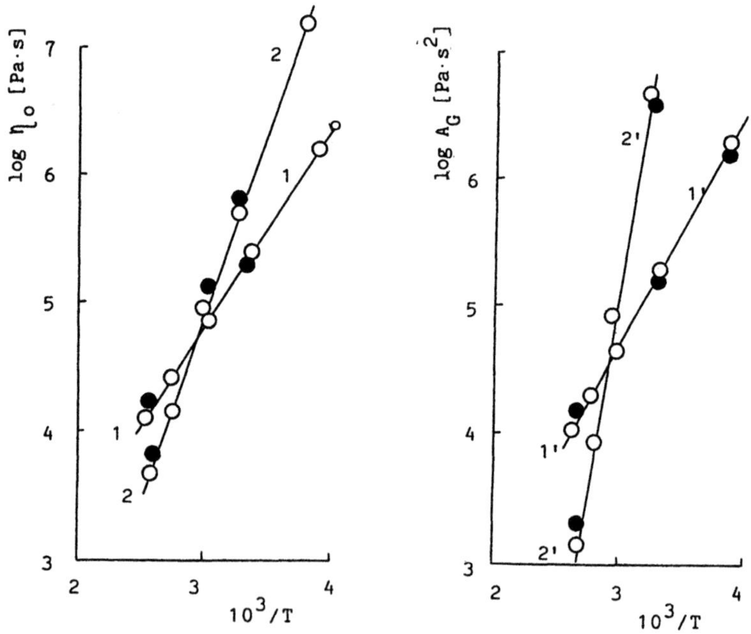

FIG. 2.14. Temperature dependences of η_0 and A_G^0 calculated from plots of G' and G'' versus T (at $\log \omega = 1{\cdot}0$) (open circles), and from plots of G' and G'' vs ω (filled circles): 1,4-PB (curves 1 and 1′); 1,2-PB (curves 2 and 2′).

2.3 VISCOELASTIC BEHAVIOR IN CONTINUOUS SHEAR DEFORMATION

2.3.1 The Critical Regimes and Parameters of Deformation

The relaxation transition from the fluid to rubber-like state can be studied in simple experiments with the capillary constant pressure viscometers [25]. This can be seen from Fig. 2.15 which shows the behavior of various monodisperse 1,4-PB polymers with $M/M_e > 10$ for the temperatures above the glass transition point ($T \gg T_g$) [26–29]. Up to certain critical values of shear stress τ_{sp} the polymers flow as Newtonian liquids. This is

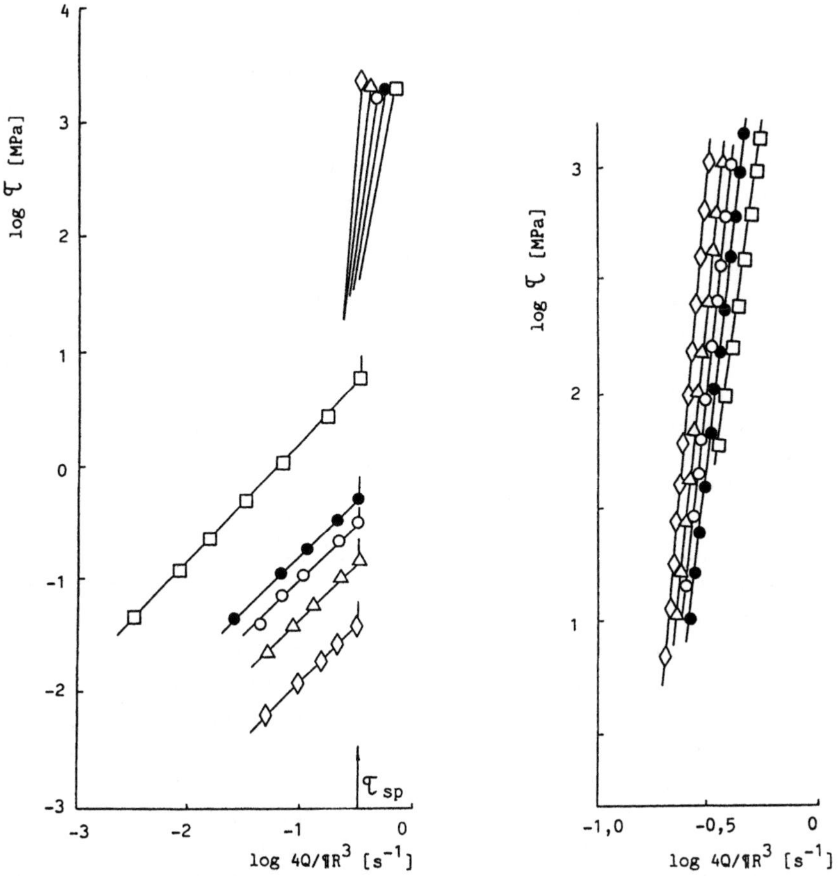

FIG. 2.15. Flow rate vs shear stress at 293 K at a capillary diameter of 1·0 mm for 1,4-PB with narrow MWD and various values of M_w: □ $1{\cdot}1 \times 10^5$; ● $2{\cdot}0 \times 10^5$; ○ $2{\cdot}6 \times 10^5$; △ $3{\cdot}2 \times 10^5$; ◇ $4{\cdot}7 \times 10^5$.

explained by the fact that dynamic equilibrium between formation and destruction of entanglement nodes is not disturbed since these polymers have entanglement networks of high density and uniformity. Analysis has revealed that the polymer structure remains practically unchanged. It should be noted, however, that with decreasing temperature the dynamic equilibrium between the formation and destruction of the entanglement network shifts towards the destruction process. Therefore, the non-Newtonian nature of the flow is manifested in a stronger and stronger way. These qualitative considerations are brilliantly corroborated by direct experiments (Fig. 2.16). (Compare also Fig. 2.16 and Fig. 2.3.)

The critical shear stress τ_{sp} is an important characteristic for all high-molecular-weight polymer compositions. The critical shear stress τ_{sp} is practically temperature-independent. As the viscosity remains constant until τ_{sp} is attained, the critical values of the shear rate are inversely proportional to the initial viscosity and therefore are correspondingly dependent on temperature and molecular weight. After the transition to the rubber-like state, a polymer behaves as quasi-crosslinked, its fluidity being suppressed. As large recoverable (high-elastic) deformations are accumulated, the polymer fractures like cured rubber. A sharp jump in the flow rate, corresponding to the spurt effect, appears as a result of detachment of the polymer in the rubber-like state from the capillary walls.

In overspurt regimes, a polymer loses its continuity at the channel entrance, after which polymer fragments, densely compressed by the pressure drop in the channel, move like a solid body. The right-hand side of Fig. 2.15 shows in enlarged scale a bundle of curves that represent overspurt regimes of the polymers. The overspurt regimes are described by the straight lines with slopes from 9 to 13. For the same flow rate, the samples with higher molecular weight experience lower shear stress. This is explained by the fact that with increasing molecular weight, the dissipative losses in the rubber-like state become lower.

The spurt process does not depend on temperature (Fig. 2.17). The explanation is that the critical shear stress τ_{sp} is temperature-independent since the entanglement density varies only slightly with temperature.

It should be noted that in overspurt regimes of polymer motion, the nature of the dependence of the volume output on the shear stress is practically independent of the chemical nature of the linear flexible chain polymers. The difference in stresses required to obtain the same flow rate is small; in the limiting cases, these stresses may differ by a factor of 2 (Fig. 2.18).

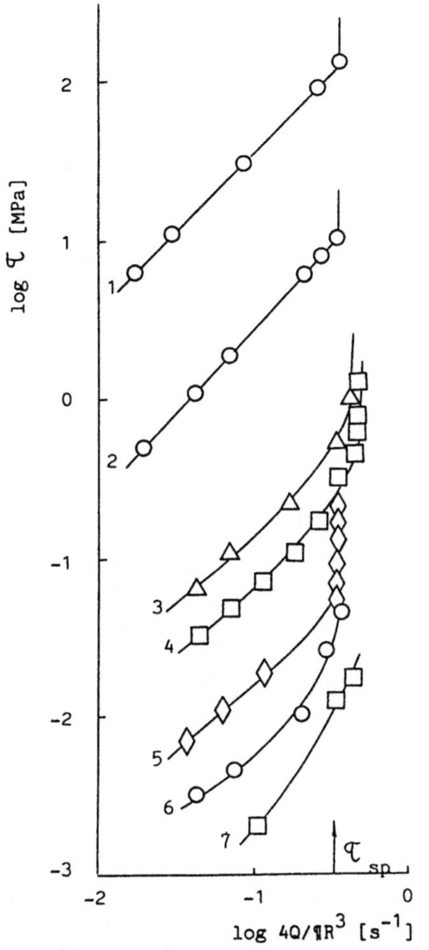

FIG. 2.16. Flow rate vs shear stress for 1,4-PB with narrow MWD at 373 K (curve 1), 295 K (2), 256 K (3), 246 K (4), 226 K (5), 224 K (6) and 222 K (7).

2.3.2 The Spurt Effect and the Transition of Polymers from the Fluid into the Rubber-Like State

Since in the steady-state flow the stress uniquely determines the unrecoverable energy loss, there exists a quantitative correspondence between the shear stress and the loss modulus. For high-molecular-weight polymers with a narrow molecular weight distribution this is confirmed

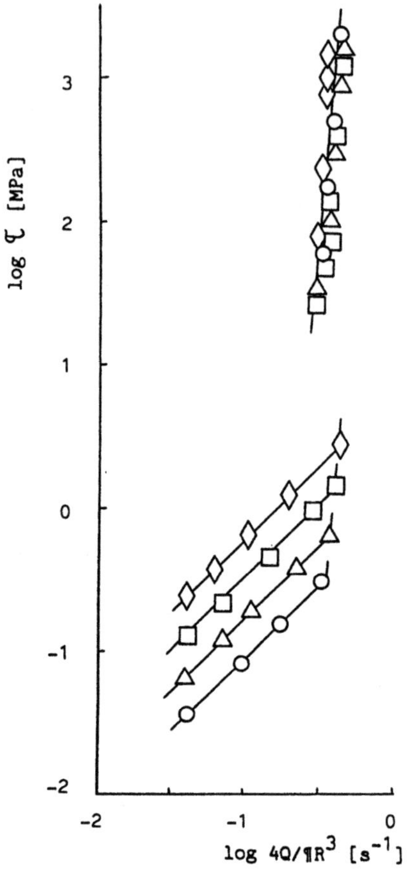

FIG. 2.17. Flow rate vs shear stress at a capillary diameter of 1·0 mm for 1,4-PB with narrow MWD and $M_w = 2{\cdot}6 \times 10^5$ at various temperatures: ○ 293 K; △ 313 K; □ 333 K; ◇ 353 K.

experimentally up to the values of the maximum loss modulus. The critical shear stress τ_{sp} is numerically equal to the loss modulus at the maximum. The correlation between these values is emphasized by the fact that the two quantities are practically independent of the temperature and molecular weight of the polymers. The parameters under consideration are important characteristics of polymer compositions. For various polymer considerations, the values of the critical shear stress, loss modulus at the maximum, and dynamic storage modulus on the plateau differ by not more than a factor of 20. The absolute value, the critical

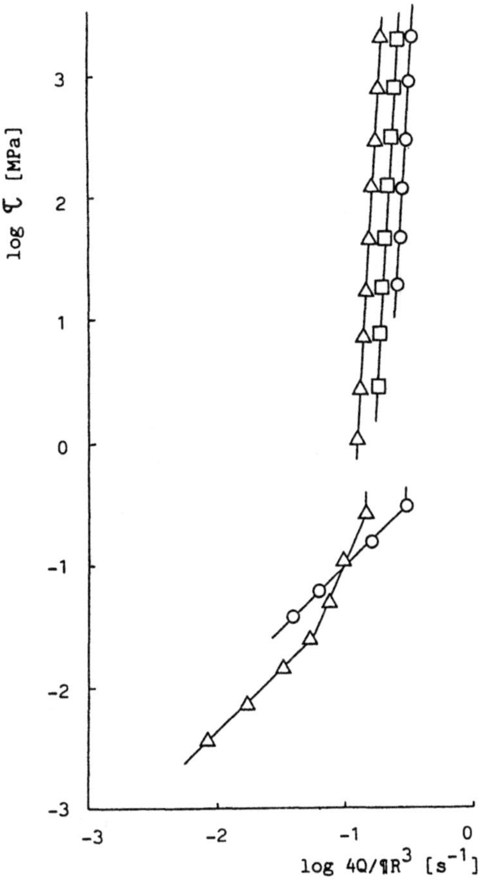

FIG. 2.18. Flow rate vs shear stress at a capillary diameter of 1·0 mm for 1,4-PB with narrow MWD and $M_w = 2{\cdot}6 \times 10^5$ at 293 K (○); for 1,4-PI with $M_w = 8{\cdot}4 \times 10^5$ at 376 K (△); and for 1,2-PB with $M_w = 1{\cdot}0 \times 10^5$ at 295 K (□).

stresses, amount to 0.1–1 MPa. These values characterize limiting stresses corresponding to steady-state flow regimes for linear polymers.

When considering the complex of phenomena observed in deformation of polymers in the spurt region, the following question arises. How does the strain affect the polymer transition to the rubber-like state? The answer to this question can be obtained from the results of dynamic tests with increasing amplitudes of deformation. With increasing amplitude, the transition from the fluid state to the rubber-like state shifts towards lower frequencies. A comparison of frequencies corresponding to critical

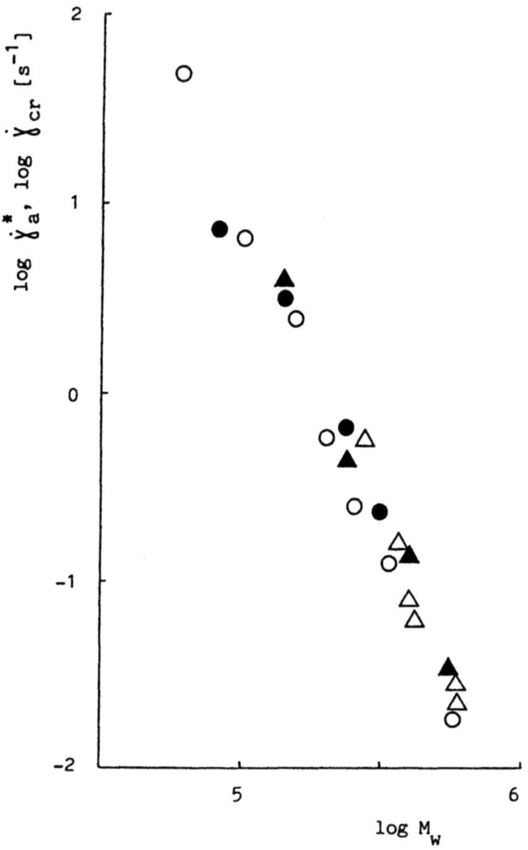

FIG. 2.19. Dependence of the critical amplitude of the deformation rate $\dot{\gamma}_a^*$ (△) and critical shear rate $\dot{\gamma}_{cr}$ (○) on the molecular weight of 1·4-PB at 293 K (open symbol) and 323 K (filled symbol).

amplitudes of deformation at which the spurt effect takes place (the spurt is registered by the abrupt drop of the torque as a result of polymer separation from the measuring surface in experiments with a working unit of the coaxial cylinders type), with the frequencies corresponding to the maximum on the loss modulus vs frequency curves in experiments with low amplitudes, shows that although large deformations affect to a certain extent the transition of the system from the fluid state to the rubber-like state, the spurt occurs as a result of limited fluidity of a polymer in the rubber-like state. Since the transition under consideration is of the relaxation type, its deformation rate characteristic, i.e. the amplitude of the rate of deformation, is most important. Experiments show that for large deformations, the transition from the fluid state to the rubber-like state occurs at a critical value of the amplitude of the deformation rate, which is constant for a given temperature. Moreover, the spurt observed in these experiments can be characterized by the same value. The magnitude of the critical amplitude of the deformation rate coincides with the magnitudes of the critical shear rate obtained in experiments with a capillary viscometer and depends in a similar way on the molecular weight and temperature (Fig. 2.19).

2.4 VISCOELASTIC BEHAVIOR IN UNIAXIAL EXTENSION

2.4.1 The Critical Regimes and Parameters of Deformation

The spurt and the failure of adhesive contact of the polymer with the wall with respect to which it moves, limit investigation on polymers in the simple shear regime. Experiments with uniaxial extension are free of these limitations. It turned out that there exists a relationship between the spurt regime and fracture conditions for uniaxial extension, and the appearance of spurt can be used to predict fracture conditions for a polymer under extension. In uniaxial extension, however, polymer fracture can be studied within the limits of many decimal orders of rates and stresses, and strength characteristics are determined over a very wide range of rates of deformation, in particular for the transition from the rubber-like to the leather-like states [30, 31].

With uniaxial extension, two limiting modes in polymer behavior can be seen clearly: steady-state flow and polymer fracture (Fig. 2.20). In the fluid state region, to the left of the vertical dashed line, theoretically unlimited strain can be accumulated and steady-state flow can be attained. With increasing rate of deformation, a transition from the fluid to the rubber-like state occurs where polymer deformation is limited, as a result of which the sample breaks, and this is a criterion for the transition to the rubber-like state. An increase in rates and stresses above their critical values rapidly reduces the ability of a polymer to accumulate unrecoverable deformation. Consequently, in a longitudinal deformation, the critical parameters must be determined by the regime of transition from unlimited to limited deformation, or, more rigorously, by the extrapolation of rates and stresses to the values corresponding to steady-state flow conditions.

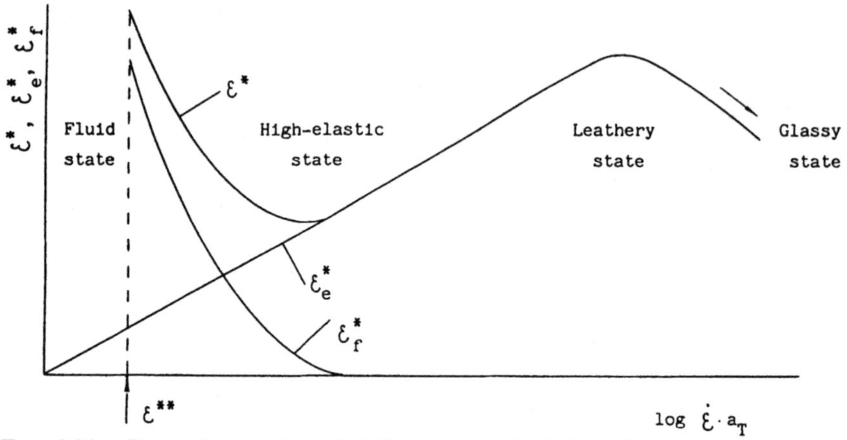

FIG. 2.20. Dependence of total (ε^*), recoverable (ε_e^*) and unrecoverable (ε_f^*) strains on strain rate.

To the critical deformation rate $\dot{\varepsilon}^{**}$, which corresponds to the vertical dashed line, is correlated the critical fracture stress σ^{**} that is in a simple relation with the critical shear stress at which the spurt occurs, namely $\sigma^{**}=3\tau_{sp}$. There exists a correlation between node formation in an entanglement network and the formation of an adhesive joint between polymer segments and the surface of solid bodies.

The critical deformation rate in extension, $\dot{\varepsilon}^{**}$, is inversely proportional to the initial viscosity and, as a function of molecular weight, varies by many decimal orders. In this case, the effect of temperature can be accounted for by the reduction temperature coefficient α_T determined by the temperature variation of the initial viscosity.

In overcritical regimes, unrecoverable deformation rapidly decreases while recoverable (high-elastic) deformation continuously increases. The result is that the total deformation passes through a minimum. To the right of the minimum, the deformation is virtually recoverable, and the curve has a maximum corresponding to the transition of a polymer first to the leather-like and then to the glass-like state. The latter state is more easily attained the closer the experimental temperature is to the glass transition point of the polymer.

It is essential that, near the maximum of the dependence of strain on the deformation rate $\varepsilon^*(\dot{\varepsilon})$, the degree of deformation of a linear polymer can amount to 2000–3000 and more per cent. These values considerably exceed recoverable fracture strains for crosslinked rubbers. The analysis of the results for a number of polymers leads to an important conclusion that the maximum of the $\varepsilon^*(\dot{\varepsilon})$ curve corresponds to the limiting degree of the straightening of the macromolecules [27, 28, 30, 31].

The polymer samples can be quickly vitrified at the strains that are close to the fracture strain values and their strength can be measured. It has been shown that a maximum of the strength corresponding to the maximum of fracture strain versus strain rate is observed. The relevant data indicate that for such vitrified amorphous polymers only a strength of up to 100 MPa can be obtained.

2.4.2 Prediction of Polymer Fracture in Uniaxial Extension on the Basis of a Destructive Test

The prediction of polymer fracture on the basis of a destructive test is of importance. As has already been shown above, the transitions of linear polymers from one physical state to another can take place in isothermal conditions. Available data [27–31] indicate that there exists a close relationship between the dependence of the total deformation on the deformation rate and the frequency dependence of the dynamic loss modulus $G''(\omega)$. The minimum on the strain vs strain rate curve corresponds to the maximum of the loss modulus, and the maximum on the same curve corresponds to the minimum of the loss modulus (Fig. 2.21). This correlation has far-reaching consequences, since it becomes possible to predict the change in the ultimate fracture strains for a linear polymer using the data on non-fracture low-amplitude deformation for this polymer.

The correlation between $\varepsilon^*(\dot{\varepsilon})$ and $G''(\omega)$ is based on the relations between the energy stored and dissipated during deformation of a polymer sample. According to this, in the linear viscoelastic approximation, the following relation should exist for ε^* and G'':

$$3(\varepsilon^*)^2 G'' = W_{\mathrm{d}} \qquad (2.2)$$

where W_{d} is the energy dissipated upon extension of the sample up to the failure. It is assumed that W_{d} is a constant or a slightly varying function of ε. In Fig. 2.21(a) the dashed line corresponds to the value of $\varepsilon^*(\dot{\varepsilon})$ calculated by formula (2.2). It can be seen that a good quantitative agreement between experimental and theoretical results is attained. Here $W_{\mathrm{d}} = 1$ MPa for 1,2-PB. For a better coincidence between theoretical values of $\varepsilon^*(\dot{\varepsilon})$ with the experimental results of 1, 2-PB, when going over from the minimum to the maximum on the curve, W_{d} should be increased by 50%.

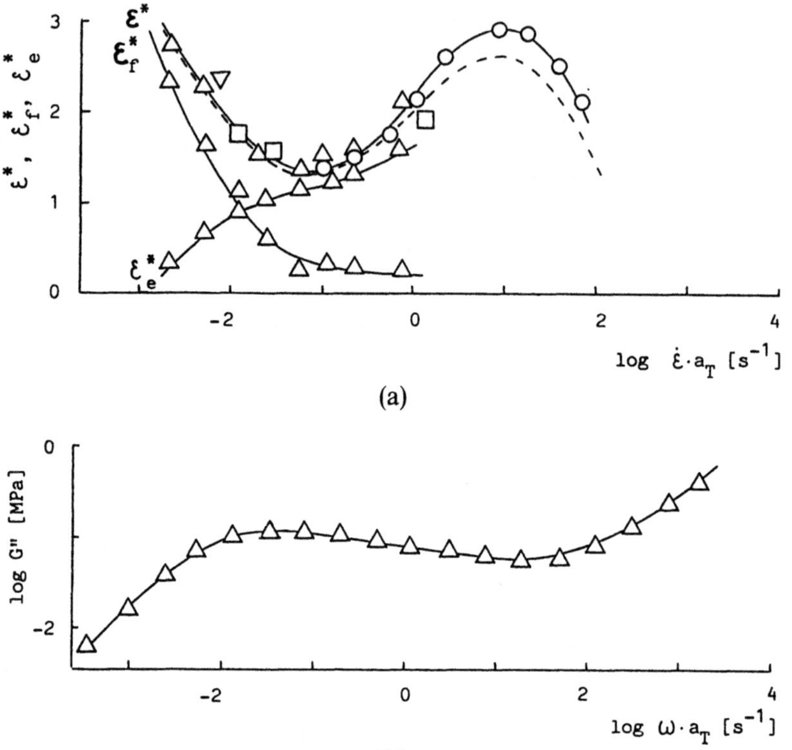

FIG. 2.21. (a) Total fracture strain (ε^*) and its recoverable (ε_e^*) and unrecoverable (ε_f^*) components vs strain rate, and (b) loss modulus vs frequency, for 1,2-PB. Reduction temperature 297 K; experimental temperatures: ○ 273 K; □ 283 K; △ 297 K; ▽ 323 K. Dashed line calculated from eqn (2.2).

Available data allow us to directly express the minimum ε_{min}^* and the maximum ε_{max}^* of the function $\varepsilon^*(\dot{\varepsilon})$ in terms of such important parameters as M_c and the storage modulus on the high-elasticity plateau, G_p', which does not depend on the molecular mass. It is well known that $G_{max}''=0{\cdot}46G_p'$, hence, using formula (2.2), we get

$$\varepsilon_{min}^*=0{\cdot}91(W_d/G_p')^{0{\cdot}5} \tag{2.3}$$

According to Ferry [7], the loss angle tangent corresponding to G_{min}'' is

$$(\tan\delta)_{min}=G_{min}''/G_p'=1{\cdot}04(M/M_c)^{-0{\cdot}8} \tag{2.4}$$

Substituting eqn (2.4) into formula (2.2), we have

$$\varepsilon^*_{max} = 0{\cdot}556(W_d/G'_p)^{0{\cdot}5}(M/M_c)^{0{\cdot}4} \tag{2.5}$$

2.4.3 Fracture Envelope and its Generalization for Polymer Melts. Universal Characteristic of Polymer Fracture

It is well known that polymer fracture is time dependent and assumes a certain relation existing between the breaking strain σ^* and breaking stress. For cured rubbers, this relation is established by the well-known Smith fracture envelope. Figure 2.22 shows the fracture envelope for 1,2-PB in the true breaking stress vs total fracture strain coordinates (dark circles). In the same figure, there are deformation curves obtained under conditions of a constant deformation rate ($\varepsilon = \text{const.}$). It can be seen that an increase in the deformation rate or a decrease in temperature results in the displacement of breaking points along the envelope towards higher stresses. The σ^* vs ε^* curve is qualitatively similar to the deformation rate curve for linear polymers. Experiments show that the upper part of the curve, convex with respect to the coordinate axis, is invariant relative to deformation conditions. As in the case of cured rubbers, its extreme nature is explained by the polymer transition to the leather-like and then to the glass-like state associated with increasing stress or decreasing temperature. The lower part of the envelope, concave with respect to the coordinate axis, is determined by the transition of a linear polymer from the fluid to the high-elastic state, which is typical only of uncured polymers. Hence, the present work generalizes the results obtained by Smith for uncured polymers.

In the highly elastic state, the polymer fluidity is suppressed. In this connection, two questions arise: to what extent is linear polymer fracture above glass transition and melting point determined by the storage of recoverable deformation? Does the mode of storage of recoverable deformation affect fracture characteristics in the fracture regime of deformation?

Figure 2.23 shows the dependence of the true breaking stress on the recoverable component of breaking strain for PI with different molecular weights and for 1,2-PB. The solid lines were obtained for different loading regimes and different temperatures, and the dashed lines correspond to the dependence of stress on highly elastic strain for a steady-state deformation regime. For linear polymers, there exists a universal critical value of recoverable strain determined by the intersection of the solid and the dashed lines. When the strain exceeds ε_e^{**}, the polymer

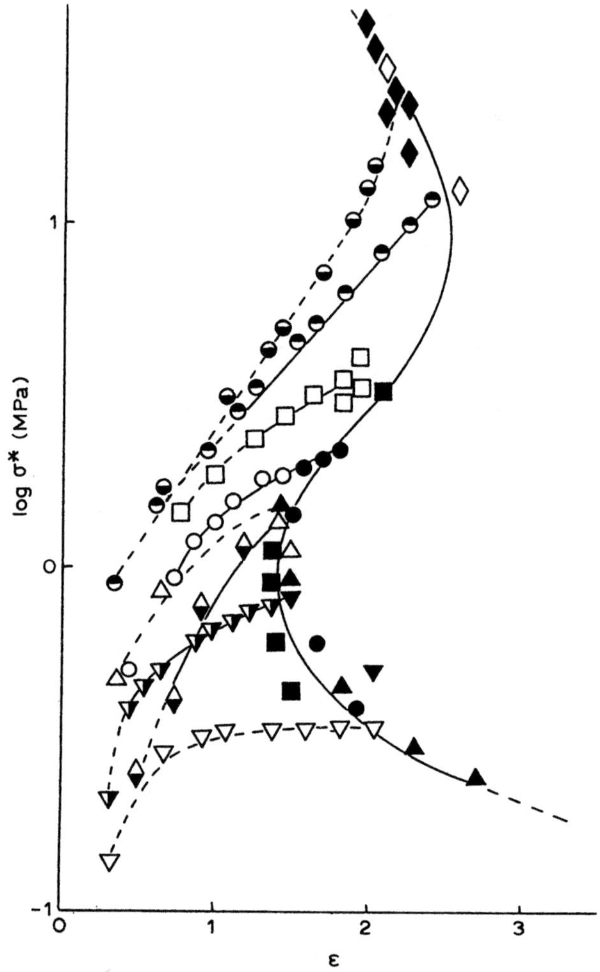

FIG. 2.22. Fracture envelope to 1,2-polybutadiene, measured at σ = const. and at various temperatures: ▼ 323 K; ▲ 297 K; ■ 283 K; ● 273 K; ◆ 263 K. Dashed lines σ–ε prior to rupture at constant strain rates (s^{-1}) and various temperatures: ▽ 5×10^{-2}; ▼ 2×10^{-1} (323 K); △ 1×10^{-1} (298 K); ○ 1×10^{-2}; ◑ 2×10^{-1}; ◐ 4×10^{-1} (273 K); □ 5×10^{-2} (283 K).

fractures. For polymer homologous series under study, $\varepsilon_e^{**} \approx 0{\cdot}5$, which corresponds to the extension ratio of 1·65. Along with the critical values of σ^{**} and $\dot{\varepsilon}^{**}$, the value of ε_e^{**} is also a critical parameter determining the transition of a polymer from the fluid to a highly elastic state [17, 27].

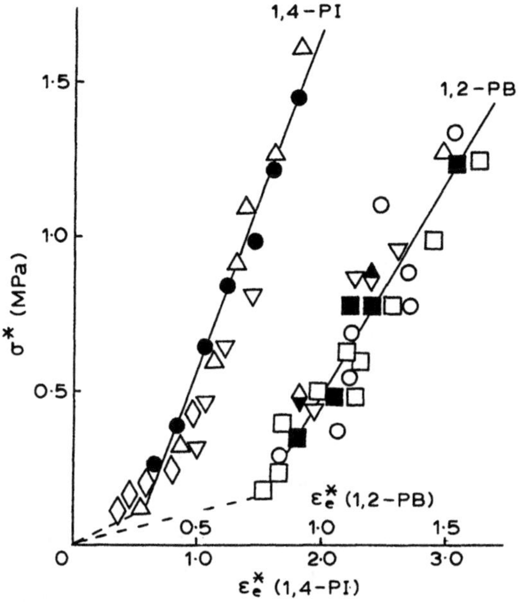

FIG. 2.23. True breaking stress vs recoverable component of breaking strain at various deformation modes for 1,4-PI and 1,2-PB. Deformation conditions: 1,2-PB, σ=constant (■ 283 K, □ 295 K, ⬘ 323 K); $\dot{\varepsilon}$=constant (⬙ 283 K, ○ 298 K); constant force (△ 298 K); stepwise loading conditions at 298 K (▽ deformation at $\dot{\varepsilon}_1 = 2\times10^{-3}\,s^{-1}$ for 30 s, relaxation at constant deformation for 60 s, deformation at $\dot{\varepsilon}_2 = 5\times10^{-3}\,s^{-1}$ before fracture; ▼ deformation at $\dot{\varepsilon}_1 = 5\times10^{-3}\,s^{-1}$ for 180 s, relaxation at constant deformation for 60 s; deformation at $\dot{\varepsilon}_2 = 2\times10^{-2}\,s^{-1}$ before; 1,4-PI at 298 K, the $\dot{\varepsilon}$=constant regime (molecular weight MM: ▽ $3{\cdot}75\times10^6$, ● $5{\cdot}75\times10^6$, △ $8{\cdot}3\times10^6$); the σ=constant regime, ◇ $5{\cdot}75\times10^5$.

It is essential that breaking stresses linearly depend on respective recoverable strains. This dependence is invariant with respect to molecular weight, temperature, and deformation regimes. It follows that at stresses and strains exceeding the critical values the ratio of the rates of deformation and of relaxation processes alone cannot be considered to be a criterion of fracture of linear polymers as continuous media.

It is well known that for a quantitative description of the long-term durability, or simply the durability (i.e. the time interval between the onset of deformation and fracture of the specimen), different functions have been used. On the one hand, an exponent-type function has been widely and successfully used by Zhurkov's school since the 1950s [32].

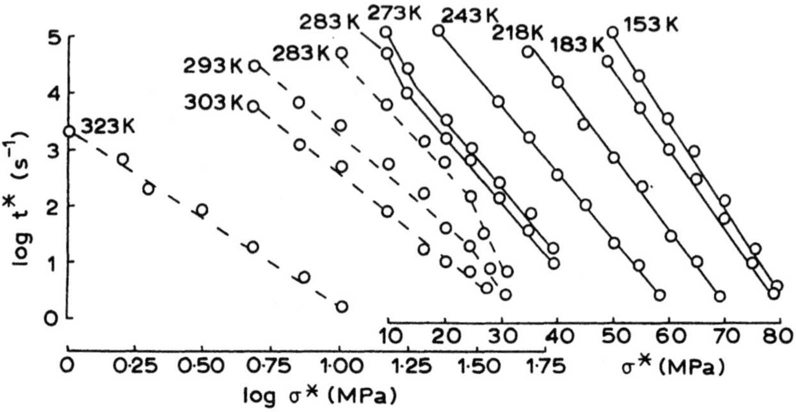

FIG. 2.24. Long-term durability vs normal stress for poly(butylmethacrylate) at different temperatures.

On the other hand, following the data that are known for plastic metals and inorganic glasses, Gul' [33] and Bartenev and Zuev [34] suggested a power function for cured rubbers. Vinogradov and others used successfully the power function for a description of the durability of uncured elastomers [30, 31, 35]. A question arises where the borderline lies between the two above-mentioned functions. The answer to this question was given by Kurbanaliev *et al.* [36]. The corresponding data are given in Figure 2.24 for poly(butylmethacrylate) [27, 28, 36]: here $M_w = 4{\cdot}7 \times 10^4$, and T_g is close to 293 K.

2.5 RELAXATION TRANSITION OF POLYMERS IN THE TRIAXIAL STRESSED STATE

The following generalization of the concept of induced relaxation transitions of polymers from the fluid to the rubber-like state is made for the case of the triaxial state of stress. A simple method for studying polymers in these conditions employs the contact of thin films between two parallel plates. Polymers under study were stretched at varying loading rate with a constant force acting perpendicular to the polymer surface. The behavior of a material in failure depends on its state; cohesive fracture is typical of polymer deformation regimes for the transition from the fluid to the rubber-like state. Further transition of the polymer to the rubber-like state is accompanied by a more clearly pronounced adhesive fracture.

A qualitative picture of this process was obtained for the contact of PB ($M_w = 6{\cdot}4 \times 10^5$) with solids of different natures [27, 37]. The results are presented in Fig. 2.25 (curve 1 corresponds to polytetrafluoroethylene, 2 to polished steel, and 3 to polymethyl methacrylate).

Let us turn now to the long-term durability of 1, 4-PB in contact in the triaxial state of stress. Figure 2.26 shows the dependence of long-term durability t^* on the breaking stress σ^* for 1, 4-PB in contact with steel and teflon at 293 K. It can be seen that the t^* vs σ^* curve for steel has two regions of stresses and long-term durability which correspond to different types of fracture of 1, 4-PB films. For cohesive fracture, the durability decreases with the molecular mass of the adhesive. With increasing stress, cohesion–adhesion fracture is observed. In this case, only a certain part of the metal surface was covered by a polymer during its separation. Above 0·9 MN m^{-2}, the fracture is of an adhesive nature, but the spurt occurs so rapidly that it is impossible to assess long-term durability ($t^* > 0{\cdot}1$ s).

The nature of the dependence of long-term durability on the stress,

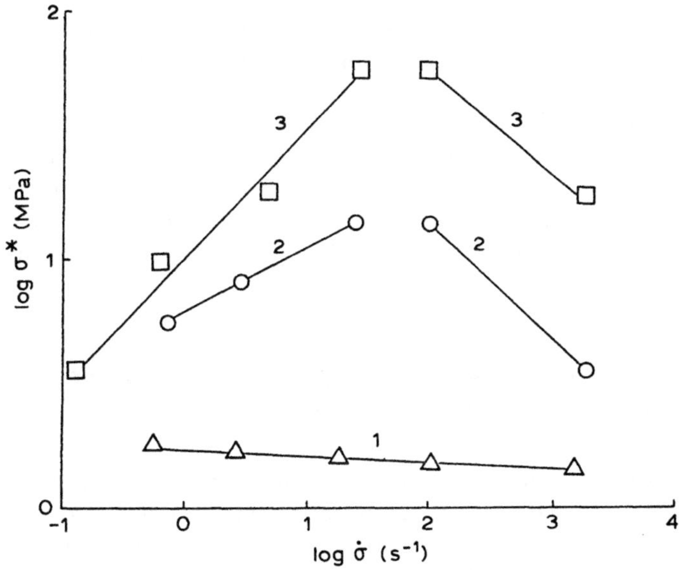

FIG. 2.25. Dependence of true breaking stress σ^* on loading rate $\dot{\sigma}$ for the contact 1,4-PB (with narrow MWD and $M_w = 6{\cdot}4 \times 10^5$) with solids of different natures at 293 K: polytetrafluoroethylene (curve 1), polished steel (curve 2) and polymethyl methacrylate (curve 3).

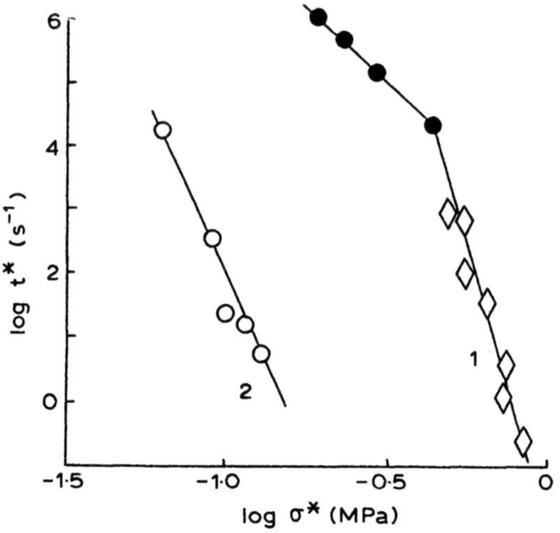

FIG. 2.26. Dependence of long-term durability t^* on the breaking stress σ^* for 1,4-PB (with narrow MWD and $M_w = 6{\cdot}4 \times 10^5$) in contact with steel (curve 1) and polytetrafluoroethylene (curve 2) at 293 K.

temperature, and molecular weight indicates that cohesive deformation in the triaxial state of stress is determined by the relaxation characteristics, in particular the initial viscosity.

The long-term durability for films in triaxial stress is 1–1·5 decimal orders higher, and the breaking stress 2–3 times larger, than the respective values for uniaxial extension. A comparison of stresses and long-term durabilities for triaxial and uniaxial stresses was carried out at approximately equal values of durability and breaking stress, respectively [27, 37].

2.6 RELATION BETWEEN RELAXATION PROPERTIES AND LAWS OF POLYMER FRACTURE

As shown in preceding sections, the achievements in systematization of the results obtained in continuous shear and tension are due in many ways to the use of model systems, viz. polymers with a narrow MWD. Within a broad range of rates of stress application, temperatures, and strains, regardless of the deformation conditions, such systems behave like linear viscoelastic bodies. This has made it possible, without emerging from the framework of linear approximations, to describe the entire set of observed viscoelastic effects including the critical conditions and parameters of deformation responsible for the failure of a polymer body as of a continuum.

In the general case, however, the laws of destruction of a viscoelastic material cannot be limited to the framework of a linear approximation.

2.6.1 Fracture as a Kinetic Process

The necessity of analyzing the process of polymer fracture as a phenomenon having a relaxation nature and characterized by a definite relaxation time had already been indicated in Ref. 38. It was noted in Refs 33, 39 and 40 that the diversity of forms of the kinetic units affecting the process of polymer fracture makes it necessary to characterize this process in the general form not by a single relaxation time, but by a set of times—their spectrum.

It is also very important that in the failure of a polymer body a sharply pronounced inhomogeneity of the stressed state of a system over its volume is possible and, consequently, inhomogeneity of the proceeding relaxation processes. One must also have in view that in the course of fracture of a viscoelastic body, its structure may become rearranged and, consequently, the nature of the relaxation processes will alter because of the replacement of one set of kinetic units by others. This process differs in principle from that of structural relaxation [41–43] in the nature of the kinetic units participating in the structural rearrangement.

The concept of breaking of the chemical bonds by the forces of the main chemical valences as the process determining the mechanical fracture of polymer bodies was formulated in Refs 44 and 45. But it already followed from Ref. 38 that in the fracture, for instance, of vulcanizates, in the general case the contribution of the forces of intermolecular interaction is also quite great and is frequently a factor determining the magnitude of the ultimate strength. According to Ref. 33, a polymer body may fail either because of breaking of bonds at the expense of the forces of the main chemical valences or by overcoming the forces of intermolecular interaction. This depends on the degree of orientation of the macromolecules and the ratio of the duration of action of a force to the time needed to destroy the bonds or ties counteracting this process.

The above information, and also many other results, show that the failure of viscoelastic materials in the rubbery state is a relaxation process due to the mobility of various kinetic elements, and it is expedient to describe it in terms of the relaxation time.

By characterizing the strength of a system by the time from the beginning of load application to the instant of failure t_f, according to Ref. 7 we can obtain a relation of the following kind for all bodies including polymer ones:

$$t_f = t_0 \exp \frac{U - \gamma\sigma_u}{kT} \tag{2.6}$$

where t_0 is the period of vibrations of the kinetic units participating in the process of failure of a body, σ_u is the ultimate stress kept constant by the conditions of an experiment, γ is a 'structure-sensitive' coefficient, and U_0 is the activation energy of an elementary event of failure at $\sigma_u = 0$.

It was shown for various polymer materials (except for elastomers) that the experimentally observed relations are described by linear expressions of the kind

$$\begin{aligned} \ln t_f &= \ln t_0 + \frac{U_0 - \gamma\sigma_u}{R}\frac{1}{T} \\ \ln t_f &= \left[\ln t_0 + \frac{U_0}{R}\right] - \frac{\gamma}{RT}\sigma_u \end{aligned} \tag{2.7}$$

The deviations from the linear relation of these expressions observed for elastomers were generally explained by the change in the 'structure-sensitive' coefficient γ. Constancy of the parameter γ is due only to the features of the studied polymers—highly oriented (in the direction of action of the destroying stress) fibers. As a result, the most specific property of polymers—the ability of the macromolecules to change their shape under the effect of external forces—was not realized. When establishing the laws of elastomer destruction, these structural limitations were absent [33], hence such a feature of the chain structure of the polymer macromolecules as flexibility was realized quite completely. This should naturally also be manifest in the specific nature of the process of destruction of polymer bodies.

It is evident from the above that special attention must be given to studying the process of polymer failure exactly in the rubbery unoriented state, the experiments being conducted in the relevant temperature range. By employing just such an approach and varying the range of deformation rates within broad limits, experimentalists have observed the failure of various polymer bodies. Analysis of ciné films of polymer fracture obtained in polarized and unpolarized light shows that the distribution of the strains in the volume of samples is not uniform.

In the general case, this may be a very serious obstacle in the way of a correct description of the mechanism and laws of polymer fracture. In the cases described in the preceding sections, inhomogeneity in the distribution of strain in the bulk of a polymer body was practically excluded by the use of model objects—monodispersed polymers with a high homogeneity of the physical entanglement network.

Having in view that the destruction of a polymer body being deformed at temperatures above T_g occurs in the region of the rubbery or leather-like states when its fluidity is suppressed, a slightly crosslinked elastomer with a quite regular network of chemical bonds can be adopted as an acceptable model for studying the fracture process. Attention was given to this circumstance in Refs 46 and 47.

2.6.2 The Strength Equation and Its Analysis

Analysis of the process of failure of various unoriented polymers within a broad range of temperatures (by the results of high-speed filming) showed [33] that in uniaxial tension of a polymer in the rubbery state, the ultimate stress is proportional to the rate of breaking of the bonds that prevent the separation of a sample into parts. Processing of the information obtained by high-speed filming of the failure of unoriented polymers in ordinary and polarized light [33] resulted in a quantitative expression relating the magnitude of the ultimate stress σ_u, the rate of bond breaking v, and the temperature T:

$$\sigma_u = Kv \exp\left(\frac{U}{RT}\right) \tag{2.8}$$

Statistical processing of the experimental results showed that for crosslinked systems the values of v and rate of tension V are related by the expression $v = AV^n$, where A and n are the parameters of the approximation equation. Accordingly, eqn (2.8) can be written as

$$\sigma_u = K_0 V^n \exp\left(\frac{U}{RT}\right) \tag{2.9}$$

where $K_0 = KA$.

It was noted in Ref. 48 that eqns (2.8) and (2.9) describe the state of a system in which the molecular mobility of the polymer chains is high, therefore these equations are inapplicable to polymers in the glassy state. Considering the possibility of elastic properties being exhibited by glassy polymers (a forced rubbery state), this prohibition seems to be unjustified.

It would be more correct, in our opinion, to limit the region of application of these equations to conditions under which the ability of macromolecules to change their shape is sufficiently manifest. Study of durability conducted in Ref. 48 for a variety of polymers within a broad temperature range revealed that the temperature–time dependence of the strength of elastomers has the form:

$$t_f = C\sigma_u^{-b} \tag{2.10}$$

where C and b are parameters of the equation.

It also follows from Ref. 48 that the time dependence of the strength of elastomers is not described by formula (2.6), whereas expressions (2.8), (2.9), and (2.10) are related and reflect the laws of elastomer failure in a single way.

The parameters of eqns (2.8) and (2.9) have a definite physical meaning, namely, that U is the activation energy of the failure of a polymer body in the rubbery state or one close to it with uniaxial deformation, and n is a parameter showing the influence of the mobility of the kinetic units on the ultimate strength.

Equation (2.10) can be written in the form [48]:

$$\sigma_u = AV^{1/m} \exp\left(\frac{U'}{mkT}\right) \tag{2.11}$$

When $m = 1 + b$, eqns (2.9) and (2.10) are qualitatively identical.

Since the equation of strength (2.9) is qualitatively similar to the equation describing the viscous flow of a pseudoelastic liquid (the Ostwald de Vill equation), the exponent in the latter case characterizing the rearrangement of the internal structure of a system, it is natural to assume that the exponent in eqn (2.9) also reflects the structural changes occurring in the process of fracture and is connected with the mobility of the kinetic units participating in this process. Indeed, experiments show that the value of n in eqn (2.9) grows with an increase in the flexibility of the polymer macromolecules. For example, in the series including linear polyethylene, a mixture of high- and medium-pressure polyethylenes, polyurethane, and polyethylene terephthalate, the values of n in eqn (2.9) are arranged as follows: 0·12, 0·09, 0·06, and 0·02.

When determining the relaxation properties of polymer bodies and comparing the results of this determination with the parameters of a fracture process, at least two circumstances must be taken into account:

(1) the structure and relaxation characteristics of an elastomer change in the course of fracture, and in the final stage of failure they should

appreciably differ from the initial values (orientation of the macromolecules plays the most substantial role in exactly this stage);

(2) nonequilibrium distribution of the stresses in the bulk of a sample when it is being deformed may cause a considerable difference in the parameters characterizing the mobility of the kinetic units at various points of a polymer body in the prefailure region.

2.6.3 Determination of the Relaxation Characteristics of a Material Being Deformed Including the Fracture Stage

We have already indicated that for polymers with a high molecular weight at temperatures above the glass transition temperature, their viscoelastic behavior is a result of two groups of relaxation processes—slow and rapid ones. The group of slow processes is significant in the flow of polymer solutions and melts, determines the laws of the appearance of large reversible strains, and is associated with the behavior of the polymers when they are being processed. The rapid processes are realized in polymer materials at an adequate intensity of the external action. Deformation in this case causes failure of the system because under these conditions the material can exhibit limited reversible strains.

The change in the structure of a polymer body in the course of its deformation up to failure is naturally attended by a change in the relaxation characteristics of the material. The most complete idea of this can be obtained by studying the change in the spectra of relaxation times or retardation times at various stages of fracture of a polymer body, including the stage of separation of the body into fragments.

Since the existing ways of calculating the spectra of relaxation or retardation times are based on the assumption of linearity of a viscoelastic body, it will be interesting to find a way of correctly appraising the relaxation function in the nonlinear region of deformation up to failure of a material [46].

The changes experienced by the 'initial' relaxation spectra of polymer melts in their shear deformation were described in Ref. 1. The initial spectra implied the distribution function of the relaxation times calculated for the region of low strain values.

The change in the viscoelastic behavior of polymer materials at various values of the static deformations were considered in Refs 46 and 47. Two circumstances were considered. Firstly, since the failure of viscoelastic materials in the general case is beyond the framework of linear approximations [6, 7], one employs the method of direct determination of the relaxation functions from experimental results by their analytical approximation. Secondly, since it is impossible to run relaxation experiments after the failure of a viscoelastic material, the self-contraction in failure was used to appraise the viscoelastic functions in this region.

Preliminary analysis of these two independent approaches has shown that the stress relaxation studied for various stages of deformation of a system up to the values of the strains directly adjoining the last stage, namely separation of a body into fragments, yields a considerable volume of useful information on the mechanism and kinetics of regrouping of the internal structural elements. However, extrapolation of the obtained characteristics for the stage of separation of a polymer body into fragments requires additional substantiation. The problem is that in stress relaxation two oppositely directed forces act on each section of a macromolecule confined between neighboring links of the spatial network and prevent its contraction. At the same time as a sample breaks up into fragments in the last stage of failure of a polymer body, no external forces prevent contraction of the fragments. Consequently, self-contraction reproduces the polymer failure situation in a way differing somewhat from stress relaxation. Regrouping of the kinetic units in self-contraction should proceed, other conditions being equal, more rapidly than in stress relaxation.

An analytical expression of the time dependence of the relaxation modulus E_t can be obtained with the aid of a computer. Here we approximate the experimental data in the following variants:

$$E_t = E_0 \exp(-t/\tau)$$
$$E_t = E_0 \exp(-t^C/\tau)$$
$$E_t = E_\infty + (E_0 - E_\infty) \exp(-t/\tau)$$
$$E_t = E_\infty + (E_0 - E_\infty) \exp(-t^C/\tau) \tag{2.12}$$

$$E_t = E_\infty + (E_0 - E_\infty) \exp(-t/\tau)^C \tag{2.13}$$

A function of the following kind was proposed in Ref. 47 for an analytical description of self-contraction:

$$J_t = K\sqrt{1 + A_1^2 A_2^2}\, \exp(-t/A_2) \sin(A_1 t + \tan^{-1} A_1 A_2) \tag{2.14}$$

where K, A_1, and A_2 are approximate coefficients, J_t is the compliance in self-contraction, $J_t = \varepsilon_t/\sigma_0$; here ε_t is the strain at each moment of time in self-contraction, and σ_0 is the stress at the time $t = 0$, i.e. when a sample fails or is cut up.

Expression (2.14) considers the effect of inertial forces that in addition to contraction of a sample cause it to vibrate. It was found [47] that the parameters A_1 (the frequency of vibrations in self-contraction) and $A_2=\Theta$ (the retardation time in self-contraction) are related by a simple expression, namely $A_1A_2=1$. In other words, the vibrations of samples in self-contraction occur at a frequency numerically equal to the reciprocal of the retardation time. The coefficient K was determined experimentally, while the retardation time in self-contraction was found by computerized approximation of the time dependences of the compliance J_t at various values of ε_0.

It was shown in Ref. 47 that eqns (2.12)–(2.14) give a reliable analytical description of experiments involving stream relaxation and self-contraction. In Ref. 49, these equations were used for approximating the results of experiments obtained for various materials, namely low-density polyethylene, slightly crosslinked elastomers based on nitrile rubber with a diverse homogeneous density of the crosslinks, poly(isobutylene *co* styrene), etc. The materials had different strengths; the stress–strain curves for $\varepsilon=$const. and the stress relaxation curves of σ vs. t for $\varepsilon=$const. were obtained in an Instron tensile-testing machine, and the self-contraction was filmed using a high-speed camera (the speed was $1{\cdot}5\times10^4$ frames per second).

It is convenient to select the approximation coefficient by the method of least squares. Calculations performed [49] for each relaxation curve within a wide range of ε for over 50 values of time covering the interval from 5×10^{-1} to 5×10^3 s revealed that the best approximate description of the experimental relaxation curves within a broad region of values of $\varepsilon=$const. is achieved when employing eqns (2.12) and (2.13). The latter equation enables one to normalize the time axis, i.e. balance the contribution of the experimental values with long and short times. It should be noted that eqn (2.13) for the time dependence of the relaxation modulus is known as the Kohlrausch equation in the literature [50–52]. In the present case, the equation was obtained as a result of approximation of the experimental data. Equation (2.13) was reduced to a linear form:

$$\ln t=\frac{1}{C}\ln\ln\frac{E_0-E_\infty}{E_t-E_\infty}+\ln\tau$$

Analysis of the tension curves for polymers having various strengths in the coordinates σ vs $\varepsilon/\varepsilon_u$ makes it possible to compare the strength of viscoelastic systems and their relaxation properties in the relevant states, i.e. at an equal distance from the place of failure (Fig. 2.27). The elastomer

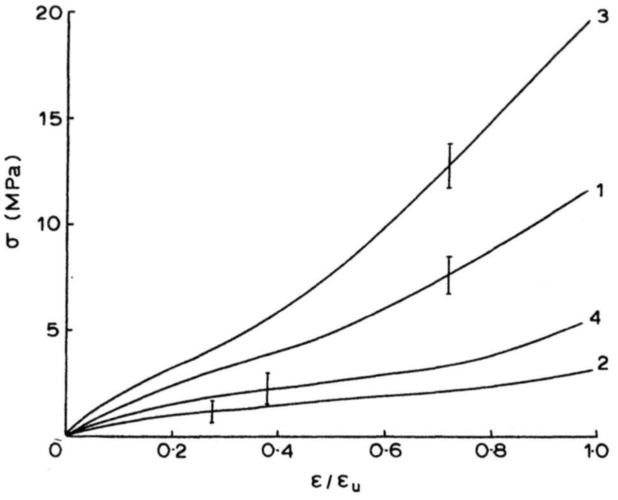

FIG. 2.27. Tension curves for viscoelastic materials of various strengths. The curve numbers correspond to the composition numbers in Table 2.3.

(Table 2.3) containing carbon black as a reinforcing filler (composition 3) had the highest strength. An elastomer crosslinked by dicumyl peroxide with magnesium methacrylate (composition 1) also had an increased strength. Magnesium methacrylate in crosslinking exhibits properties of a reinforcing filler. Compositions 2 and 4 had the lowest strength. The vertical lines on the tension curves indicate the maximum strains at which a relaxation experiment can be run. (These maximum strains will be called 'pre-ultimate' ones below.) For weaker compositions, the pre-

TABLE 2.3
Composition of Elastomer Compounds Based on Rubber SKN-26

Ingredients	*Mass fraction per 100 mass parts of rubber for composition no.*			
	1	*2*	*3*	*4*
SKN-26	100	100	100	100
Dicumyl peroxide	0·5	0·5	0·75	—
Magnesium methacrylate	20	—	—	—
Carbon black PM-75	—	—	40	—
Sulfur	—	—	—	2
Zinc oxide	—	—	—	5

ultimate values are 0·3–0·4 times ε_u, for the stronger ones 0·7–0·8 times ε_u. Figures 2.28 and 2.29 show how the parameters of eqn (2.13) change depending on the strains $\varepsilon/\varepsilon_u$ for the same viscoelastic materials. The change in the parameter $(E_0 - E_\infty)$ with an increasing strain correlates qualitatively with the tension curves shown in Fig. 2.27 (within the entire range of strains the magnitudes of $(E_0 - E_\infty)$ are higher for the stronger compositions 1 and 3). The stress at the relevant points of the tension curves is a measure of the strength of the chemical network at various strain levels. For compositions 2 and 4 whose tension curves have no section of orientation hardening, the values of $(E_0 - E_\infty)$ are constant or grow slightly up to the pre-ultimate state. For the stronger compositions, a similar change in the parameter $(E_0 - E_\infty)$ is observed only up to strains of $0{\cdot}2\varepsilon_u$. Next a sharp linear growth of the quantity $(E_0 - E_\infty)$ occurs that is connected with the orientational stretching of sections of the macromolecule chains between the network links.

The parameter C shows the degree of retardation of the relaxation processes in a real system in comparison with an ideal one modeled by a single Maxwell element for which $C = 1$. The highest value of the parameter C is observed for all materials at low values of $\varepsilon/\varepsilon_u$ (Fig. 2.28).

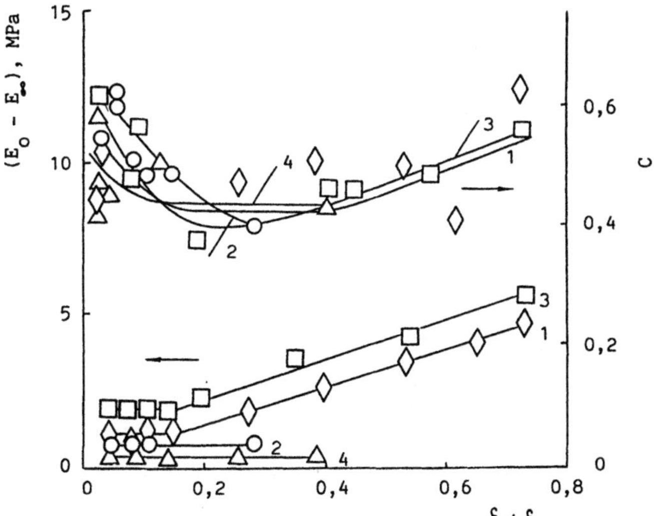

FIG. 2.28. Parameters $(E_0 - E_\infty)$ and C vs $\varepsilon/\varepsilon_u$ for materials of various strengths. The curve numbers correspond to the composition numbers in Table 2.3.

The drop in the parameter with a growth in the strain is associated with the passage from rapid 'Hookean' strains to slow conformational processes and the participation of longer lengths of the polymer chains in relaxation. For the stronger compositions 3 and 1 at $\varepsilon/\varepsilon_u > 0{\cdot}4$, the parameter C grows. This is connected with the orientation hardening of the material.

The relations τ vs $\varepsilon/\varepsilon_u$ are more involved (Fig. 2.29). The extremum at low strains witnesses a transition from purely elastic to viscoelastic strains. At values of $\varepsilon/\varepsilon_u > 0{\cdot}2$, the nature of the change in the parameter τ depends on the polymer strength.

The difference in the curves of τ vs $\varepsilon/\varepsilon_u$ for polymers varying in the chemical structure of their network reflects the features of their behavior in deformation. When a material with a strong network is stretched, sections of chains of various lengths are gradually involved in deformation, next comes their orientational hardening, and, finally, failure. In the stretching of a material with a weaker network, deformation on the chains is accompanied by rupture of both the chemical and physical fluctuation bonds.

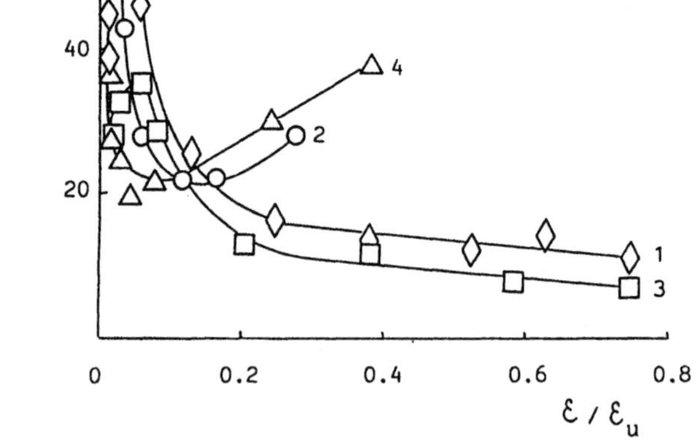

FIG. 2.29. Relaxation times τ vs $\varepsilon/\varepsilon_u$ for materials of various strengths. The curve numbers correspond to the composition numbers in Table 2.3.

In view of the foregoing, one should expect a definite relation between the relaxation time τ and the stress appearing in the process of deformation, and this is what is actually observed (Fig. 2.30). The stress changes for two reasons: (i) the strain $\varepsilon/\varepsilon_u$ increases, and (ii) the chemical structure of the network changes at $\varepsilon/\varepsilon_u = \text{const}$. It can also be seen that with a growth in the stress σ regardless of what caused this growth, the relaxation time τ diminishes, asymptotically approaching the value of $\tau = 9$–10 s. A comparison of Figs 2.29 and 2.30 shows that the plot of τ vs σ is universal because it reflects the properties of polymers with a diverse chemical structure of their network.

Consequently, analysis of how the parameters of the relaxation equation depend on the magnitude of the strain and stress points to the existence of a clear relation between the relaxation and strain–strength properties of a material, including the region of nonlinear viscoelasticity. Experiments on stress relaxation, however, allow this relation to be traced within a limited interval of ε_u, but they do not yield this relation in failure of a material. In the latter case, experiments involving self-contraction are more fruitful (Figs 2.31 and 2.32).

A comparison of experimental information on stress relaxation and self-contraction reveals that the nature of these relations coincides qualitatively. However, the magnitudes of the relaxation time τ and the retardation time Θ differ appreciably, which is connected with the features of these processes. The qualitative similarity of these laws, also

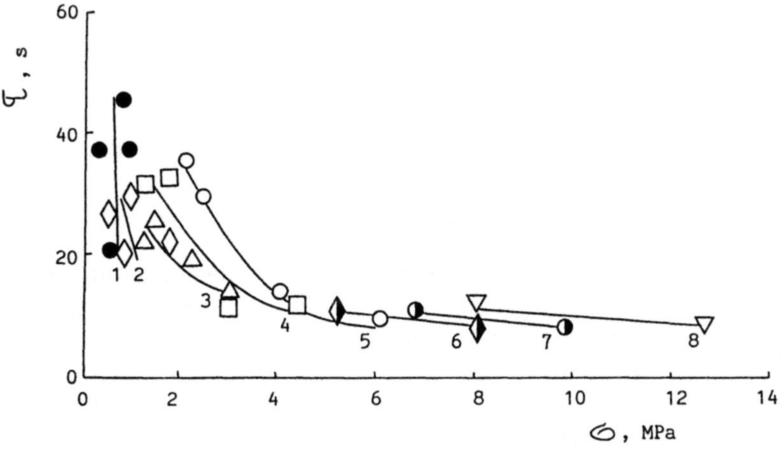

FIG. 2.30. Relaxation times τ vs σ for materials of various strengths. Curves 1–8 correspond to values of $\varepsilon/\varepsilon_u$ of 0·05, 0·1, 0·2, 0·3, 0·4, 0·5, 0·6 and 0·7, respectively.

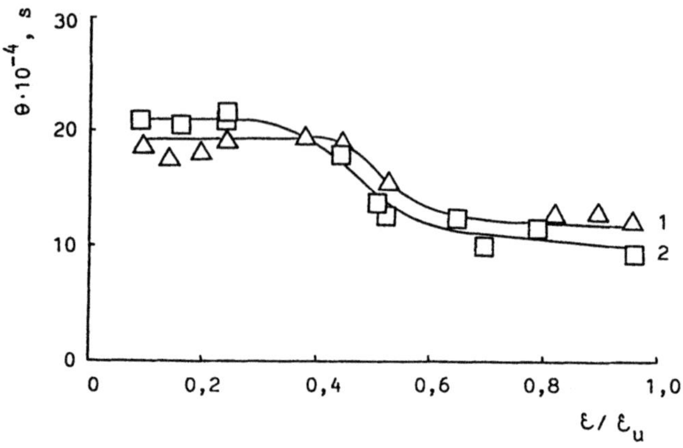

FIG. 2.31. Retardation time θ in self-contraction vs strain $\varepsilon/\varepsilon_u$ for compositions 4 (curve 1) and 3 (curve 2) in Table 2.3.

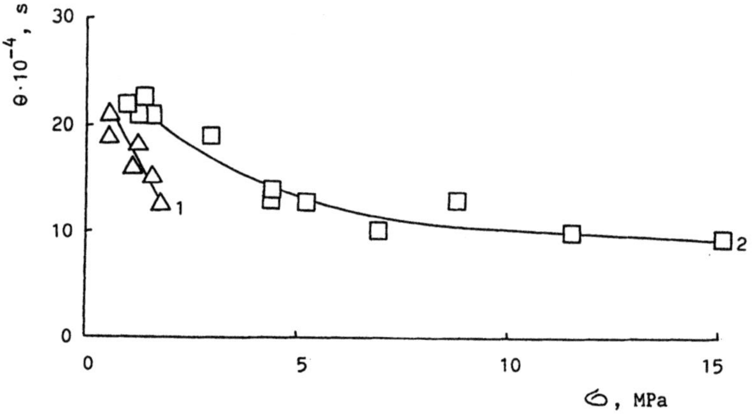

FIG. 2.32. Retardation time θ in self-contraction vs stress σ for compositions 4 (curve 1) and 3 (curve 2) in Table 2.3.

observed for other polymers, indicates their common nature. But this does not at all signify the absence of distinctions in the rates of stress relaxation and self-contraction processes.

When analyzing the obtained results, it is useful to again revert to the foregoing and widely known strength equation (2.9) relating the ultimate stress σ_u, rate of tension V, and temperature T. Taking into account the set of experimental results, including those given above, indicating a change in the 'curtailment' of the time responsible for the conformational movement of chains in the passage from a slightly strained state of a material to the pre-ultimate one, the parameter n should be presented in the form of the function $n=f(\Theta)$, where Θ is the retardation (or relaxation) time relating to the last failure stage. This allows us to write eqn (2.9) in the form

$$\sigma_u = K_0 V^{n(\Theta)} \exp(U/RT) \tag{2.15}$$

In summarizing, we can state that the fracture of polymer bodies in the general case is determined not only by the strength of the bonds at the expense of the basic chemical affinity and intermolecular interaction, but also by the relaxation characteristics at both the initial stage of loading and the final stage of failure.

We find that the most effective information on the relaxation properties of a polymer in the failure stage can be procured when analyzing the self-contraction of a sample after failure.

REFERENCES

1. Vinogradov, G.V., Yanovsky, Yu.G. and Isaev, A.I. *J. Polym. Sci.*, **8**(A2) (1970) 1239.
2. Vinogradov, G.V., Yanovsky, Yu.G. and Malkin, A.Yu. *Vysokomol. Soed.*, **14A**(11) (1972) 2425.
3. Graessley, W.W. *Advance Polymer Science*, Springer, Berlin, 1974, Chapter 16.
4. Oser, H. and Marvin, R.S. *J. Res. Nat. Bur. Standards*, **B67** (1963) 87.
5. Vinogradov, G.V., Malkin, A.Ya. and Kulichikhin, V.G. *J. Polym. Sci.*, **8**(A2) (1970) 333.
6. Gross, B. *Mathematical Structure of the Theories of Viscoelasticity*, Hermann, Paris, 1953.
7. Ferry, J.D. *Viscoelastic Properties of Polymers*, 3rd ed., Wiley, New York, 1980.
8. Graessley, W.W. *J. Chem. Phys.*, **54** (1971) 5143.
9. Vinogradov, G.V., Pokrovsky, V.N. and Yanovsky, Yu.G. *Rheol. Acta*, **11** (1972) 258.
10. Vinogradov, G.V., Yanovsky, Yu.G., Malkin, A.Ya., *et al. J. Polym. Sci.*, **10**(A2) (1972) 1061.
11. Pokrovsky, V.N., Yanovsky, Y.G. *Rheol. Acta*, **12** (1973) 280.
12. Yanovsky, Yu.G. *Int. J. Polym. Mater.*, **8** (1980) 257.
13. Yanovsky, Yu.G. and Vinogradov, G.V. *Vysokomol. Soed.*, **22A**(11) (1980) 2567.
14. Tobolsky, A.V. and McLoughlin, J.R. *J. Polym. Sci.*, **8** (1952) 543.
15. Boyer, R.E. *Eng. Polym. Sci.*, **17** (1981) 661.
16. Plazek, D.J. *J. Polym. Sci.*, **20**(9) (1982) 1533.

17. VINOGRADOV, G.V. and MALKIN, A.YA. *Rheology of Polymers*, Mir Publishers, Moscow, 1980.
18. VINOGRADOV, G.V., YANOVSKY, YU.G., *et al. Polym. Eng. Sci.*, **20**(17) (1980) 1138.
19. ONOGI, S., MASUDA, T. and KITAGAWA, K. *Macromolecules*, **3** (1970) 3.
20. ROCHEFORT, W.E., SMITH, G.G., RACHAPUDY, H., *et al. J. Polym. Sci., Polym. Phys. Ed.*, **17** (1979) 1197.
21. RAJU, V.R., SMITH, G.G., MARIN, G., *et al. J. Polym. Sci., Polym. Phys. Ed.*, **17** (1979) 1183.
22. YANOVSKY, YU.G., POKROVSKY, V.N., KOKORIN, YU.K., *et al. Polym. Sci. SSSR*, **30**(5) (1988) 1037.
23. NIELSEN, L.E. *Mechanical Properties of Polymers and Composites*, Marcel Dekker, New York, 2 vols, 1974.
24. YANOVSKY, YU.G., VINOGRADOV, G.V. and IVANOVA, L.I. *DAN SSSR*, **282**(5) (1985) 1190.
25. VINOGRADOV, G.V. *Pure and Appl. Chem.*, **26** (1971) 423.
26. VINOGRADOV, G.V., PROTASOV, V.P. and DREVAL, V.E. *Polym. Bull.*, **10** (1983) 274.
27. YANOVSKY, YU.G., VINOGRADOV, G.V. and DREVAL, V.E. *Trans. CSME*, **8**(2) (1984) 84.
28. VINOGRADOV, G.V., DREVAL, V.E. and YANOVSKY, YU.G. *Rheol. Acta*, **24** (1985) 574.
29. YANOVSKY, YU.G. *L'actualité chimique*, March–April 1991, p. 109.
30. VINOGRADOV, G.V., DREVAL, V.E., BORISENKOVA, E.K., *et al. Rheol. Acta*, **20** (1981) 433.
31. BORISENKOVA, E.K., DREVAL, V.E., VINOGRADOV, G.V., *et al. Polymer*, **23** (1982) 91.
32. REGEL, V.R., SLUTSKER, A.I. and TOMASHEVSKY, E.E. *Kineticheskaya Priroda Razrusheniya Tverdykh Tel* (Kinetic Nature of the Strength of Solid Bodies) Nauka, Moscow, 1974.
33. GUL', V.E. *Struktura i Svoiistva Polimerov* (Structure and Strength of Polymers) Khimiya, Moscow, 1978.
34. BARTENEV, G.M. and ZUEV, YU.S. *Prochnost i Razrushenie Vysokoelastichnykh Materialov* (Strength and Fracture of High Elastic Materials), Khimiya, Moscow, 1964.
35. VINOGRADOV, G.V., *et al. J. Polym. Sci., Polym. Phys. Ed.*, **13** (1975) 1721.
36. KURBANALIEV, M.K., VINOGRADOV, G.V., DREVAL, V.E., *et al. Polymer*, **23** (1982) 100.
37. VINOGRADOV, G.V., EL'KIN, A.I. and SOSIN, S.E. *Polymer*, **19** (1978) 1458.
38. GUL', V.E., SIDNEVA, N.YA. and DOGADKIN, B.A. *Kolloid. Zh.*, **13**(6) (1951) 425.
39. LIBOWITZ, H. (Ed.). *Fracture of Non-Metals and Composites*, Academic Press, New York, 1972.
40. KAUSCH, H.H. *Polymer Fracture*, Springer-Verlag, Berlin, Heidelberg, New York, 1978.
41. TANG, C.L., VIJAYAN, K. and PAE, K.D. *Polymer*, **29**(3) (1988) 410.
42. JOHARI, G.P., *et al. J. Polym. Sci., Polym. Phys., Ed.*, **268**(9) (1988) 1923.
43. ROY, S.K., KYN, T. and MANLEY, R.S. *Macromolecules*, **B21**(6) (1988) 1741.

44. BUECHE, F. *J. Appl. Phys.*, **26** (1955) 1133.
45. BUECHE, F. *J. Appl. Phys.*, **29** (1958) 1231.
46. GUL', V.E., YANOVSKY, YU.G., SHAMRAEVSKAYA, T.V., *et al. DAN SSSR*, **294**(4) (1987) 905.
47. GUL', V.E., SHAMRAEVSKAYA, T.V., YANOVSKY, YU.G., *et al. DAN SSSR*, **297**(1) (1988) 125.
48. BARTENEV, G.M. *Prochnost i Mechanizm Razrusheniya Polimerov* (Strength and Mechanism of Polymer Structure), Khimiya, Moscow, 1984.
49. GUL', V.E., YANOVSKY, YU.G., SHAMRAEVSKAYA, T.V., *et al. Mekhanika Komposit. Mater.*, **4** (1992) 242.
50. SLONOMSKY, G.L. *J. Tekhn. Phys.*, **9**(20) (1939) 1791.
51. ASKADSKII, A.A. *Deformatsiya Polimerov* (Deformation of Polymers), Khimiya, Moscow, 1973.
52. ASKADSKII, A.A. Physical properties of polymers. In: *Encyclopedia of Fluid Mechanics*, Ed. N.P. Cheremisinoff, Gulf Publishing, Houston, TX, 1990, Vol. 9, Chapter 2, p. 103.

Chapter 3

Rheological and Relaxation Properties of Polymer Blends

3.1 INTRODUCTION

Blending is one of the most available and effective technological ways of modifying the properties of polymers. It improves the workability of a given material and imparts the required properties to it. The components of a blend are generally incompatible systems. According to numerous publications, the properties of blend both when melted and when solid depend on many factors. They include the chemical nature of the polymers, the composition of the blends, the level of rheological action, and the influence of additives making the components compatible.

Depending on the deformation conditions, the blends may exhibit various types of behavior, either typical of each component or advantageously differing therefrom. It is often quite difficult to predict the possible consequences of the not always predictable anomalies of their properties. The latter are due to the phase structure (its formation is determined by the phase diagram), the preparation conditions, and the extent of completeness of phase stratification in the steps of preparation, processing, and service of a material. The technology of producing articles from blend compositions is very important in forming the physical and mechanical properties of the latter. When producing films from a solution, the phase structure is greatly affected both by the chemical nature of the solvent and by the viscoelastic properties of the fluid system as a whole. When producing blends from a melt, a major role is played by technological parameters such as the blending and processing temperatures, shear stress in the blending and processing equipment, and the rheological parameters of the components and their ratio.

The prognosis of the physical, mechanical, structural, and morphological properties of blends thus requires deep scientific analysis. It includes, in particular, (i) plotting a phase diagram of the blend and studying the kinetic characteristics determining the phase transition process (diffusivities), and (ii) studying the features of the rheological behavior. On this basis, one can trace the relation between the structural and morphological features of a system, on the one hand, and its properties on the other. Unfortunately, only a few such publications have appeared (see, for example, Refs 1–3). Most authors, however, have studied either the structure of a specific composite and its properties, or the thermodynamic parameters of a blend, without analyzing its morphology, kinetic stability, and properties. Such an approach did not yield the required degree of generalization and reveal the relation between the thermodynamic characteristics, the structure, and the rheological and relaxation properties. In the present section, as will be seen below, a definite attempt is being made to fill this gap.

A strict quantitative description of the flow of blend composition requires a high level of working out of the theoretical ideas combining the thermodynamic, structural, and relaxation aspects of the problem. The thermodynamic appraisals must predict the limits of mutual solubility of the components, the interphase tension, and distribution of the components at the interface; the structural appraisals must predict the size and shape of the phases under the given conditions of blend preparation; and the relaxation appraisals must embrace the features of viscoelastic behavior in connection with the set of the first two factors.

Since the theoretical problems in this field are still being solved (even for the flow of the very simple two-phase blends), all one can perform practically is a qualitative or semi-quantitative analysis of rheological data based on comparing characteristics of blends (including thermodynamic ones) with their viscosity and, more rarely, with the parameters determining the elastic properties. When explaining the discovered laws, one generally has to concentrate attention on only one of the acting factors without considering its relation to the others. For example, when considering the thermodynamic interaction of polymers on the one hand, and the viscosity of their blends on the other, in the best case one can state that with a higher affinity of the components the viscosity of the blend exceeds the additive values, and with a lower one, it is smaller than these values. In practice, however, in most cases an

S-shaped viscosity–composition relation is observed with a varying deviation from the additive values [4]. The overwhelming majority of authors explain this only by the change in the magnitude of the interaction between the molecules in a system. Such a unilateral approach fails to consider the different extents of deformation of phases with different viscosities, the influence of the shape and size of the particles of the dispersed phase, the interface, and also phase inversion [5]. Simplification of the complicated picture of the blend flow process makes one doubt the correctness of generalizing the laws established in specific cases.

We can thus state that, although the observed experimental property–composition relations do reflect in a definite way the changes in the internal structure of a system, intermolecular interaction, and so on, a strict physical interpretation of the processes is possible only on the basis of a sufficiently substantiated theoretical model.

If we revert to the 'purely rheological problems' when studying the flow of blend compositions, the following must be noted.

3.2 GENERAL PROBLEMS OF DESCRIPTION OF POLYMER BLEND PROPERTIES

An enormous number of publications differing in both their tasks and their ideology have been devoted to the rheology of polymer blends. Here the influence of diverse theoretical ideas and directions of rheological schools is clear. The close attention to these systems was caused primarily by the interests of practical work, namely, by the search for a rational technology of polymer processing and ways of directed variation of the properties of articles fabricated from polymer materials. This is why publications on applied rheology of polymer blends prevail. The typical objects of study include the most diverse variants of blend compositions whose choice was determined by specific goals.

When analyzing these publications as a whole, it is quite difficult to systematize them in detail, and moreover to establish the general laws of the viscoelastic behavior of objects within a broad range of concentrations of the component, temperatures, rates, conditions of deformation, etc.

It was exactly the need to predict sufficiently general rules of the change in the rheological properties of melts of polymer blends that led to the running of experimental studies aimed at obtaining a number of empirical relations between the change in some rheological constants on the one hand, and the individual features of the blend components and their concentration on the other.

However, the most complete system of generalizing rheological data must be based on definite theoretical ideas whose experimental verification requires the running of special studies meeting the demands of a model approach to the greatest possible extent. The latter necessitates the use of well-characterized model objects that can also exhibit an anomaly of their viscoelastic behavior in deformation like that of high-molecular-weight compounds. Moreover, a relatively complete description of the relaxation properties of such systems can be obtained on the basis of experiments within a sufficiently broad range of temperatures and frequencies.

The procuring of the essential combination of rheological information and the 'minimal' set of viscoelastic constants for describing the most typical features of blend composition behavior is important in all cases, whether applied studies or a model approach. This requires, for example, an analysis of the characteristics obtained (i) under conditions of continuous deformation in steady flow, and (ii) with low-amplitude periodic deformation; in other words, an analysis (i) of how the tangential and first difference of the normal stresses depend on the deformation rate and stress, and their relation, and (ii) how the dynamic moduli of storage and loss depend on the deformation frequency. Studies under continuous and periodic deformation conditions yield mutually supplementary information and are consistent in interpretation of the results. It is also good practice to calculate the initial constants of a system, viz. the initial viscosity and elasticity and the characteristic values of the viscosity and elasticity (their determination for polymer blends will be introduced below).

Rheological studies can determine the integral characteristics of a system that undoubtedly depend on the individual features of its components. Nevertheless, when considering below the rheological properties of single- or two-phase polymer systems, or mutually soluble and insoluble, one must have in view that these concepts concern primarily not the thermodynamic, but the kinetic aspects of the problem because they are analyzed from the viewpoint of manifestation of the main relaxation process—glass transition. In our opinion, these approaches must not be opposed as is sometimes encountered in the literature, and no discussions should be opened as to which of them is more physical or which criteria are better. The problem consists in finding mutual correlation between the thermodynamic and kinetic appraisals. It is awaiting solution, and the author hopes that the present publication will facilitate this.

3.3 QUALITATIVE APPRAISALS OF STRUCTURAL–MORPHOLOGICAL PROPERTIES OF BLENDS: CLASSIFICATION

To discuss rheological data covering the most typical kinds of polymer blends, let us revert to their classification [6]. A rough classification of blends divides them into homogeneous (miscible or 'compatible') and heterogeneous (immiscible of 'incompatible') ones. The former give birth to single-phase systems in which the individual components are mutually soluble in one another. The properties of such objects generally obey the law of mixtures, though their physical and mechanical properties are sometimes superior to those observed in the individual components. Heterogeneous blends have two different types of dispersed state, viz. (1) one component in the form of a continuous phase and the other in the form of discrete one, and (2) individual components each in the form of a continuous phase initiating the mutually penetrating state of the dispersion.

To date there are no sufficiently reliable rheological criteria and theories that would predict which of the two components will be present in the form of the continuous or discrete phase, though definite attempts have been made in this direction [7, 8].

Of major significance for the rheology of heterogeneous blends, when the decisive role is played by their morphology, are, firstly, the method of preparing blend samples, mixing intensity, temperature, stresses, and other factors, and, secondly, the conducting of a comprehensive rheological analysis, e.g. the combination of continuous and low-amplitude periodic deformation. It is also doubtless that data on the morphology must also be invoked in addition to rheological information for interpretation of the results.

Limitedly miscible (or limited compatible) blend compositions are of separate interest. Depending on the concentration of the components or the temperature in a melt or solution, they exhibit properties of either a single- or a two-phase system. Attention will also be devoted below to the rheophysical aspects of the behavior of such compositions.

The compatibility of polymers is sometimes determined by the behavior of their solutions, in particular by the ability of the latter to stratify.

The decisive role is played here not only by the nature of the components, but also by their ratio in the blends. Of major interest is the region with a low content of one component. It will be shown below that here is exactly where one can expect the sharpest change in the properties of a blend composition. Klykova *et al.* [1] and Yanovsky *et al.* [9] showed, in particular, that with a low content of one component there may be no stratification at all. The influence of the molecular weight on the ability of a mixture to stratify has also been considered [2].

Stratification may also be observed in the absence of a solvent, however. This can be exemplified by a blend of fractions of the same homological series differing considerably in molecular weight and exhibiting features of a two-phase structure [10, 11]. But here in a definite range of concentrations, such a system can be considered as a solution in which the component with the higher content by volume (or weight) is the solvent.

Proceeding from the above rough (though generally adopted, from the rheological viewpoint) classification of the types of blend compositions [6, 8], we shall analyze schematically their rheological behavior.

Blends of fractions of various molecular weights (MW) of the same polymer are a typical example of compatible, i.e. homogeneous, blend compositions. Rheological studies of such systems have been described in sufficient detail in the literature (see, for example, Refs 6, 8 and 9). The typical rheological relations obtained in this case have the form shown schematically in Fig. 3.1. They include curves determined with continuous deformation under steady flow conditions, i.e. plots of the apparent viscosity $\eta = \tau_{12}/\dot{\gamma}$ and the first difference of the normal stresses $(\tau_{11} - \tau_{22})$ versus the strain rate $\dot{\gamma}$; and characteristics measured with periodic low-amplitude (the linear region) deformation, i.e. plots of the dynamic viscosity $\eta' = G''/\omega$ and the modulus of storage G' versus the cyclic frequency ω. All the curves are monotonic. The curves for blends are between those for the initial components (Fig. 3.1(a) and (b)).

Inspection of Fig. 3.1(a) and (b) reveals that at low values of $\dot{\gamma}$ and ω (the region of Newtonian flow), $\eta_0 = \eta'_0 = \text{const}$. With an increase in $\dot{\gamma}$ and ω, the curves of η' and G' vs ω are lower than η and $(\tau_{11} - \tau_{22})$ vs $\dot{\gamma}$. Such a law is typical of single-phase polymer systems (melts, solutions) and is explained by the specific nature of manifestation of an anomaly in the viscoelastic behavior when they are deformed under these two different conditions (continuous and periodic). The correlation of rheological data measured with continuous and periodic low-amplitude deformation has been treated in sufficient detail elsewhere [12].

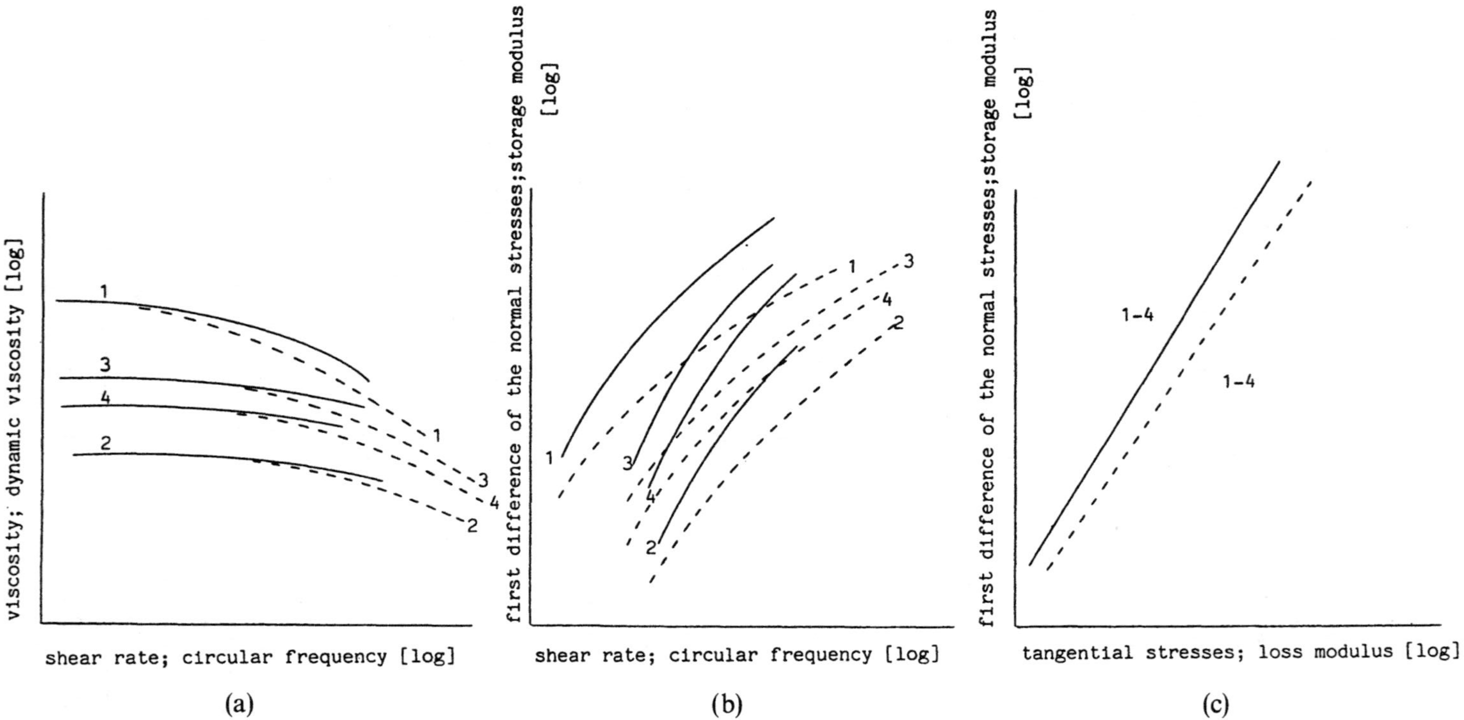

FIG. 3.1. Typical rheological relations observed when studying binary blends of compatible polymers with continuous (solid lines) and periodic (dashed lines) deformation. Curves 1 and 2 depict the initial components, 3 and 4 their blends in various proportions.

According to the scheme in Fig. 3.1(b), useful generalized information on the ratio of the elastic and viscous (or elastic and dissipative) components in the deformation of a viscoelastic medium can be obtained from plots of $(\tau_{11}-\tau_{22})$ vs τ_{12} and G' vs G''. For single-phase melts (or solutions) of polymer blends within a definite interval of $\dot{\gamma}$ and ω, these plots are linear for both homopolymers and mixtures thereof regardless of the composition. Moreover, the straight lines $(\tau_{11}-\tau_{22})=f(\tau_{12})$ and $G'=f(G'')$ (dashed line) uniquely describe the properties of melts of both individual components and blends thereof, i.e. they are invariant relative to the blend composition. It is also significant that these plots are invariant with respect to the temperature too. To mention quantitative consistency, the values of $(\tau_{11}-\tau_{22})$ are higher than G' (see Fig. 3.1(b)). Chuang and Han [6] indicate that for blends of high-density polyethylene fractions this difference within a definite interval of $\dot{\gamma}$ and ω is a double or triple one. It should be noted that such results are well consistent with the conclusions following from the linear theory of viscoelasticity and its generalizations by which [12]

$$\lim_{\dot{\gamma}\to 0}[(\tau_{11}-\tau_{22})/\dot{\gamma}^2]=\lim_{\omega\to 0} 2G'/\omega^2$$

If we turn to the temperature dependences of the loss tangent tan δ for such objects, they are characterized first of all by one main peak corresponding to the glass transition temperature of the system, T_g.

Let us next consider schematically the rheological behavior of melts of limitedly compatible blends (Fig. 3.2). As in the preceding case (Fig. 3.1), Fig. 3.2 presents characteristics of single-phase systems, i.e. corresponding to regions of the component composition in which the blends are still compatible. A comparison of Figs 3.1 and 3.2 shows that the pictures of the change in the elastic and viscous properties of the system versus $\dot{\gamma}$ and ω are similar only qualitatively, while quantitatively they differ quite clearly. This is manifest not only in that for the blends even at low values of $\dot{\gamma}$ and ω (the region of Newtonian flow) we have $\eta_0 \neq \eta'_0$, but also in that the anomaly in the viscoelastic behavior on the plots of η' vs ω stands out much more than on η vs $\dot{\gamma}$. Although for these blends too the plots of $(\tau_{11}-\tau_{22})/\dot{\gamma}^2$ vs τ_{12} and G' vs G'' are of a linear nature, they do not coincide with similar relations for the individual components, i.e. here the invariance relative to the composition is not retained (though the invariance relative to the temperature is observed as previously). Correlation of the elastic properties registered in continuous and periodic deformation is also not observed, i.e. $\lim_{\dot{\gamma}\to 0}(\tau_{11}-\tau_{22})/\dot{\gamma}^2 \neq$

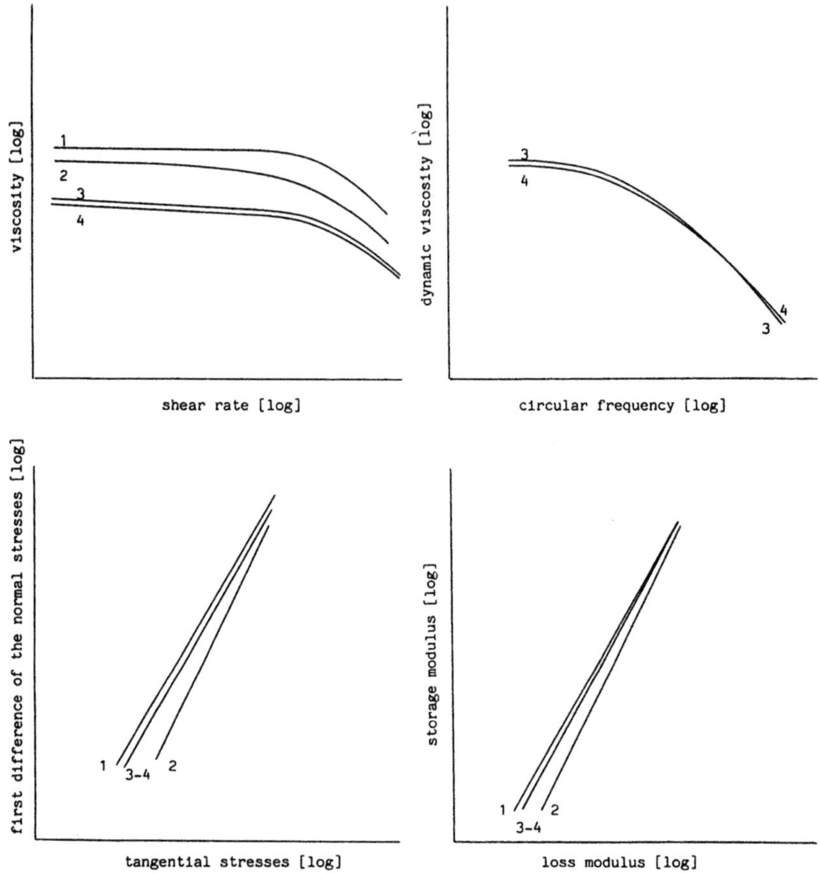

FIG. 3.2. Typical rheological relations observed when studying binary blends of limitedly compatible polymers with continuous and periodic deformation. The designations of the curves are the same as in Fig. 3.1.

$\lim_{\omega \to 0} 2G'/\omega^2$. The lack of correlation in continuous and periodic deformation for such systems is a consequence of the microheterogeneity of the melt [12, 13]. It is significant that, as shown by the scheme, it is quite difficult to reach such a conclusion on the basis of the results of only one of the rheological experiment types considered above.

The plot of tan δ vs T for limitedly miscible systems in the region of the compositions being considered has only one peak connected with glass transition. It occupies an intermediate position between the T_g values of the initial components. It is important here that the glass transition temperature of the blend $(T_g)_{bl}$ does not follow the additivity rule, i.e. $(T_g)_{bl} \neq W_1 T_{g,1} + W_2 T_{g,2}$, where W_1 and W_2 are the weight fractions of the components.

An example of a limitedly consistent blend is that of poly(methyl methacrylate) (PMMA) and poly(vinylidene fluoride) (PVDF). Their rheological properties have been described quite well in the literature (see, in particular, Ref. 6, while Privalko *et al.* [14] analyze their thermodynamic and relaxation characteristics. Chuang and Han [6] indicate that the viscosities of PMMA and PVDF blends are lower than those of the individual components (i.e. the rule of logarithmic additivity is not observed). This may be related to the features of their internal structure. These matters will be considered below in greater detail.

When dealing with incompatible, i.e. heterogeneous, melts of blends, one must first of all note the diverse types of their rheological behavior. Figure 3.3 shows what is in our opinion the most characteristic type. For blends, the region of Newtonian flow is reached with much more difficulty at lower values of $\dot{\gamma}$ and ω than for the individual components. The rheological curves may have breaks and inflections corresponding to the manifestation of the individual features of the components in a certain region of temperatures and rates (or frequencies) of deformation.

There is no correlation of the plots of η vs $\dot{\gamma}$ and η' vs ω, as well as of $(\tau_{11} - \tau_{22})$ vs τ_{12} and G' vs G'' for the blends. The rheological curves for the blend compositions depending on the component concentration may change places (inversion). This points to the complicated nature of their change as a function of the composition. For heterogeneous blends, there is no invariance of the plots of $(\tau_{11} - \tau_{22})$ vs τ_{12} and G' vs G'' relative to the composition, but invariance is retained relative to the temperature.

Hence, for heterogeneous blends, the observed type of their rheological behavior depends not only on the initial morphology, but also on the kind of rheological tests. With periodic low-amplitude (nondestructive) deformation, the properties of a system unchanged by external action are registered, which tells on both the quantitative and qualitative appraisal.

When analyzing relaxation transitions in heterogeneous systems, we must note first of all the presence of several main relaxation peaks on the curves of $\tan\delta$ vs T corresponding to temperatures close to T_g of the initial components. Depending on the nature of the components, their concentration, frequency of testing, etc., there may be various laws of the change in these glass transition temperatures [15].

We can conclude from the above schemes and from an analysis of published information that the rheological characteristics of blends

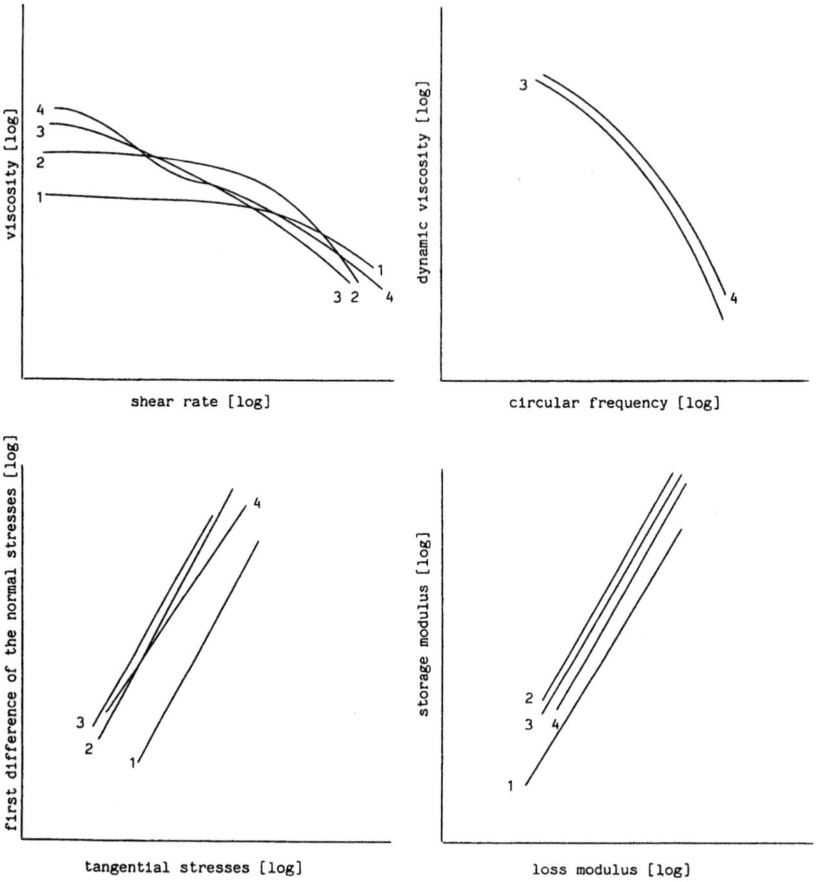

FIG. 3.3. Typical rheological relations observed when studying binary blends of incompatible polymers with continuous and periodic deformation. The designations of the curves are the same as in Fig. 3.1.

change in quite a complicated way depending on the rate (or frequency) and stress (or amplitude) of the deformation, temperature, blend composition, and many other factors, which hinders an analytical description of the obtained laws. The picture becomes somewhat simpler if we analyze only the initial rheological parameters of a system. The literature contains quite a few empirical expressions describing the changes in the initial viscosity of a blend composition depending on the content of the components, and procured by processing a large amount of experimental results.

The first empirical expression describing the dependence of the viscosity of liquid mixtures on their composition was already proposed by Arrhenius, namely,

$$\ln \eta_{\text{mix}} = N_1 \ln \eta_1 + N_2 \ln \eta_2 \tag{3.1}$$

where N_1 and N_2 are the mole fractions of the components, and η_1 and η_2 are their viscosities. Horio *et al.* [16] have proposed an empirical expression for the initial (maximum) Newtonian viscosity of a two-component mixture depending on the mass fractions of the components W_1 and W_2:

$$\log(\eta_0)_{\text{mix}} = W_1 \log \eta_{0,1} + W_2 \log \eta_{0,2} \tag{3.2}$$

Expressions (3.1) and (3.2) presume a logarithmic additive dependence of the viscosity. Generally, it is quite rare for such a relation to be observed experimentally, e.g. for the flow of blends of fractions of a single polymer [11]. In practice for the flow of compatible (single-phase) blends, one may encounter various cases of a change in η_0 with the composition of the blend components either exceeding the additive values or being lower than them. The causes may be quite diverse [5, 8, 14].

According to the literature (see, for example, Refs 5 and 8, the composition dependence of the initial viscosity of single-phase blends of polymers can be described by various empirical expressions. For instance, an expression has been proposed [17, 18] for the viscosity of a blend composition consisting of fractions of the same polymer differing in molecular weight. According to Kuleznev *et al.* [17]:

$$(\eta_0)_{\text{bl}} = (W_1 \eta_{0,1}^{1/\alpha} + W_2 \eta_{0,2}^{1/\alpha})^{\alpha} \tag{3.3}$$

while according to Ninomija and Ferry [18]:

$$(\eta_0)_{\text{bl}} = \frac{D \eta_{0,1}^{V_1} \eta_{0,2}^{V_2} [1 + V_2(N_{21} - 1)]}{(N_{21})^{V_2}} \tag{3.4}$$

where $(\eta_0)_{\text{bl}}, \eta_{0,1}$ and $\eta_{0,2}$ are the viscosities of the blend and the two initial components; V_1 and V_2, or W_1 and W_2, are the bulk or mass fractions of the components, respectively; D and N_{21} are dimensionless numbers [18]; and α is a constant [17]. Having in view that for compatible blends in the region of Newtonian flow $(\eta_0)_{\text{bl}} = (\eta'_0)_{\text{bl}}$, we can see that expressions (3.2)–(3.4) also hold for calculating results obtained by dynamic experiments, i.e. under conditions of periodic low-amplitude deformation.

If we turn to the relations describing the change in the Newtonian viscosity of heterogeneous blend compositions, here a still greater diversity of forms of the empirical expression is observed. However, attempts to extend these expressions beyond the bounds of the specific examples for which they had been established are generally not successful. If we consider the specific expressions for calculating the initial viscosity of heterogeneous compositions, they differ from (3.1)–(3.4) primarily in their considerable 'unwieldiness', indefinite concentration and temperature limits of capacity for work, low efficiency with a change in the blend composition, the nature of the components, etc. An example is the expression obtained by Uemura and Takayanagi [19] for calculating the magnitude of the complex dynamic viscosity of a two-phase blend composition:

$$|\eta^*_{\text{bl}}| = |\eta^*_1| \frac{3|\eta^*_1| + 2|\eta^*_2| - 3(|\eta^*_1| - |\eta^*_2|)V}{3|\eta^*_1| + 2|\eta^*_2| + 2(|\eta^*_1| - |\eta^*_2|)V} \tag{3.5}$$

where $|\eta^*_{\text{bl}}|, |\eta^*_1|$ and $|\eta^*_2|$ are the magnitudes of the complex dynamic viscosity of a blend and its initial components, and V is the volume content of the dispersed phase.

When treating the difficulties of describing the concentration dependence of even the Newtonian viscosity of blend compositions, we can readily imagine how complicated this problem becomes for the apparent viscosity, when a substantial role begins to be played by the rate and stress of deformation (see, for example, the scheme in Fig. 3.4(a)). Still more involved problems appear when describing the concentration dependences of the elastic properties of melts of polymer blends (Fig. 3.4(b)). Here to date it has been difficult to obtain a good generalization of experimental results for various types of blend compositions, not only because of the more involved nature of the change in the elastic parameters of the system in comparison with the viscous ones, but also because of the difficulty of correctly determining them by an experimental method [20].

In summarizing the above, we can state that universal relations describing the change in the rheological parameters of polymer blends can be obtained only by a strictly model approach.

Figures 3.1–3.4 illustrate only some quite general and in a definite sense 'idealized' features of behavior of very simple blend compositions. This qualitative analysis has failed to touch on a very important region of blend behavior, namely, the precritical and critical one when a system for

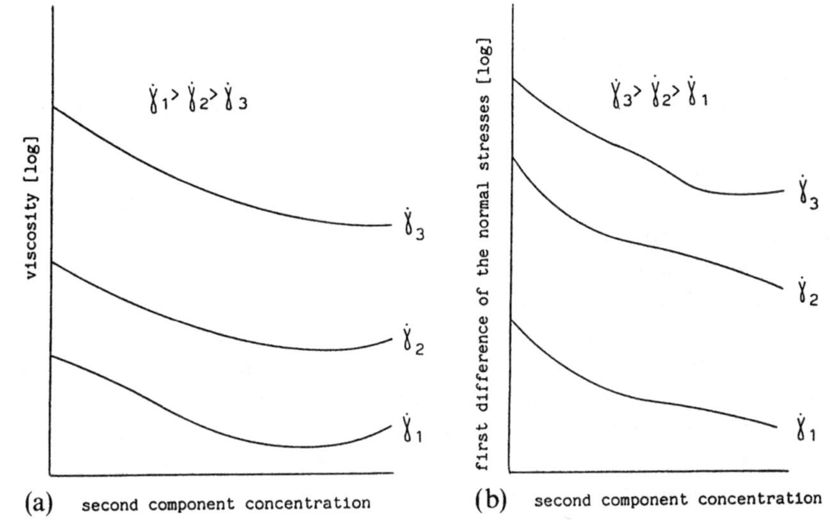

FIG. 3.4. Concentration dependences of the apparent viscosities and first difference of the normal stresses observed when studying binary blends of incompatible polymers. Curves 1–3 show various deformation rates.

some reason or other passes over from a stable-phase state to an unstable, and then two-phase state. Moreover, an actually observed experimental picture of the rheological behavior of flowing blend compositions will be much more complicated and 'richer' for interpretation, especially if it is considered in close relation to the results of thermodynamic analysis and morphology. To prove this, we must consider several specific examples of blend composition behavior reflecting, in our opinion, the diversity of the forms of manifestation of their viscoelastic and relaxation nature. These examples will relate to limitedly compatible and incompatible systems.

3.4 RHEOLOGICAL PROPERTIES OF LIMITEDLY COMPATIBLE BLENDS

3.4.1 Polymer–Solvent Systems

Study of the viscoelastic behavior of limitedly miscible polymer systems, e.g. solutions and melts of polymer blends, is of considerable interest in connection with the features of their anomalous behavior due to the specific state of the system in critical regions of compositions and temperatures. Solutions of polymer blends are a thermodynamically stable single-phase system only in the region of quite low concentration. A growth in the latter is attended by stratification of the system. A complicated set of phenomena at the molecular and supermolecular levels occurs in the region of phase stratification of polymers, while the systems which appear are of major theoretical interest and have a practical value in connection with the features of their physical and mechanical properties.

For example, there has been noted the presence of a peak of asymmetry of scattered light [21, 22], which is due to critical opalescence. A strong and constant compression of the polymer coils is observed [23] within a certain temperature interval (near the critical temperature, T_{cr}). On the other hand, Lirova *et al.* [24] indicate that there is intermolecular aggregation in the transition from a homogeneous solution to stratification and further settling out of the polymer that increases the duration of relaxation of dipole polarization.

Semenchenko [25] has shown that for flowing systems of two liquids the viscosity of a blend in the critical region passes through a maximum and the sign of its temperature coefficient changes. The presence of a peak on the viscosity polytherm has been observed in concentrated solutions [26]. The height of the peak grew with an increasing concentration of the polystyrene solution in cyclohexane. Such a peak was not observed in dilute solutions.

The specific nature of polymer chain structure determines the various types of structure formation processes under conditions of phase separation [27]. For example, for flexible-chain polymers these processes are due to concentration and density fluctuations, while for rigid-chain polymers orientation fluctuations also play a substantial role [27]. Kuleznev *et al.* [28] established an increase in the viscosity of a decalin solution of PS with an increase in the polyisoprene content in the region close to stratification, and also discovered lowering of the viscosity in stratification. The effect of a sharp drop in the viscosity of polymer blend solutions and melts was observed only in the region of a transition of a system from the single-phase state to a two-phase one [28]. The drop in the viscosity and modulus of elasticity [29] at the moment of stratification may be due either to the addition of a less viscous polymer to a more viscous one, or, conversely, to the addition of a more viscous polymer to a less viscous one [30].

The transition through the binodal curve in the phase diagram is attended by the appearance of an interface carrying excess energy. If the interphase tension is great, an ordinary, unstable, rapidly stratifying emulsion forms. With a low interphase tension at the interface of the polymer solutions (hundredths to thousandths of an erg cm^{-2}) in the region of compositions directly adjoining the binodal curve, thermodynamically stable two-phase colloidal systems appear. They form spontaneously under conditions of low values of the interphase stresses σ_{in} [31, 32].

3.4.2 Polymer–Polymer–Solvent Systems

Thermodynamically stable colloidal systems may also appear at low values of σ_{in} provided that the increment of the free surface energy in the formation of a droplet of the emulsion ($\propto \sigma_{in} r^2$) is compensated by the gain in the free energy at the expense of the growth in the system's entropy because of the formed droplet participating in Brownian motion, i.e. the condition of existence of such systems is $\sigma_{in} r^2 < kT$.

Thermodynamically stable emulsions have been obtained experimentally (i) in polymer–polymer–solvent systems [33], (ii) in mixtures of low-molecular-weight substances when the interphase tension drops to thousandths of an erg cm^{-2} [32], and (iii) on separation from a metastable solution of acetyl cellulose in ethyl ether with a change in the temperature [34].

It was discovered that solutions of polymer blends in addition to the turbidity concentration, i.e. the limit of their molecular solubility, have a second higher concentration limit associated with the beginning of the appearance of a second layer. Highly dispersed emulsions can form independently in the interval between these concentrations [33].

The above examples show that the behavior of polymer–polymer–solvent systems is quite complicated. This can also be concluded when determining the changes in their rheological properties. As an example, the data given below [1] were obtained for a polymer–polymer–solvent system in the stratification region ($\propto$6·5% toluene solution of blends of polyisoprene rubber (PI) of Grade 'Kariflex' with $M = 9{\cdot}0 \times 10^5$ and block polystyrene (PS) with $M = 3{\cdot}7 \times 10^5$). The transition from the single-phase to the two-phase state is achieved here (as in many other systems) in two ways, viz. (1) by an increase in the content of one of the components in the blend at a constant temperature, or (2) by a change in the temperature at a constant composition of the blend close to stratification. In the example considered, a change in the concentration of PS relative to PI from 0·1 to 0·5 wt.%, on the one hand, and in the temperature in the interval of 293–352 K, on the other, revealed a phase transition in the blend within the indicated range of concentrations and temperatures. The most graphic rheological results are observed in nondestructive experiments run with low-amplitude periodic deformation [35].

Precision rheological tests point to a lack of equilibrium of already the initial PI solution in toluene itself. This is readily detected by the appearance of a sort of hysteresis on the temperature dependence of the dynamic modulus of storage G' and loss modulus G'' (Fig. 3.5(a)). The appearance of hysteresis is a result of rearrangement of the solution structure that is not completed during an experiment. The moduli grow and the structure forms unusually intensely with lowering of the temperature, on the one hand, and lowering of the rate of its change on the other. A comparison of the plots of G' and G'' vs T obtained for various rates of change of the heating–cooling temperature conditions reveals an increase in the hysteresis with diminishing of these rates. The latter indicates a sufficiently long lifetime of the supermolecular structures in the solution.

The addition of 0·1% of PS to the PI solution changes the nature of the relations being considered (Fig. 3.5(b)). No indications of stratification or critical opalescence are observed in the solution. Although structurizing also occurs in this case (the heating–cooling curves do not coincide), the relaxation processes have time to be completed (the initial and final experimental points for 343 and 353 K coincide, respectively). This is caused by the second component, whose introduction even in insignificant amounts lowers the intermolecular interaction in the system (the polymers are incompatible), which facilitates the progress of long-term relaxation processes and ensures an equilibrium change in the properties during a cycle. With a PS content in a blend close to the stratification limit, the hysteresis reaches its maximal value (Fig. 3.5(c), curves 1′ and 2′); here G'' passes through a minimum, which is usually observed in the stratification region [29].

Hence, rheological data of such a type reflect the 'long-term' structural changes that are a result of relaxation processes in solution with a polymer concentration much lower than is needed to form a three-dimensional fluctuation network. The distinctive 'plasticizing' effect observed when small amounts (0·1–0·2%) of a second polymer are introduced is manifested primarily in accelerating the relaxation processes, which prevents stratification of the system. There is observed an effect, as it were, of a growth in the compatibility of the components in the blend.

Important information can be procured in such experiments by analyzing how the loss modulus depends on the content of the second

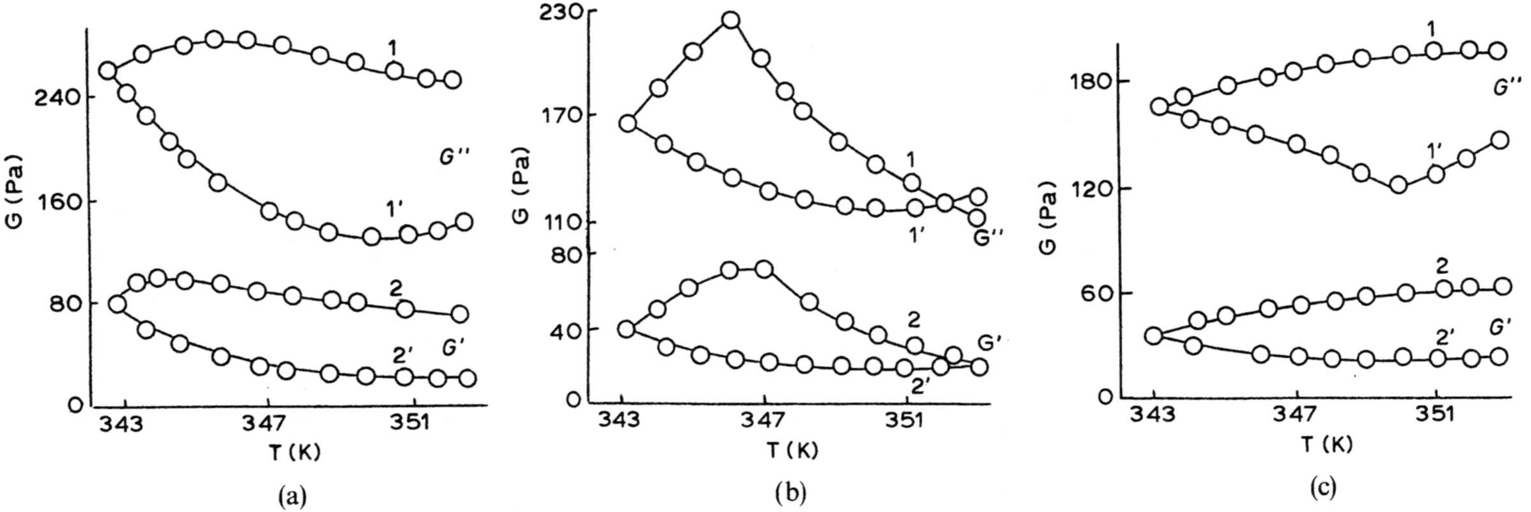

FIG. 3.5. Temperature dependences of the moduli G' and G'': (a) 6·5% solution of PI in toluene; (b) the same PI solution with a 0·1% content of PS; (c) the same solution with a 0·5% content of PS. The rate of temperature change is 1·2 K min^{-1}. The deformation frequency is $\omega = 1{\cdot}4\ s^{-1}$. Curves 1′ and 2′ show heating, 1 and 2 cooling.

component, in the given case PS (Fig. 3.6). The value of G'' is lowered considerably here only with slow cooling of the single-phase solution (Fig. 3.6, curve 3) or slow heating of a two-phase one (curve 2) as a result of the appearance and preservation of metastable or highly dispersed colloidal systems. With a rapid change in the temperature, the system 'jumps' over the near-critical region and becomes single-phase, or, conversely, unstable (two-phase), and now no diminishing of G'' (or the viscosity) is observed (Fig. 3.6, curve 1).

The size of the particles determined by the light scattering techniques (according to the turbidity spectrum) up to the moment when macro-stratification began was about 1000 Å and remained virtually unchanged during many months. This witnessed the spontaneous formation of a highly dispersed emulsion [5]. Such emulsions are characterized by a low surface tension at the interface of the polymer solutions in the common

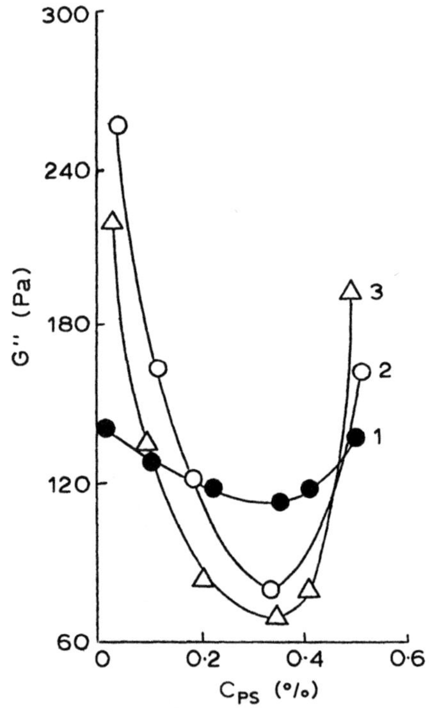

FIG. 3.6. Concentration dependence of modulus G''. Curve 1 corresponds to 353 K (rapid heating), curve 2 to 353 K (slow heating), and curve 3 to 343 K (slow cooling from 353 K). The deformation frequency is $\omega = 1{\cdot}4\ \mathrm{s}^{-1}$.

solvent, and also by a sharp drop in the surface tension in the near-critical region of composition [32]. Owing to the low particle size, the emulsion has a highly developed interphase surface with reduced polymer interaction in the interphase layer because of the incompatibility of the polymers [5]. This is why relaxation processes in such a system are facilitated by the presence of microadditions of a second polymer (0·2–0·4% by weight). At the moment of stratification, these microadditions to a definite extent are also a modifier, because they noticeably alter both the structure of the blend and a number of properties of the system as a whole (e.g. they diminish the dissipative losses). An appreciable role in stratification processes is undoubtedly played by the time factor. This draws special attention to the relaxation behavior of such substances.

3.4.3 Polymer–Polymer Systems

If we turn to the sufficiently general behavioral features of limitedly compatible blend compositions of the polymer–polymer type, the picture is still complicated, as was shown, in particular, by Klykova *et al.* [2]. The structure of polymer blends in the region of the transition from a single-phase state to a two-phase one can be characterized by the formation of both highly dispersed stable colloidal emulsions (compare with polymer–polymer–solvent systems) and heterophase fluctuations. This as a whole underlies the specific nature of the change in a number of physical properties of polymer blends in the stratification region. Hence, if we trace the dynamics of the change in a number of properties in the stratification region in polymer–solvent [26], polymer–polymer–solvent [2], and polymer–polymer systems, we can find an obvious trend of a growth in the anomaly of many composition–property laws.

The existence of a region of highly dispersed stable emulsions in polymer systems has been established experimentally by Kuleznev *et al.* [28]. It has been shown [31, 36] that if the interphase tension is less than $10^{-4}-10^{-5}\,N\,m^{-1}$, the decrease in the free energy of the system because of the growth in the entropy when a highly dispersed emulsion forms exceeds in magnitude the increase in the free energy of the system when a phase interface forms. The free energy of the system changes with increasing particle size along a curve with a minimum, and this is just what underlies the thermodynamic stability of the emulsion. The surface tension at a polymer–polymer interface is about $10^{-3}\,N\,m^{-1}$ [33]. Moreover, at the moment of stratification, the value of the surface tension decreases by two or three orders of magnitude because of 'blurring' of the phase interface [37]. This is exactly what underlies the sharp drop in the increment of the free energy when an interface forms near the binodal curve.

By the theory of 'heterophase fluctuation and pretransition phenomena' developed by Frenkel [38], near the binodal curve there should be an especially sharply pronounced formation of 'heterophase' fluctuations, i.e. fluctuations that extend beyond the limits compatible with the initial state of the system and are nuclei of a new phase. In the region of a stable single-phase state, these nuclei reach an insignificant size and decompose. In the metastable region, the nuclei of a new phase after reaching a critical size corresponding to the maximum value of the potential acquire a trend of unlimited growth. In the region of an absolutely unstable state (after the spinodal), all the formed nuclei grow. It has also been shown that the presence of heterophase fluctuations in the transition region should produce an anomalous growth in the heat capacity and thermal expansion coefficient.

Semenchenko [39] showed an anomalous decrease in the stability coefficients of a system in the transition region (the thermal coefficient $(\mathrm{d}T/\mathrm{d}S)_p = T/C_p$ and the mechanical coefficient $(\mathrm{d}p/\mathrm{d}V)_T$. The values of the stability coefficients are inversely proportional to the size of the fluctuations, consequently the maximum development of heterophase fluctuations corresponds to the passage of the stability coefficients through a minimum. Diminishing of the stability coefficients should be manifested in a lower value of T/C_p (or in an increase in the heat capacity C_p itself), and in a decrease in the reciprocal of the permittivity and modulus of elasticity. Moreover, kinetic parameters such as the thermal diffusivity, speed of elastic waves, and fluidity (the reciprocal of the viscosity) are proportional to the stability coefficients to a first approximation. Hence, in the transition region, these quantities must have their minimum values.

The conclusions reached in the above theoretical and experimental publications are confirmed quite convincingly by studies of recent years. Let us consider as an example the set of some properties of blends of polystyrene (PS) and butadiene–styrene copolymer (SKMS-30) in the stratification region [2], which we deem to be typical.

The phase diagrams of these blend compositions show that they relate to systems with an upper critical dissolution temperature. The plots of the radius of the PS dispersed phase particles and their concentration versus the PS content in the blend (Fig. 3.7) show the composition to be featured by an extended region after the binodal curve ($C_{PS} = 3$–7%) in which the radius of the PS dispersed phase particles changes slightly and is about

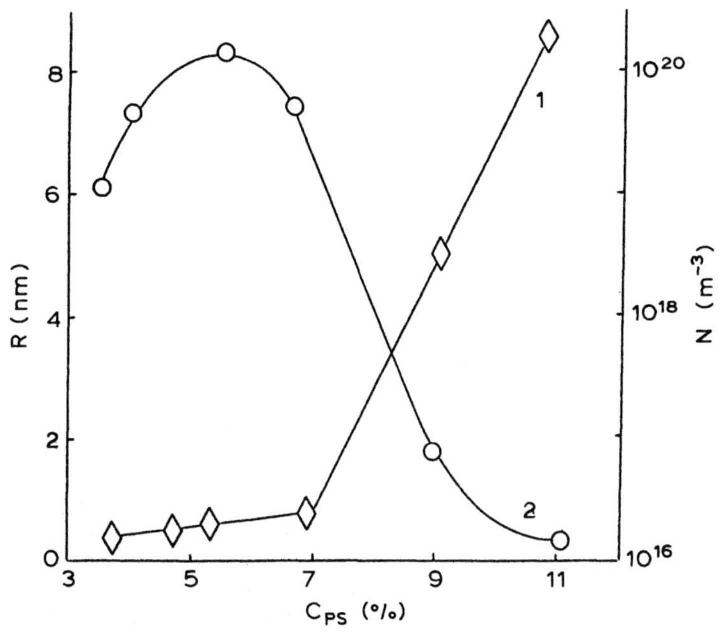

FIG. 3.7. Size (1) and concentration (2) of particles of the PS dispersed phase in the transition region in a blend of SKMS-30–PS.

60 nm. The number of particles per m^3 grows with increasing C_{PS} up to 10^{20}. With a further growth in the PS concentration, the size of the dispersed phase particles grows rapidly owing to coalescence, while their number diminishes by several orders of magnitude. Observations in the region of $C_{PS}=3-7\%$ show that the structure of the blend does not change during a prolonged period, but at $C_{PS} \geqslant 9\%$, the size of the dispersed phase particles grows in time, while their number decreases. The physical picture of the structural changes becomes clear when we involve data on rheological and heat capacity measurement.

The change in the complex dynamic viscosity $|\eta^*|=|G^*|\omega^{-1}$ (Fig. 3.8) shows that when $C_{PS} \simeq 3\%$ (this corresponds to the solubility limit), the viscosity grows by 20% in comparison with that of the initial SKMS-30. With a further growth in the PS concentration to 5%, the viscosity drops, and then grows again because of the higher viscosity of the PS at the measurement temperatures. With elevation of the temperature, the anomalous change in the viscosity diminishes, and at 353 K it degenerates completely, which is associated with the increase in PS solubility when

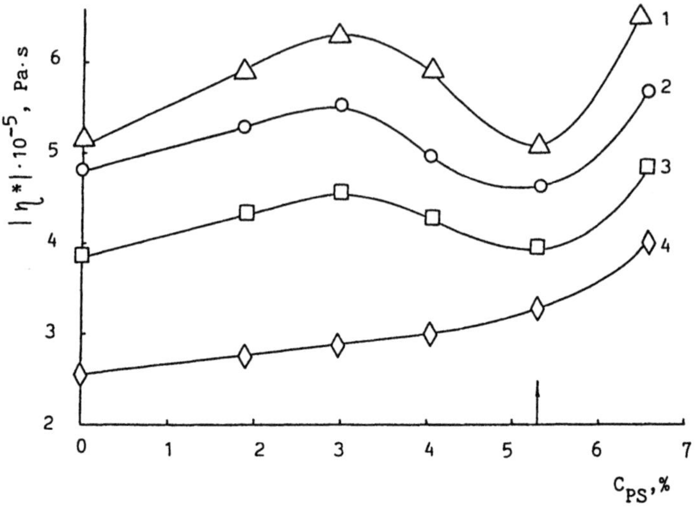

FIG. 3.8. Magnitude of complex dynamic shear viscosity of SKMS-30–PS blend vs PS content at 293 K (curve 1), 313 K (2), 333 K (3), and 353 K (4). The deformation frequency is $\omega = 0{\cdot}6\ s^{-1}$. The arrow shows the PS concentration at which the maximum number of particles in unit volume is observed (Fig. 3.7, curve 2).

heated. With an increase in the frequency of mechanical action from 0·6 to $6\ s^{-1}$, the deviations in the change in the viscoelastic characteristics G', G'', and η^* from monotonic reduce considerably, while at $\omega = 60\ s^{-1}$ there are no extrema on their plots vs the PS content in the blend. Hence, the influence of a growth in the frequency of mechanical action on a blend of polymers in the stratification region is similar to an increase in the temperature. One can presume that with a growth in the frequency of action there is a tendency to destruction of the highly dispersed colloidal emulsions.

A comparison of Figs 3.7 and 3.8 reveals that a maximum of the viscosity (and of the heat capacity) is observed in the region of composition corresponding to the binodal curve, while a minimum is observed in the region of stable highly dispersed colloidal emulsions having a small dispersed phase particle size with a maximum particle concentration. The surface area of the phase interface in such an emulsion calculated by the data of Fig. 3.7 is $6 \times 10^6\ m^2$ per m^3 of blend. Because of blurring of the phase interface, the region of contact of the different species of macromolecules should be much larger than is determined by the size of the surface. The lower interaction of the different macromolecules leads to a reduced density of the cohesion energy of the blend as a whole [40] and to an increased fraction of the free volume in the interphase layer [41]. This is just what lowers the viscosity of the system.

3.5 RHEOLOGICAL PROPERTIES OF INCOMPATIBLE BLENDS

We have already discussed above some possible schemes of the change in the viscous and elastic properties of very simple blend composition. Let us now consider some specific examples of the features of rheological behavior of heterogeneous blends. They undoubtedly do not cover all the variants of the diversity observed in practice, but do illustrate the main relation between the thermodynamic, structural, and rheological characteristics, namely, the values of the interaction parameter, the mean size of the dispersed phase particles, and the apparent viscosity.

Let us consider the results obtained by Vershinin *et al.* [3] who quite clearly, in our opinion, established such a relation using melts of two-phase blends—the rigid-chain polymer poly(methylmethacrylate) (PMMA) with elastomers (see Table 3.1). Samples of marketed materials were employed as the objects (see the table), namely, suspension PMMA ($M_\eta = 7{\cdot}5 \times 10^4$), butadiene nitrile copolymer SKN-40 ($M_\eta = 3{\cdot}1 \times 10^5$), chlorosulfonated polyethylene CSPE ($M_\eta = 1{\cdot}6 \times 10^5$), polyurethane PU ($M_\eta = 9{\cdot}0 \times 10^4$), and the ethylene–propylene copolymer SKEP-50 ($M_\eta = 6{\cdot}8 \times 10^4$). Blends of PMMA with the indicated elastomers model compositions with a high impact strength. A glance at the table reveals that the components of the blends are characterized by various thermodynamic affinities, which are indicated by the value of the interaction parameter.

TABLE 3.1
Interaction Parameter of PMMA with Elastomers at 293 K

Elastomer	*Interaction parameter χ calculated by*		
	Scott's method	*Solubility limit*	*Hildebrand's equation*
SKN-40	0·011	0·002 2	0·11
CSPE	0·022	0·003 5	0·001 1
PU	0·032	0·005 5	0·018
SKEP-50	0·055	0·006 9	0·21

The difference of about a decimal order of magnitude in the absolute values determined by Scott's method and by the limits of solubility of elastomers with PMMA is due to the known drawbacks of the classical Flory–Scott theory. The results of calculating χ by Hildebrand's equation (we used the tabulated values of the solubility parameters, while for PU they were calculated by Smoll's rule) are not consistent with experimental results, not only quantitatively but also qualitatively. It should be noted that such a discrepancy was also observed previously [5].

To obtain quantitatively coinciding characteristics of the interaction parameter determined in various ways, one must consider the noncombinatorial blending entropy component χ_S for both a solution and a block [42, 43]. To date this task has been quite involved because the presence of the component χ_S results in χ depending on the ratio of the components and their molecular mass. For the same reason, the plot of χ vs T does not obey the laws predicted by the Flory–Scott theory.

A comparison of the rheological relation (flow curves) determined from experiments on a capillary viscometer [3] has shown the possibility of their linearization in logarithmic coordinates in a definite range of deformation rates $\dot{\gamma}$ and stresses τ_{12}. This made it possible, firstly, to determine the change in the exponent in the Ostwald de Vill equation (it altered from 2·1 to 2·9 as a function of the mixture composition) and, secondly, to establish how the value of the apparent viscosity $\eta = \tau_{12}/\dot{\gamma}$ depends on the PMMA concentration (ϕ_{PMMA}) in a blend (Fig. 3.9). The cited example shows well that the nature of the curve of η vs ϕ_{PMMA} depends substantially on the compatibility of the polymers. The apparent viscosity of melts of blends with weak PMMA–SKEP interaction is much lower than the additive values, while for blends with relatively strong PMMA–SKN interaction it is appreciably higher than the additive values.

The involvement of the results of structural studies has shown that with a growth in the dispersed phase particle size the values of η diminish, i.e. the fluidity of the melt of a blend increases (naturally, within identical temperature and rate conditions of comparison). As regards quantitative criteria, in the given specific case a change in the mean dispersed phase particle size of 2·0–2·5 times is attended by an approximately four-fold increase in the fluidity of a melt [3]. This is explained by the features of the morphology and velocity profile of the flow, i.e. a stratified (or lamellar) structure of the blend and the nature of its flow by layers. The rate of deformation in the phase of the low-viscous component is higher than in that of the high-viscous one [5, 8]. With an increase in the length of the layers (the size of the dispersed phase), the nature of the flow by

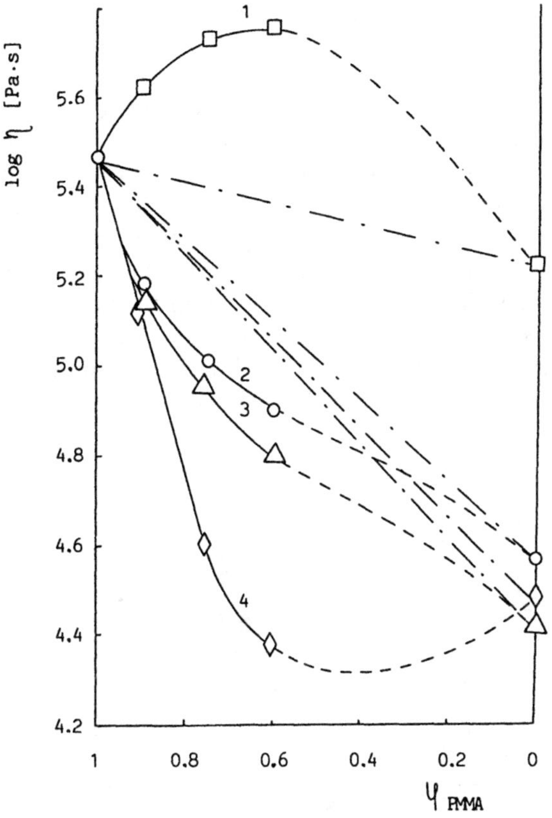

FIG. 3.9. Apparent viscosity of melts of blends of PMMA with SKN-40 (curve 1), PU (2), CSPE (3), and SKEP (4) vs PMMA content.

layers grows, and the contribution due to the deformation of the dispersed phase of the low viscosity component increases. This lowers the value of the apparent viscosity of the blend melt.

The anomaly in the viscoelastic behavior observed in this specific case can be related to the typical features of flow of heterophase blends. In the general case, these features are due to the following causes: (i) the specific nature of the flow structure (close to a stratified one), the flow having a high velocity in the phase of the low-viscosity component, (ii) the interaction of the components at the phase interface, and (iii) the growth in the dissipative losses because of collisions of the dispersed phase particles in the flow.

To determine the individual role of each of these factors, model blends have been studied [12] in which the component phases were alternated. This was done by alternately filling the viscometer tank with layers made from the blend components [12]. In the flow of such compositions, the thickness of each phase changed from the wall of a capillary to its axis. It was found (Fig. 3.10) that in the arrangement of the component phases by layers the viscosity of the mixture melt is much lower than that of a blend in which the components are mutually dispersed (e.g. when preparing a blend composition in rolls). This underlay the prevailing role of the component phase having a lower viscosity in the flow process.

It is interesting to note that the viscosity of a blend with a layered structure is even lower than that of the less viscous individual component (PU in Fig. 3.10). The cause of this has already been noted—the rate of

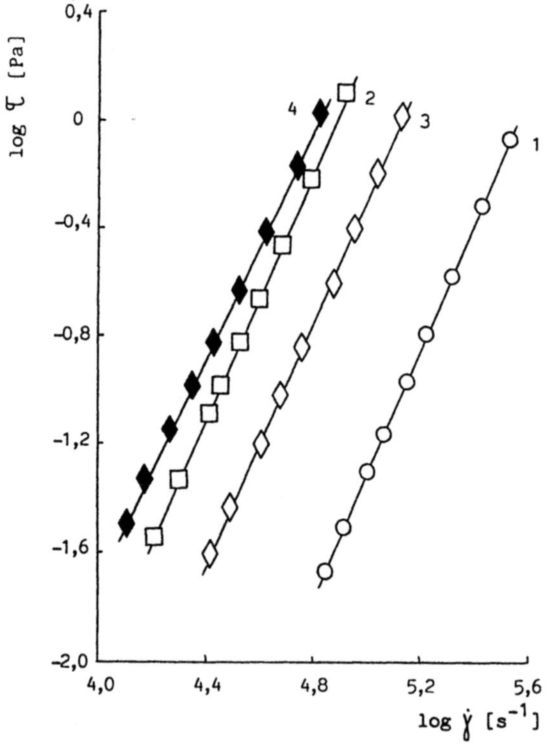

FIG. 3.10. Flow curves of PMMA (curve 1), PU (2), and blends thereof (1:1) obtained in rolls (3) and with layered arrangement of components (4). The temperature was 453 K.

deformation in the phase of the less viscous component of the blend in deformation of the blend composition is higher than under the corresponding conditions in an individual component. Because of the sharply expressed anomaly of the viscosity in the flow of the system as a whole, lower values of η correspond to a higher $\dot{\gamma}$. Hence, the value of the apparent viscosity of a blend composition often decreases in comparison with η of a low-viscosity component not because of sliding along the phase interface, but because of the feature of deformation of the two-phase system.

The influence of the size of the blend component particles on the level of its viscoelastic characteristics can be seen from the data shown in Fig. 3.11. It can be seen that the loss modulus and, consequently, the values of the viscosity of the coarsely dispersed blend are lower than those of the finely dispersed one. This is associated with a larger number of particles and a thinner interlayer between them in the highly dispersed blend, on the one hand, and the larger phase interface and volume of the interphase layer on the other. In the case being analyzed (20% by volume of the dispersed phase and a cubic lattice model), diminishing of the size from 100 to 0·1 μm causes the number of particles to grow from 4×10^5 cm^{-3} to 4×10^{14} cm^{-3}, the thickness of the interlayer between them to reduce from 35 to 0·035 μm, and the surface of the phase interface to increase from 0·04 to 40 m^2 per cm^3 of the blend.

It should be noted that the theory of coalescence in a flow [45] predicts the absence of a dependence of the number of dispersed phase particle collisions on their size ($\phi=$const.) Moreover, the theoretical ideas on the flow of an emulsion with consideration of the deformation and interaction of the particles in a flow point to the viscosity being independent of the number of particles [46], hence a growth in the dissipative losses due to friction of the dispersed phase particles is only slightly probable with an increase in the degree of dispersion.

The values of the viscosity calculated by the equation [46]

$$\eta_{\mathrm{b1}}=\eta_{\mathrm{med}}\left[1+\frac{5K+2}{2K+2}\phi+\frac{5(5K+2)^2}{8(K+1)^2}\phi^2\right]$$

where $K=\eta_{\mathrm{ph}}/\eta_{\mathrm{med}}$ is the ratio of the viscosities of the dispersed phase and the dispersion medium, are practically identical for all the studied blends (when $\phi=0{\cdot}1$, $\log\eta=5{\cdot}54$, and when $\phi=0{\cdot}2$, $\log\eta=5{\cdot}6$ Pa s) and exceed the additive values, but are considerably lower than the experimentally determined values for blends of PMMA with SKN-40 (see Fig. 3.9).

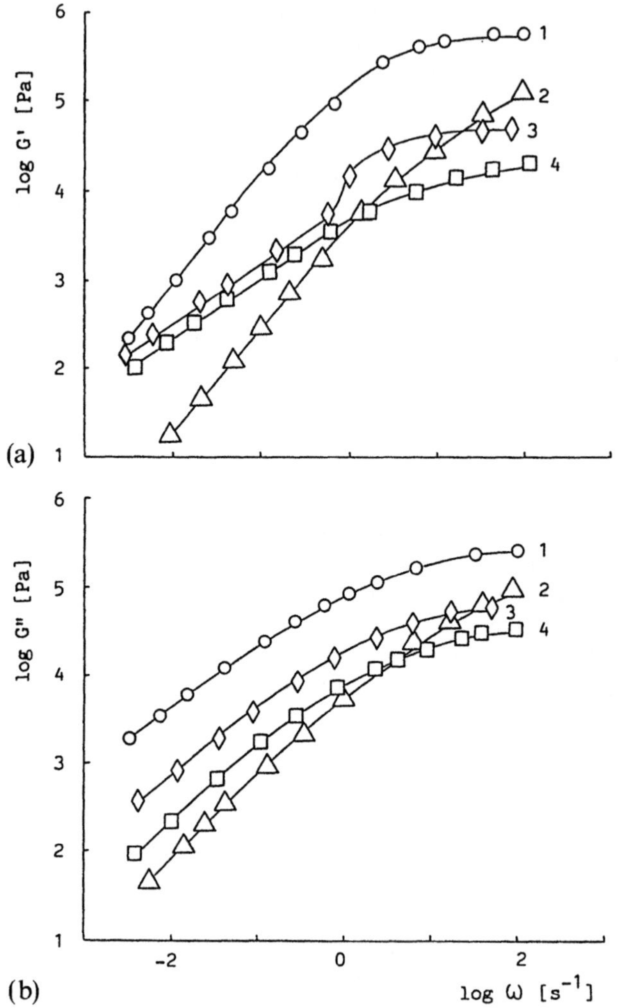

FIG. 3.11. (a) Storage G' and (b) loss G'' moduli vs deformation frequency of PMMA (curve 1), PU (2), and blends thereof (2:8) with particle sizes of 0·1–0·3 μm (3) and 70–100 μm (4). The temperature was 453 K.

In the general case, the growth in the viscosity of melts of heterogeneous polymer blends observed with an increase in the degree of dispersion of the particles distributed therein and, consequently, in the phase interface surface, can also be explained by purely relaxation reasons. Indeed, at the phase interface of mutually insoluble polymers, the conformational set of macromolecules is limited [47], and this should lower their mobility [48]. With a growth in the unit phase interface area, the fraction of macromolecules inhibited in this way grows, and this naturally increases the viscosity of the blend.

The difference in the frequency dependences of the moduli of storage of coarsely or finely dispersed melts may also be caused by the difference in the size of the interphase layer in a specific case (Fig. 3.11). In the studied interval of frequencies, the second component, PU, is in the plastic state (Fig. 3.11, curve 2), whereas PMMA when $\log \omega > 1$ already passes over into a rubber-like state, which is indicated by the plateau on the curve of G' vs ω (Fig. 3.11, curve 1). The plot of $\log G'$ vs $\log \omega$ for a coarsely dispersed blend (curve 4) is qualitatively similar to that observed for PU. On the relevant plot for a finely dispersed blend (curve 3) at $\log \omega \approx 1\ s^{-1}$ the modulus of storage grows sharply, after which there is a clearly expressed transition to a rubber-like state identical to that observed for PMMA (curve 1). Hence, the sharply pronounced transition of PMMA (the dispersed phase) in a highly dispersed blend from a fluid state to a rubber-like one substantially affects the rigidity of a composition, which is not observed for a coarsely dispersed blend.

In summarizing, we can conclude that for melts of blends of mutually insoluble polymers, the viscosity grows with an increase in the thermodynamic interaction of the components and a decrease in the dispersed phase particle size. The viscosity of melts of blends with weak interaction of the components is chiefly determined by the size and shape of the low-viscosity dispersed phase particles because this phase deforms predominantly in flow of the blend. The viscosity of blends with strong interaction is determined by the influence of the phase interface.

3.6 MODEL APPROACHES

3.6.1 Compatible Blends

Rheological studies of polymer blends can sometimes give very convenient information for verifying theoretical models describing the features of the viscoelastic behaviour of polymers. For example, a theoretical concept has been developed on the basis of a one-molecule approach [49–51]. This approach was employed in order to propose a model [50, 51] by which the motion of macromolecules among their like is considered to be equivalent not in a viscous, but in a viscoelastic medium. The characteristics of this apparent viscoelastic medium (microviscoelasticity) do not coincide with those of the viscoelasticity for the system as a whole (macroviscoelasticity). We can see that already such an approach is based on definite ideas on the possible microheterogeneity of a viscoelastic continuum.

An experimental example of the reality of this concept is its successful application for describing the viscoelastic behavior of polymer blends. Indeed, when adopting a single-molecule approximation, one must naturally consider that one of the blend components with respect to the macromolecules of the other component (or components) is a medium that should be considered as viscoelastic. A very simple example of such a system is a blend of two different linear polymers or a blend of two fractions of the same linear polymer with different molecular weights.

The region of low concentrations of one of the components is obviously of the greatest interest here.

Let us consider specific results of such an approach enabling one to determine to a definite extent the degree of microheterogeneity of the polymer medium (with different molecular weight) and its influence on the formation of viscoelastic properties of the system as a whole.

It is convenient to consider as an example the results procured when studying the viscoelastic properties of blends of linear flexible-chain polymers having various molecular weights with a narrow MWD in the region of low concentrations of additives of one of the polymers [9]. For this end, we shall consult the results of dynamic experiments [9] obtained for samples of blends of PB (polybutadiene of various MW with $M_w/M_n \leqslant 1{\cdot}2$) within a relatively broad range of change of the component concentrations (see Table 3.2). They are shown in Figs 3.12 and 3.13. Since the concentration of one of the components in the considered blends prevails, we can assume that this component plays the role of a matrix. It is significant that the MW ratios of the polymer matrix and additives in the series A, B, C are 10, 50, and 100, while the initial viscosities of the first and second components differed $1{\cdot}2 \times 10^4$, 7×10^5, and $1{\cdot}25 \times 10^7$ times. In the series D and E, the MW ratios of the first and second components are 5 and 10, respectively, while the initial viscosities differed $5{\cdot}8 \times 10$ and $1{\cdot}05 \times 10^4$ times.

Attention is primarily attracted in the figures by the plots of G' and G'' vs ω for the matrix. For series A, B and C, it is a polymer with MW = 1×10^4. Within the entire range of studied particles, the plots of G' and G'' vs ω for this PB do not reveal anomalies in the viscoelastic behavior (the curves ascend virtually monotonically). Such a behavior is typical of polymers with a low MW with weak rubber-like properties. Yanovsky

TABLE 3.2
Characteristics of PB and Compositions of Binary Blends Based Thereon

Additive concentration (%)	*Mixtures*				
	$M_V^m = 1\times10^4$			$M_V^m = 1\times10^5$	
	$M_V^a = 1\times10^5$	$M_V^a = 5\times10^5$	$M_V^a = 1\times10^6$	$M_V^a = 5\times10^5$	$M_V^a = 1\times10^6$
0·125	A_1	—	B_1	—	D_1
0·25	A_2	—	B_2	—	D_2
0·5	A_3	B_1	C_3	E_1	D_3
1·0	A_4	B_2	C_4	E_2	D_4
3·0	—	—	C_5	—	D_5
5·0	A_5	B_3	C_6	E_3	D_6
7·0	—	B_4	—	—	—
10·0	A_6	B_5	C_7	—	—
20·0	A_7	B_6	—	—	—

Note: M_V^a = MW of additive; M_V^m = MW matrix. When $M_V^m = 1\times10^4$ and 1×10^5, $\eta_0 = 2{\cdot}52\times10$ and 3×10^5 Pa s, respectively.

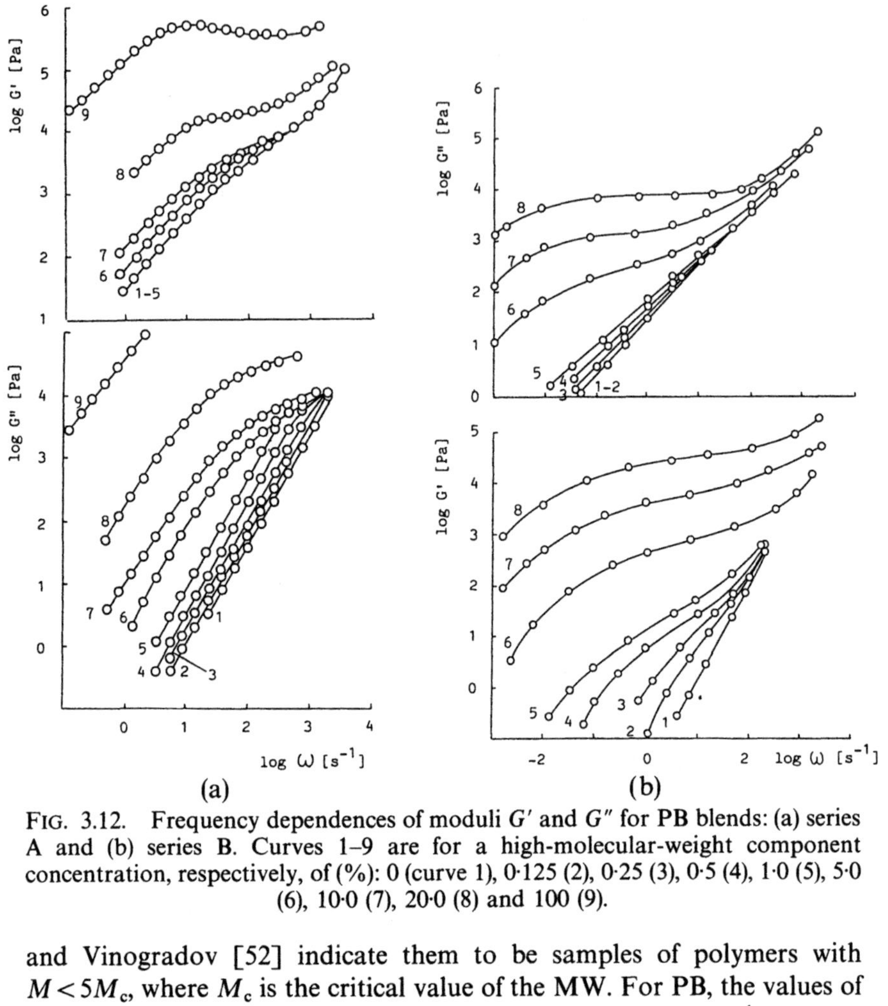

FIG. 3.12. Frequency dependences of moduli G' and G'' for PB blends: (a) series A and (b) series B. Curves 1–9 are for a high-molecular-weight component concentration, respectively, of (%): 0 (curve 1), 0·125 (2), 0·25 (3), 0·5 (4), 1·0 (5), 5·0 (6), 10·0 (7), 20·0 (8) and 100 (9).

and Vinogradov [52] indicate them to be samples of polymers with $M < 5M_c$, where M_c is the critical value of the MW. For PB, the values of $M_c \sim 5600$, and, consequently, for a sample with $MW = 1 \times 10^4$ the ratio $M/M_c < 2$. For the series D, the first component—the matrix—has $MW = 1 \times 10^5$, while the ratio $M/M_c = 16$. This sample exhibits all the features of the rheological behavior of a high-molecular-weight polymer, i.e. first of all it has a clearly expressed plateau on the curve of G' vs ω and a maximum and minimum of G'' vs ω, which indicates the achievement of a rubber-like state (Fig. 3.13). The introduction of the second component in various concentrations quite sharply affects the characteristics being considered, and to a greater extent with a larger difference between the

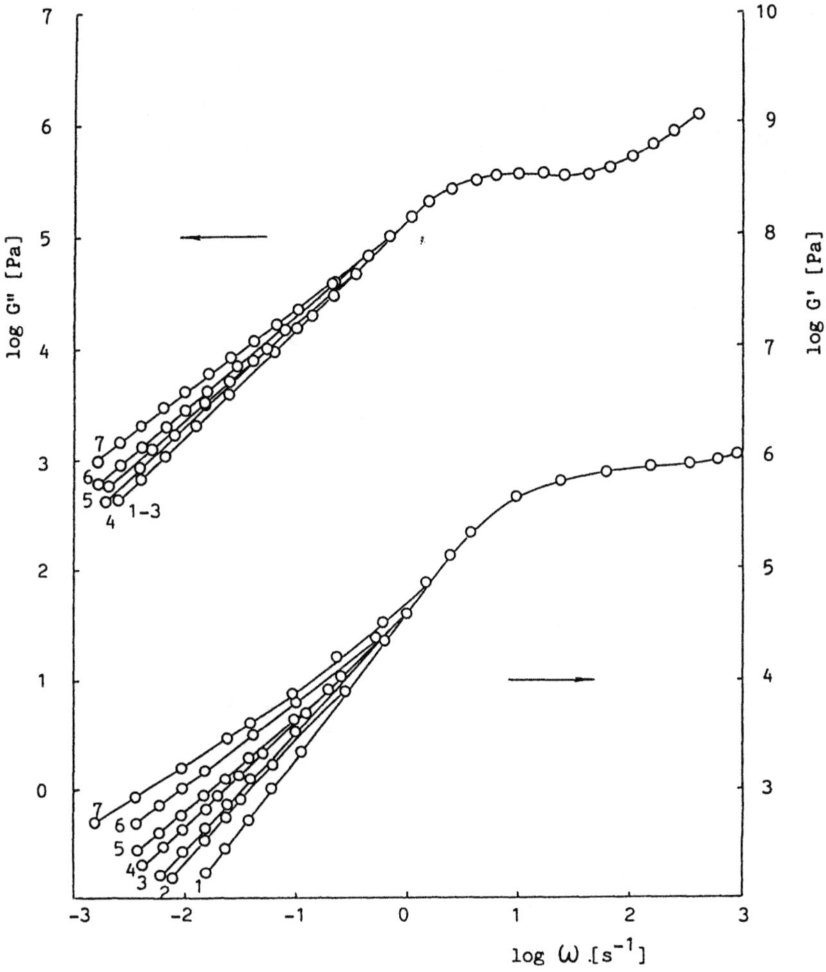

(a)

FIG. 3.13. Frequency dependences of moduli G' and G'' (a) for PB blends of series D at different concentrations of the high-molecular-weight component, (b) for various PB solutions with $M = 1 \times 10^6$, and (c) for systems with various matrix MW. The curve designations are, respectively: (a) 0 (curve 1), 0·125 (2), 0·25 (3), 0·5 (4), 1·0 (5), 3·0 (6) and 5·0% (7); (b) 2% solution in cetane (curve 1), 2% solution of α-MN (2), 5% solution in cetane (3) and 5% solution in α-MN (4); (c) cetane (curve 1), α-MN (2), PB with $M = 1 \times 10^4$ (3) and PB with $M = 1 \times 10^5$ (4).

molecular weights of the first and second components. The region of low concentrations of the additives, where their influence tells the most sharply, is also of major significance. This is especially clear on the plots of G' vs ω, on which both the quantitative and qualitative changes are obvious. At low frequencies, the final zone is reached (the flow region). Here $G'' \propto \omega$, and $G' \propto \omega^2$. It is simple to calculate the values of the initial

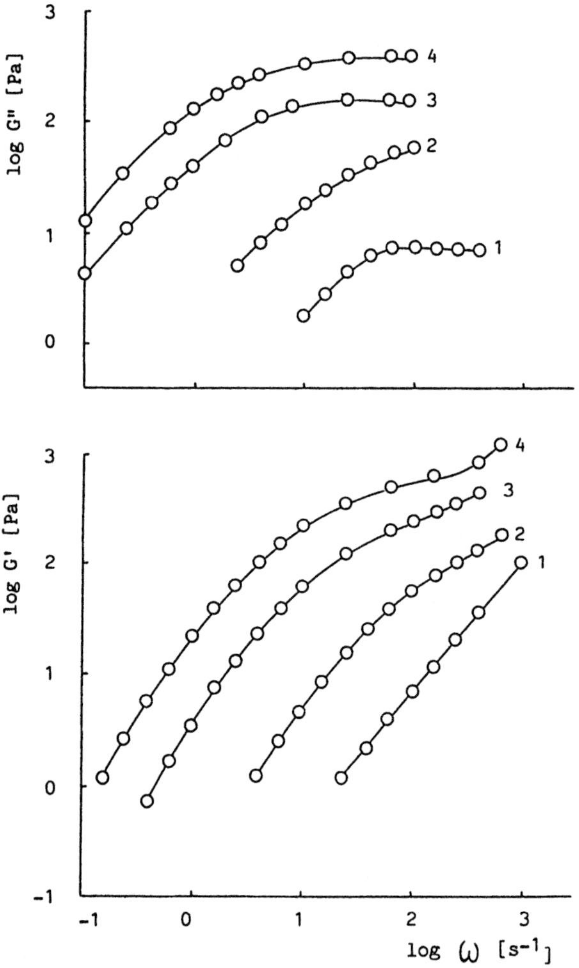

(b)

FIG. 3.13. *Cont.*

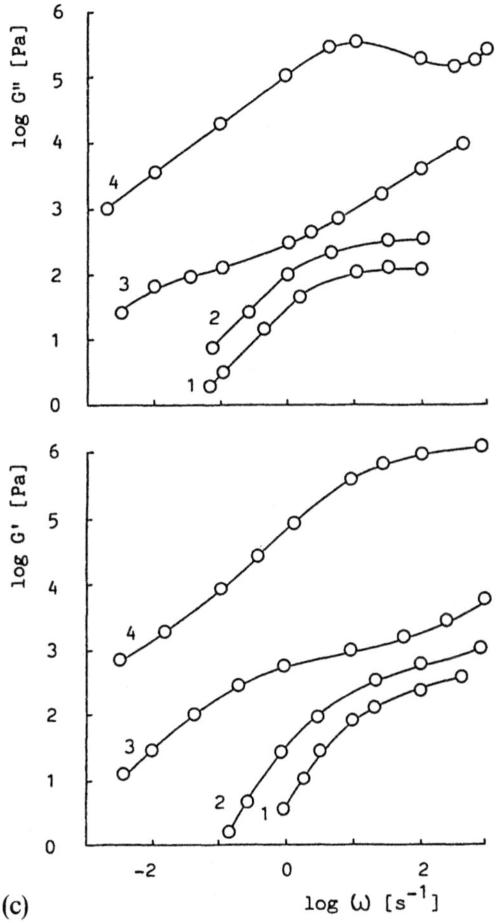

(c)

FIG. 3.13. *Cont.*

viscosity of the blends

$$\eta_0 = \lim_{\omega \to 0} (G''/\omega)$$

and the initial high elasticity coefficient

$$A_G^0 = \lim_{\omega \to 0} (G''/\omega^2)$$

for this region. When the values of the initial viscosities of the components differ more, the values of G' and G'' in the final zone change more with the concentration. An especially sharp jump in the properties is observed in the region of quite small concentrations (~1%) of the high-molecular-weight component additive. The values of G' for the blends A, B, and C with a change in the concentrations of the high-molecular-weight additive from 0 to 1% change 1·2, 40, and 25 000 times, respectively. The dissipative component (G'') changes here much less—1·2, 2·5, and 4 times, respectively.

With an increase in the concentration of the second, high-molecular-weight component and its MW, the slope of the curves of G' vs ω begins to change. First inflections, and then a plateau of the rubber-like state appear on them. The ascending right-hand branches of the curves of G' and G'' vs ω are connected with a transition of the system to a 'leather-like' state.

A comparison of the results of Figs 3.12 and 3.13 shows that the relative change in the viscoelastic properties of the characteristics is much greater in the first case (see Fig. 3.12). It can also be seen quite clearly that the joint influence of a number of factors (MW, the component concentrations, ratio of their initial viscosities) hampers the interpretation of the rheological information.

To find the individual role that the MW of each component, i.e. the matrix and additive, plays in forming the viscoelastic properties of binary blends and the distinctions in the behavior of these systems from polymer solutions, Yanovsky *et al.* [9] studied the viscoelastic characteristics of PB solutions with $MW = 1 \times 10^6$ in good and poor solvents, i.e. α-methylnaphthalene (α-MN) ($\eta = 2{\cdot}5 \times 10^{-3}$ Pa s) and cetane ($\eta = 3{\cdot}95 \times 10^{-3}$ Pa s) respectively (see Fig. 3.13(b)). The results procured for solutions in the different solvents were qualitatively similar (compare curves 1, 2 and 3, 4). Quantitatively in a poor solvent, the initial viscosities are three times lower for a 5% solution and 10 times lower for a 2% one. A comparison of the plots of G' and G'' vs ω for systems with different matrices, but an identical (5%) content of the high-molecular-weight additive PB with $MW = 1 \times 10^6$ (Fig. 3.13(c)), indicates that the MW of the matrix has a determining effect on the nature of the relations, the greatest changes in the magnitudes of the dynamic characteristics being observed for the modulus of storage. This is associated with the features of intermolecular interaction in these systems or with the ratio of the micro- and macroviscoelasticities by which this interaction is determined. For a purely viscous matrix, the curves remind one of the Rouse relation. With a viscoelastic matrix of a sufficiently low MW, a system behaves like a high-molecular-weight polymer, namely, it has a region of a fluid state, a plateau, and a transition to a leather-like state. With a growth in the MW of the matrix, the curves are similar to those for a blend of polymers—a break in the curves of G' and G'' vs ω in the region of low frequencies, and a maximum and minimum coinciding with the extrema for PB with $MW = 1 \times 10^5$ in the region of high frequencies.

The greatest distinctions in the nature of the curves observed for G' vs ω manifest themselves more when the MW of the additive is higher.

In summarizing the features of behavior of the above types of linear polymer blends in the region of low concentrations of additives of one of the components, we must indicate the enormous qualitative and quantitative differences in the viscoelastic characteristics, especially in the final zone. These differences are determined primarily by the characteristic of the blend matrix, and then by the MW concentration of the additives. It should be noted that a change in the above parameters affects first the frequency dependence of the modulus of storage. It is well known from the literature [10, 52] that in the final zone (i.e. in the region of low frequencies or in that of prolonged relaxation times), with a change in the MW, the elastic characteristic of a material, in particular the initial coefficient of high elasticity A_G^0, changes very sharply with the molecular weight (in proportion to the seventh or eighth power), i.e. much more sharply than the initial viscosity. By comparing the plots of η_0 and A_G^0 vs C, we can see that the elastic characteristic of a material in the final zone is also much more sensitive than the viscosity to the molecular inhomogeneity of a system.

The laws of the change in the rheological properties of blends can be generalized to a definite extent when plotting the concentration dependences of the initial parameters. This is shown in Fig. 3.14 (a) and (b) for the initial viscosity and elasticity coefficients, i.e. η_0 and A_G^0 vs C in comparison with solutions. For the latter, the initial viscosity changes in proportion to the square of the concentration [53]. Two groups of laws can be seen for blends. The first relates to blends for which the molecular weights of the components and accordingly the initial viscosities differ quite sharply (e.g. series A, B, and C in Fig. 3.14), and the second to blends for which this difference is much smaller. The first group is noted for a sharp change in the initial rheological parameters in a relatively narrow range of concentrations. For the second group, the initial rheological parameters change much less with the composition, e.g. the slope of the curve of $\log \eta_0 = f(C)$ with an increase in the MW of the second component changes from 0·7 to 3·0 (curves 3, 4, and 5 in Fig. 3.14). It should be noted that for concentrated polymer solutions η_0 changes in proportion to the fifth power of the concentration [53]. For blend D, η_0 up to values of $C=1\%$ is practically constant, and then grows in proportion to $C^{0\cdot3}$. Accordingly, $A_G^0 \propto C^{1\cdot5}$ for blend A, A_G^0 changes especially sharply for blend B, while the plot of A_G^0 vs C has a break at $C \approx 0\cdot5\%$. When $C>0$, $A_G^0 \propto C^{2\beta}$, where β is the exponent in the relation $\eta_0 = f(C)$. It is important that for all blends the coefficient A_G^0 changes with C to the power 2β. It should be noted that according to Vinogradov *et al.* [53] for concentrated polymer solutions the value of $A_G^0 \propto C^{10}$.

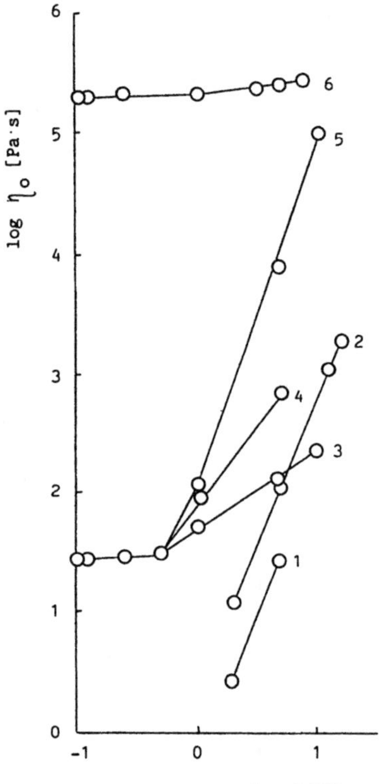

(a)

FIG. 3.14. (a) Maximum Newtonian viscosity η_0, and (b) initial coefficient of high elasticity A_G^0 vs concentration of the high-molecular-weight component. The curves signify, respectively: (a) PB with $M=1\times10^6$ in cetane (curve 1), the same in α-MN (2), and series A, B, C, and D, respectively (3–6); (b) series A, C, and D, respectively (curves 1–3).

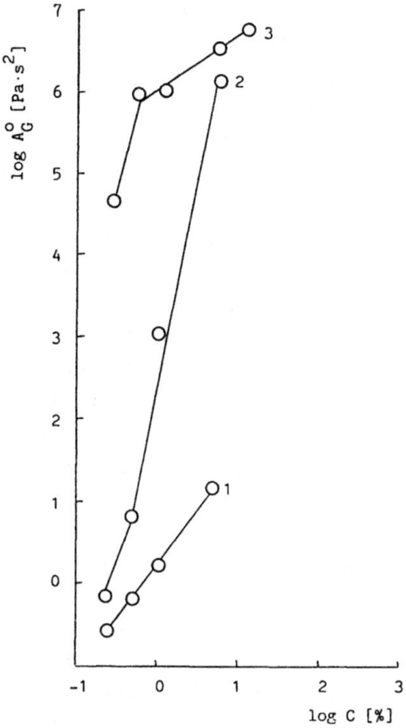

(b)

FIG. 3.14. *Cont.*

A comparison of the plots of A_G^0 and η_0 vs C shows that when $C<0{\cdot}5\%$, η_0 is practically insensitive to changes in the concentration, whereas the coefficient A_G^0 changes sharply in this region.

The observed features of the curves of A_G^0 and η_0 vs C for blends can apparently be explained by the greater influence of the matrix in comparison with conventional solvents, i.e. by the specific nature of component interaction.

Calculation of the characteristic values of the viscosity and elasticity coefficient by the expressions

$$[\eta]=\lim_{C\to 0}\frac{\eta_2-\eta_1}{\eta_1 C} \qquad [A_G]=\lim_{C\to 0}\frac{A_{G,2}-A_{G,1}}{A_{G,1}C}$$

where the subscripts 1 and 2 stand for the individual components, shows their complicated dependence on the molecular weights of the components [54], $[\eta] \propto M_1^{-\alpha} M_2^{\beta}$; here $\alpha = \beta = 0{\cdot}8$. But when $M_2 > M_1^2$, we have $[\eta] \propto M_2^{0{\cdot}5}$, which corresponds to the laws observed by Vinogradov *et al.* [53]. This indicates that the behavior of the high-molecular-weight component in a blend is the same as in a good solution. In other words, in blends of amorphous polymers of an indentical nature, the low-molecular-weight components should play a role like that of theta solvents: $[A_g] \propto M_1^{-\gamma} M_2^{\delta}$, where $3 > \delta > 1$, while $\gamma = 2$.

Of interest is the question of the relation between the values of η_0 and A_G^0 in blends, which can be appraised by the change in the equilibrium compliance of the system $J_e = A_G^0/\eta_0^2$. Philippoff [55] had already indicated that dilute solutions of high-molecular-weight polymers (polyisobutylenes—PIB) can reveal very high reversible deformation (up to 20 000%). Graessley [56] showed that solutions of high-molecular-weight polymers have clearly expressed peaks on the plots of the equilibrium compliance vs the concentration in the region of low concentrations. With a view to these ideas, it is interesting to compare the behavior of the systems considered above.

The concentration dependence of the equilibrium compliance (Fig. 3.15) shows that for blend A, a peak is observed on the plot of J_e^0 vs C at $C \approx 0{\cdot}5\%$. For blends A and D, the peak is blurred, which is connected with the lower values of the MW of the components being blended.

The existence of various laws of the viscoelastic behavior of polymer blends in various regions of concentrations of the high-molecular-weight component signifies the presence of 'critical' values of the concentrations above which the viscoelastic behavior of a system changes sharply. The critical values of the concentrations corresponding to the beginning of overlapping of the polymer coil spheres can be determined approximately after Debye: $C_{cr} = 1{\cdot}08/[\eta]$, where η is the characteristic viscosity of the high-molecular-weight component. Calculations show that the values of C_{cr} for blends A, B, and C correspond to 0·5, 0·7, and 1% respectively, i.e. to about the same values of the concentrations at which the peak is discovered on the plot of J_e^0 vs C.

We can consider from the determination of the critical concentrations that the plots in Fig. 3.12 (curves 1–5) up to concentrations of 0·5% describe the behavior of macromolecules with a high MW that do not interact with one another. Their viscoelastic behavior is featured by interaction with the surrounding polymer matrix, which they perceive as a viscoelastic medium.

With this in view, it is expedient to compare the viscoelastic behavior of the polymer blends being considered with that of extremely dilute

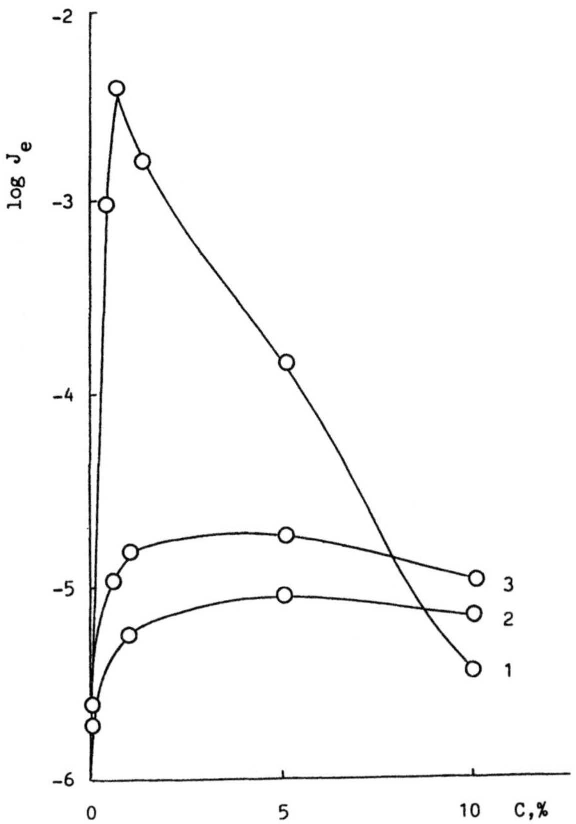

FIG. 3.15. Equilibrium compliance of blends J_e^0 vs concentration of high-molecular-weight component in blends A (curve 1) and D (3).

solutions of linear polymers. For this purpose, it is convenient to use the results of Johnson *et al.* [57], whose conclusions are consistent with the predictions of the Rouse–Zimm theory for good solvents with consideration of a hydrodynamic interaction of a moderate nature (the frequency dependences of the dynamic characteristic of extremely dilute solutions have a slope of 1/2). The characteristic values of the modulus of storage

$$[G'] = \lim_{C \to 0} (G'/C)$$

calculated after Johnson *et al.* [57] (Fig. 3.16) show that for a blend of type A the plot of G'/C vs C is approximated well by a straight line only when $C < 0{\cdot}5\%$, while when $C \geqslant 0{\cdot}5\%$ the values of G'/C grow sharply. Moreover, the curve of $[G']$ vs ω is close to G' vs ω for the component playing the role of the matrix.

Hence, generalization of the well-known theory of polymer solutions by extrapolation to zero values of the concentrations of one of the components and calculation of the characteristic values of the moduli of storage, viscosity and elasticity coefficients are highly informative when analyzing the finest nuances of the viscoelastic behavior of blend compositions.

The results considered above convincingly show that even for thermodynamically compatible blends whose components are fractions of poly-

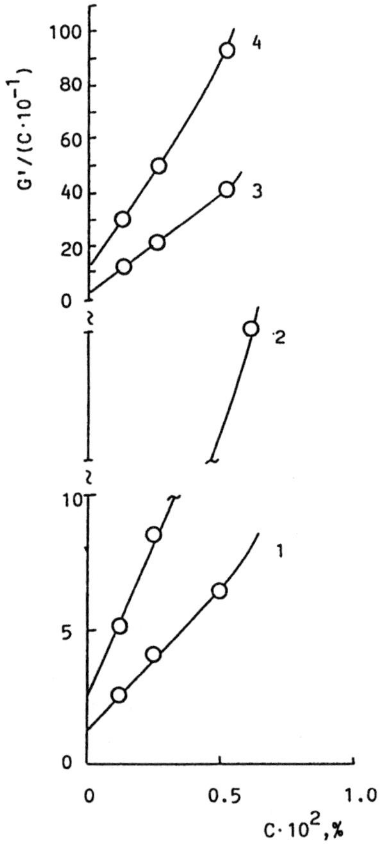

FIG. 3.16. Characteristic values of modulus of storage G' for blends of series A. Curves 1–4 show the values of the circular frequency $\omega(s^{-1})$: 1·0 (curve 1), 1·6 (2), 2·5 (3) and 4·0 (4).

mers of the same homologous series, precision rheological measurements on the one hand, and an analysis of the concentration dependences of the initial and characteristic parameters of a system on the other, confirm the existence of microheterogeneity in such systems even from the lowest concentrations of one of the components. Privalko *et al.* [14] have arrived at a similar conclusion for thermodynamically compatible blends of PVDF whose viscoelastic behavior has already been discussed above.

Our analysis of the rheological behavior of model blends of linear polymer fractions would be incomplete if we considered only the region of concentrations of one of the components from about 0·1 to 20% (see Table 3.2). Therefore we shall go over below to very simple binary blends in which both components are of high molecular weight ($M/M_c \geqslant 10$) with concentrations ranging from 20 to 80%. A number of publications have appeared giving the results of measuring the viscoelastic properties of blends of the type of PS [58–62] and PMMA [63, 64]. However, in our opinion the results obtained by Vinogradov *et al.* [10] for PB blends are clearer. They analyzed the rheological and relaxation characteristics of these blends within a broad range of temperatures and frequencies, the blends containing 20, 50, and 80% of one of the components.

For simplification and better illustration of the picture of relaxation behavior, we compare the frequency relations of the loss moduli of the initial components and their blends at two temperatures, as shown in Fig. 3.17. We first consider the case of relatively low content of the low-molecular-weight component. At high temperatures, two peaks of the loss modulus are observed, though they are expressed only slightly. With lowering of the temperature, a plateau appears on the frequency dependence of the loss modulus. It covers the entire region of frequencies between the peaks of both components. This signifies that with lowering of the temperature, the individual features of the components in the blends are less manifest. This may occur for the same reason why at low temperatures in experiments on a capillary viscometer the sharpness of the transition from flow to spurt vanishes [10].

With equal fractions of the components, a typical point of inflection appears on the curves of G'' vs ω that corresponds to a frequency close to the frequency of the peak of the loss modulus for the higher-molecular-weight component and indicates its transition to the rubber-like state. At all temperatures in the blends considered, a peak of the loss modulus is observed that coincides with the relevant peak of the lower-molecular-weight component.

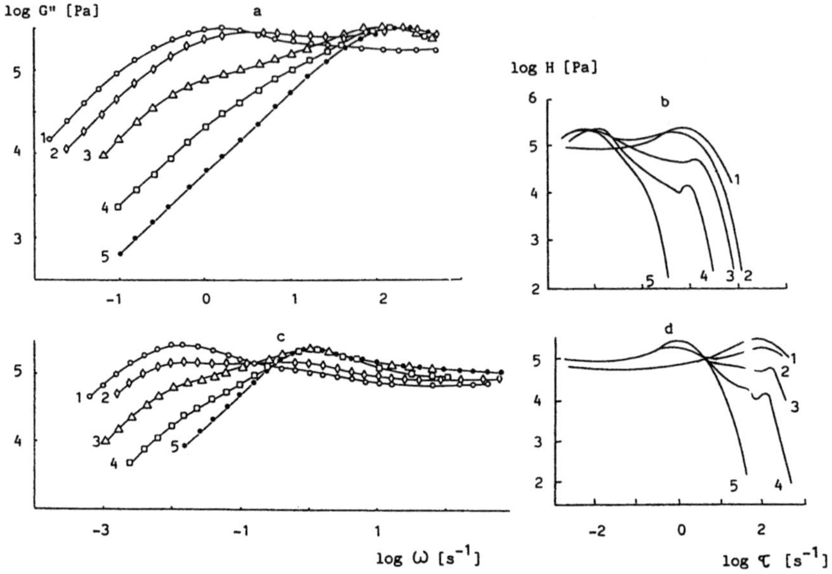

FIG. 3.17. Viscoelastic characteristics of PB and their binary blends ($\bar{M}_w = 8{\cdot}5 \times 10^4$; $\bar{M}_w/\bar{M}_n = 1{\cdot}4$; $\bar{M}_w = 3{\cdot}2 \times 10^5$; $\bar{M}_w/\bar{M}_n = 1{\cdot}05$). The high-molecular-weight component includes 0 (curve 1), 20 (2), 50 (3), 80 (4), and 100% (5) of the low-molecular-weight component. (a) Loss modulus G'' at 333 K; (b) relaxation spectrum of H at 333 K; (c) G'' at 293 K; (d) spectrum of H at 293 K.

In blends with a low content of the higher-molecular-weight component, its presence is detected by the nature of the change in the curves of G'' vs ω in the region of frequencies at which this component passes over into the rubber-like state.

Important conclusions follow from the consideration of the dynamic characteristics of binary blends of this type.

(1) Within the broad range of binary blend compositions, a more or less sharp transition of the high-molecular-weight component to the rubber-like state is discovered.
(2) Within the broad range of frequencies, the viscoelastic properties of polymer blends (polydisperse systems) cannot be described by theories introducing averaged parameters, such as the average characteristic relaxation time.
(3) The individual influence of the components on the properties of blends increases with elevation of the temperature.

A comparison of the spectra of relaxation time distribution H vs θ for blends and the initial components (Fig. 3.17) points to the existence of two characteristic regions (regions of peaks on the plots of H vs θ). It should be noted that according to Ninomiya [65], these peaks do not lend themselves to calculation by the techniques proposed for calculating the relaxation spectra of blends by Bogue [66] and Prest [67]. The presence of two peaks on the curves of G' and G'' vs ω and of H vs θ, each of which corresponds to a definite relaxation process in a binary system observed in a definite range of frequencies of deformation and temperature, indicates the kinetic 'incompatibility' of the blend components. It is important that this incompatibility is registered clearly only by the results of dynamic (nondestructive) experiments under frequency scanning conditions. When studying the temperature dependences of the dynamic characteristics, it is much more difficult to obtain a clear picture of two relaxation peaks. The reason, as can be seen from Fig. 3.17, is that a change in the temperature may lead to an unpredictable growth in the role of one of the blend components and leveling out of the role of the other one. The structural glass transition temperature T_g, both for the initial components of PB and for blends thereof, is registered quite reliably and remains practically unchanged with the composition.

The above illustrative material convincingly proves the possibility of inadequate manifestation of relaxation properties in various temperature–frequency (rate) fields even in very simple binary blends. This indicates the fundamental necessity of developing the theory enabling one to calculate the relaxation spectra of blends, especially within a broad range of a change in the MW of the components. It is also obvious that without such a comprehensive determination of the viscoelastic and relaxation behavior of a blend composition, it is impossible to draw an unambiguous conclusion regarding its compatibility or incompatibility.

3.6.2 Incompatible Blends

When discussing the results of rheological experiments run with polymer blends, calculation of their relaxation characteristics clearly reveals the definite restrictions imposed by the specific nature of the behavior of such systems in their transition from one physical state to another. This specificity is determined by the relaxation nature of the observed phenomena. It is natural to ask whether preference should be given to a particular region of the physical state (in the sequence from fluid through rubber-like and leather-like to glassy) when procuring information on the relaxation behavior of blend compositions.

We have already shown the importance of determining the initial or characteristic rheological parameters in the terminal zone or, for example, analyzing the relaxation time spectra in the region of the rubber-like state. Special attention must be given to the fact that the realization of a physical state in the indicated experiments was achieved under conditions of a change in the frequency (or rate) of deformation at a constant temperature, i.e. with a change in the 'intensity' of the mechanical field. According to Yanovsky and Vinogradov [52], physical states realized under such conditions are interpreted as 'forced' ones. On the other hand, a change (elevation) of temperature can also result in various changes in the physical state of a system, from glassy to fluid. It is obvious that from an experimental viewpoint it is simpler to run the latter studies. Only a comparison of the relaxation characteristics of a system registered independently under the conditions of both approaches will enable one, in our opinion, to answer the above question sufficiently convincingly (see Chapter 2).

To illustrate this conclusion, let us turn to the results obtained by Yanovsky *et al.* [11]. They studied PB of various microstructures, namely 1,2- and 1,4-polybutadienes with close molecular weights having a narrow MWD and blends thereof (the samples are characterized in Table 3.3). The blends were prepared by the joint dissolving of the components in benzene at room temperature. The temperature dependence of the dynamic characteristics of these blends indicates (Fig. 3.18) the involved

TABLE 3.3
Characteristics of the Investigated Polybutadienes

Characteristics	*1,4-PB*	*1,2-PB*
Molecular structure	—(CH_2—CH=CH—CH_2)—	—(CH_2—CH)— with side group —CH=CH_2 on CH
Microstructure, %		
1,2-links	8·3	84
1,4-*trans*	46·7	7·8
1,4-*cis*	46·0	8·2
Molecular weight $M_w \times 10^{-5}$	1·1	1·35
T_g, K	177	238
[η] 25°C, toluene	1·32	1·28
M_w/M_n	1·15	1·2

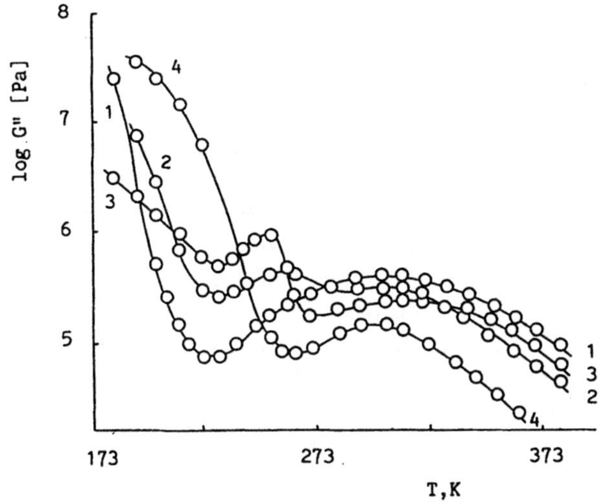

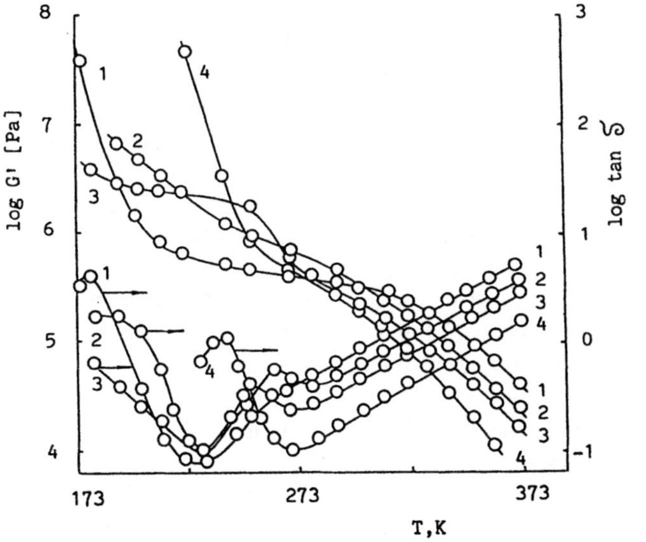

FIG. 3.18. Moduli of storage G' and loss G'', and loss tangent $\tan\delta$, vs temperature T (K): 1,4-PB (curve 1), 20% of 1,2-PB (2), 40% of 1,2-PB (3) and 1,2-PB (4). The deformation frequency was $\log\omega = 0{\cdot}8\ s^{-1}$.

nature of the change in their relaxation characteristics. The descending left-hand branches correspond to the transition from the glassy to the rubber-like state, while descending right-hand branches are connected with reaching the region of a fluid state. The peaks on the curves of $\tan\delta$ vs T at 177 and 238 K (Fig. 3.18, curves 1 and 4) characterize the glass transition temperatures of the initial 1,4- and 1,2-PB, respectively. Two peaks exist for the blends and their position approximately corresponds to the glass transition temperatures of the initial samples. Such a picture is typical of incompatible polymers [68]. But in the region of the fluid state, a change in the composition of a blend is felt weakly—the curves of $\tan\delta$ vs T for elevated temperatures change their slope only insignificantly (curves 1–4 are virtually parallel). The plots of G' and G'' vs T within a broad temperature range have a complicated nature, viz. a plateau and blurred peaks corresponding to the rubber-like, transition, and fluid states, a sharp drop in the moduli G' and G'' at elevated temperatures in the region of the fluid state, and their growth with lowering of the temperature until the attainment of a leather-like state (transition to the glassy state). At temperatures much higher than the glass transition temperature of a component, the viscosity for incompatible blends is practically independent of the composition, and by this parameter a system can be considered as compatible. The curves of G' vs T for the fluid state region were used to calculate the values of η_0 and A_G^0 vs T (Fig. 3.19). The plots of these values are straight lines with different slopes varying from 1·6 to 3·0. The straight lines of η_0 and A_G^0 vs T for blends occupy an intermediate position with respect to the initial polybutadienes. The activation energy of viscous flow for 1,4- and 1,2-PB is 8 and 14 kcal mol^{-1}, respectively, which points to the quite strong influence of the microstructure on the process.

When $T > T_g + 50$, the viscosity is an additive function of the blend composition, $\log\eta_{b1} = C\log\eta_1 + (1-C)\log\eta_2$, where C is the weight content of the added component, and η_1 and η_2 are the viscosities of the components.

The features of the viscoelastic behavior of the blends being considered in various regions of 'forced' states (fluid—rubber-like—transition to glassy—leather-like) that are manifested when considering the curves of G' and G'' vs ω show that, depending on the temperature, the dynamic characteristics reflect the specific nature of the relaxation behavior of blends in various ways. At 253 K, they enable one to characterize the behavior of blends in the region of the fluid state (the descending branches at the lowest frequencies), the rubber-like state (the plateau on

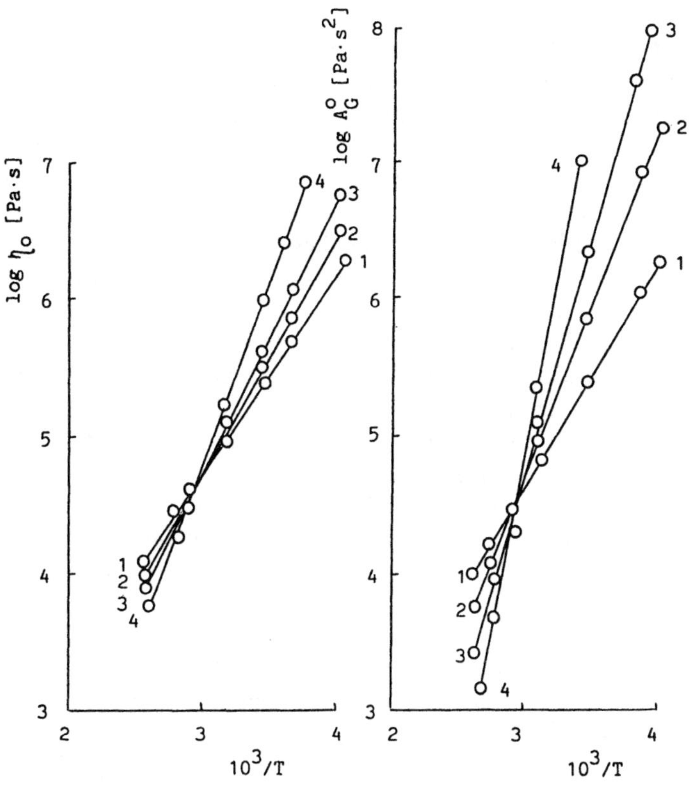

(a) (b)

FIG. 3.19. Temperature dependence of (a) the maximum Newtonian viscosity η_0 and (b) the initial coefficient of high elasticity A_G^0. For the designations of the curves see Fig. 3.18.

G' vs ω and the peaks on G'' vs ω), and the transition to a leather-like state (the ascending branches of the curves for high frequencies). For the initial samples, the features corresponding to a transition from one state to another are quite pronounced, while for blends they are weaker. The greatest difference between the behavior of samples and that of their blends is observed in the region of the transition from the fluid to the rubber-like state. It should be noted that with elevation of the temperature, the curves for the initial samples and their blends differ only slightly, in practically all states. Owing to the difference in the temperature coefficent of viscosity of the initial components, the temperature dependence of the moduli G' and G'' vs ω for 1,2-PB is greater than that for

1,4-PB. A consequence is the change in the sequence of the curves in the region of elevated temperatures (compare curves 1–4 and 1′–4′) This is well consistent with the temperature dependences of G' and G'' for the indicated substances in the regions of lowered and elevated temperatures (Fig. 3.20). For 383 K, the curves of G' and G'' vs ω for the initial PB and their blends differ insignificantly. The nature of the curves indicates the weakly expressed region of the rubber-like state in the realized frequency range. It should be expected that at these temperatures, the systems with a growth in the frequency change almost immediately from a liquid state to a forced glassy one [52].

Hence, the individual features of the initial components in the region of elevated temperatures for systems of a single polymer homologous series with a varying microstructure are leveled out. Of interest is the quantitative evaluation of the change in the values of η_0 and A_G^0 with the concentration of the components.

Calculations show (see Fig. 3.20, curves 1–3 for 253, 293, and 383 K, respectively) that the rheological constants change in a complicated

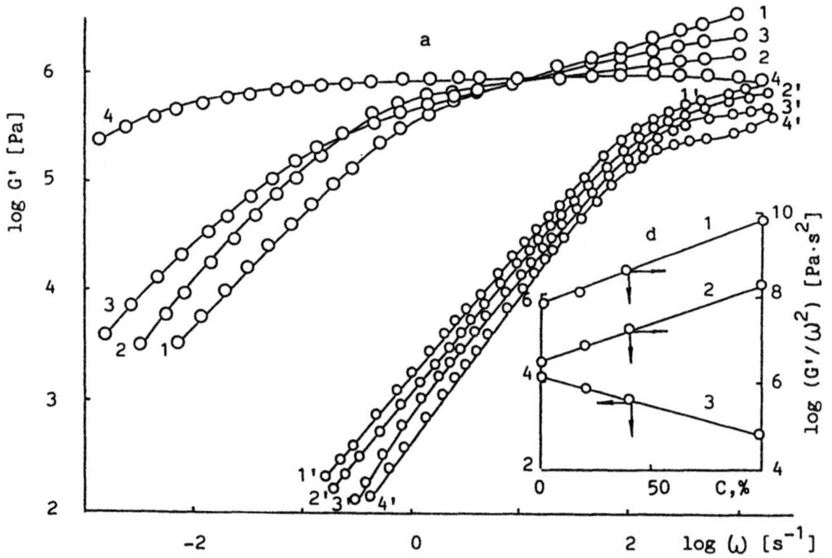

FIG. 3.20. Frequency dependences of (a) G' and (b) G'' at 353 K (curves 1–4) and 383 K (curves 1′–4′). For the designations of the curves see Fig. 3.18. The inserts show (c) the viscosity G''/ω and (d) the elasticity coefficient G'/ω^2 vs the concentration C (%) in a blend of 1, 2-PB.

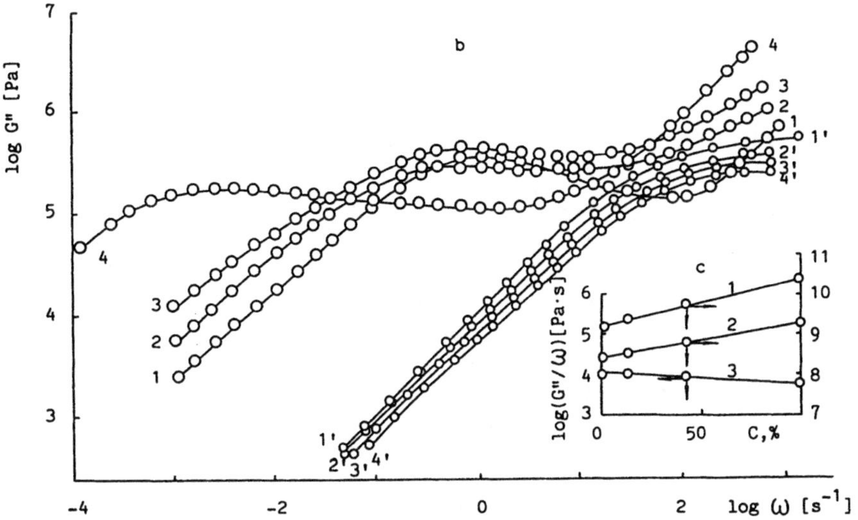

FIG. 3.20 *Cont.*

manner depending on the concentration of the blend components. At $T \leqslant T_g + 50$, we have $\eta_0 \propto C^{0 \cdot 3}$; for higher temperatures, the influence of the concentration is much lower, and $\eta_0 \propto C^{-0 \cdot 1}$. Within the temperature range 253–283 K, $A_G^0 \propto C^{0 \cdot 6}$, while at 383 K $A_G^0 \propto C^{0 \cdot 3}$. It is important that at 383 K the slope of the curves of η_0 and A_G^0 vs C changes, which is apparently a result of the different contributions of the blend components to the flow process.

When comparing the plots of G' and G'' vs ω for one blend (Fig. 3.21) procured for different temperatures, we can observe a curious trend, viz. a change in the region of the rubber-like state with the temperature. With lowering of the temperature, the length and height of the rubber-like plateau grow sharply, while the relaxation transition characterized by the peak on the curve of G'' vs ω is shifted to the region of low frequencies (by over two orders of magnitude).

Hence, the information considered above (Figs 3.18–3.21) reveals that the temperature dependences of the viscoelastic characteristics identify the relaxation properties of a blend composition quite reliably in the region of the glassy, leather-like, and rubber-like states, but are poorly informative visually in the region of the fluid state. True, the plots of G' and G'' vs T for the region of the fluid state can be used to calculate the initial rheological constants [69]. At the same time, the frequency

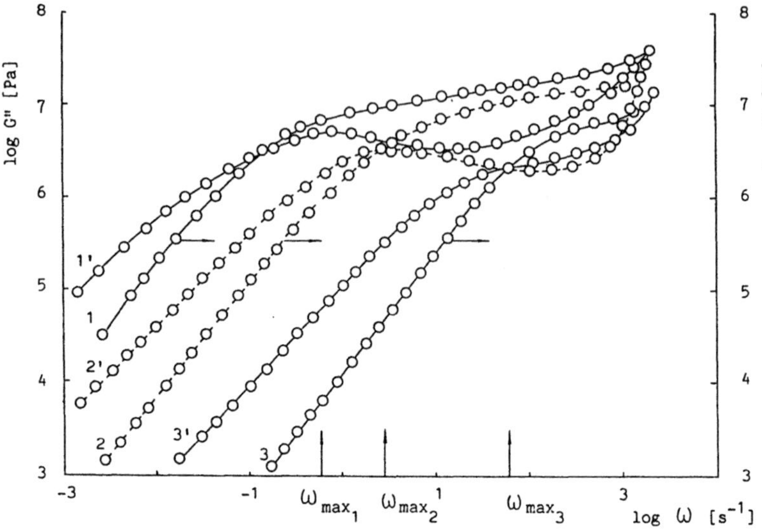

FIG. 3.21. Moduli G' and G'' vs the deformation frequency ω for a blend with 20% of 1, 2-PB at 253 K (curves 1 and 1′), 293 K (2 and 2′) and 383 K (3 and 3′).

dependences of the viscoelastc characteristics yield extensive information on the relaxation behavior of blends in the region of the fluid and rubber-like states. It is exactly by analyzing these relations that the most important features of behavior of blends of linear polymers varying in microstructure are established. In the given specific case, they have shown that when $T \leqslant T_g + 50$ the relaxation characteristics of the system are determined by the properties of the component with the higher glass transition temperature. When $T > T_g + 50$, the individual properties of the blend components are not evident for practical purposes.

3.7 CONCLUSION

The rheological behavior of blend compositions is exceedingly diverse and involved. Even blends of linear flexible-chain polymers, which have been studied well enough to date and are the simplest from the rheological and structural viewpoints, exhibit a diverse spectrum of viscoelastic anomalies depending on the concentration, nature and MW of the components, temperature, deformation conditions, etc. In different regions of component concentrations, they may exhibit properties typical of solutions, blend compositions, or highly plasticized systems with weakly-manifested rubber-like properties.

The most sensitive characteristic of a blend composition is its elasticity coefficient. The latter can change tens of thousands of times more than the viscosity with a change in the concentration composition of the components. This is especially obvious when the content of the high-molecular-weight component is low. This illustrates one way of monitoring the viscoelastic properties of a composite material, primarily by its elasticity.

At present it is difficult to recommend universal empirical relations that predict quantitatively the changes in the elastic properties of a system, because the latter are determined to a considerable extent by the intermolecular interaction of the components. The latter can be considered only by a strictly model approach.

We can assert that even compatible systems have an internal micro-heterogeneity due to the presence of diversified (in their nature or even in the value of the MW) macromolecules. This is supported quite reliably by precision rheological techniques (in particular, by the dynamic one). Of special significance here is analysis of the 'limiting' rheological parameters, namely, the initial and characteristic values of the viscosity and the elasticity coefficient.

A very important role in interpreting the mechanism of viscoelastic and relaxation behavior of blend compositions is played by determination of the thermodynamic properties and morphology, especially when speaking of analysis in the critical regions of compositions and temperatures.

A new approach to determining the compatibility of polymer blends in the fluid state has been undertaken recently [70, 71]. Analysis of the literature and our own experimental results revealed definite common features of the rheological behavior and structure of various blends of incompatible polymers regardless of the nature of the polymer in the dispersed phase. In logarithmic coordinates, the plots of the viscosity of equiconcentrated blends, normalized with respect to the viscosity of the matrix (η_{b1}/η_m) or the dispersed phase (η_{b1}/η_d), against the ratio of the viscosities of the initial components (η_m/η_d), were found to be linear and universal for different pairs of polymers. This indicates that the given reduction parameters are related by a power law. It was shown, with quite a high correlation coefficient, that with an increase in the ratio (η_m/η_d) the viscosity of a blend normalized according to the matrix viscosity diminishes linearly, while the viscosity of a blend normalized according to the dispersed phase viscosity grows. These relations were plotted for various polymer pairs. Either ordinary thermosetting polymers [72–75] or liquid-crystalline [76–81] and mesophase polymers [82] were employed as the dispersed phase. The ratio (η_m/η_d) in the blends being considered was varied either by changing the molecular weight or by altering the viscosity of the initial components by monitoring the shear rate and temperature. The proposed generalization clearly shows the location of the regions of various rheological behavior with respect to the parameter η_m/η_d. Coincidence of the logarithmic plots of η_{b1}/η_m vs η_m/η_d and of η_{b1}/η_d vs η_m/η_d corresponds to the condition of equality of η_m and η_d. When the parameter η_m/η_d equals 0·1 or 7, the mixture flows with the viscosity of the more fluid component, i.e. here $\eta_{b1} = \eta_m$ and $\eta_{b1} = \eta_d$. It has also been shown [70, 71] that different morphological flow patterns correspond to the typical rheological regions indicated when blends flow in channels.

At very low values of $(\eta_m/\eta_d) < 0{\cdot}1$, the structure of the flow reminds one of that of a filled polymer. When $(\eta_m/\eta_d) \approx 0{\cdot}1$, the best fibrillar structure forms in the flow. Such a self-reinforced system has the best strength indices. A further growth of η_m/η_d lowers η_{b1}. When $\eta_m/\eta_d \approx$ 0·1–7, conditions are produced for the formation of a shell of low-viscosity component at the periphery of the flow. When $\eta_m/\eta_d > 10$, layers of fibers form, and the viscosity of the blend continues to decrease because of the unique macrostructural plasticization of the system.

The proposed generalization has a predictive strength. Knowing the viscosity properties of the initial components, one can estimate the viscosity of a blend of a given composition and the structure of its extrudates beforehand.

The above method can be used, in our opinion, for determining not only structural but also relaxation inhomogeneity. The latter is urgent, for instance, for blends of narrow fractions of monodispersed polymers. We employed the above normalization procedure to process the rheological data obtained for blends of a fraction of monodispersed polybutadiene and polyisoprene (Fig. 3.22). It can be seen that even for blends of compatible polymers (e.g. fractions of the same polymer) the rule of logarithmic additivity is not observed. In the region of $(\eta_m/\eta_d) > 0$, the viscosity of a blend is always lower than that of the matrix, while for $(\eta_m/\eta_d) < 0$, the viscosity of a blend is higher than that of the matrix, i.e. here the blend behaves like one filled with a viscoelastic filler, the role of which is played by the high-molecular-weight additive. The above curve passes through the origin, where the viscosity of the blend and the equimolecular components are the same. Only at this point do we have a system absolutely compatible in every way.

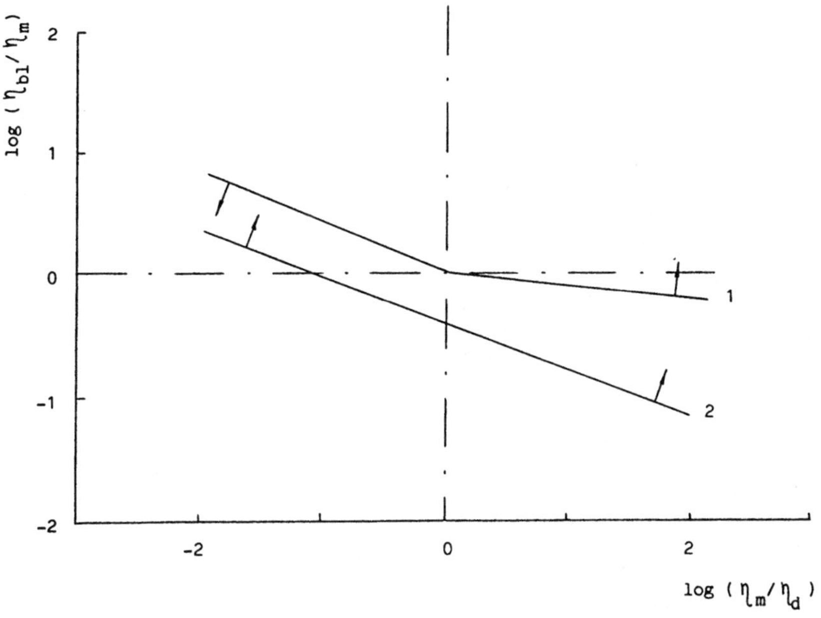

FIG. 3.22. Normalized viscosity of blend (according to matrix) vs ratio of the initial component viscosities for compatible polymer (a blend of narrow fractions of polybutadiene and polyisoprene) (curve 1) and incompatible polymers (curve 2) (after Ref. 70). The concentration is 30% by weight. The arrows show the evolution of the change in the relations with decreasing concentration.

REFERENCES

1. KLYKOVA, V.D., YANOVSKY, YU.G., KULEZNEV, V.N., *et al. Vysokomol. Soed.*, **22B**(9) (1980) 684.
2. KLYKOVA, V.D., CHALYKH, A.E., VERSHININ, L.V., *et al. Vysokomol. Soed.*, **27A**(4) (1985) 724.
3. VERSHININ, L.V., YANOVSKY, YU.G., KULEZNEV, V.N., *et al. Vysokomol. Soed.*, **30A**(5) (1988) 831.
4. UTRACKI, L.A. *Polym. Eng. Sci.*, **23**(11) (1983) 602.
5. KULEZNEV, V.N. *Smesi Polimerov* (Polymer Blends), Khimiya, Moscow, 1980.
6. CHUANG, H.-K. and HAN, C.D. *J. Appl. Polym. Sci.*, **29** (1984) 2205.
7. VAN OENE, H. *J. Coll. Interface Sci.*, **40** (1972) 448.
8. HAN, C.D. *Multiphase Flow in Polymer Processing*, Academic Press, New York, 1981.

9. YANOVSKY, YU.G., VINOGRADOV, G.V. and IVANOVA, L.I. *Vysokomol. Soed.*, **24A**(5) (1983) 1057.
10. VINOGRADOV, G.V., YANOVSKY, YU.G., MALKIN, A.YA., *et al. Vysokomol. Soed.*, **20A**(11) (1978) 2403.
11. YANOVSKY, YU.G., VINOGRADOV, G.V. and IVANOVA, L.I. *Inzhenerno-Fiz. Zh.*, **46**(6) (1984) 974.
12. YANOVSKY, YU.G. In: *Teoreticheskaya i Instrumental'naya Reologiya* (Theoretical and Instrumental Rheology), Izd. I-t Teplomassoobmena AN BSSR, Minsk, 1970, p. 119.
13. YANOVSKY, YU.G., VINOGRADOV, G.V., ZHDANYUK, V.K., *et al. Rheol. Acta*, **27**(3) (1988) 298.
14. PRIVALKO, V.P., NOVIKOV, V.V. and YANOVSKY, YU.G. *Osnovy Teplofiziki i Reofiziki Polimernych Materialov* (Principles of Thermophysics and Rheophysics of Polymer Materials), Naukova Dumka, Kiev, 1991.
15. PAUL, D.R. and NEWMAN, S. (eds). *Polymer Blends*. Academic Press, New York, 1978.
16. HORIO, M., FUJII, T. and ONOGI, S. *J. Phys. Chem.*, **68** (1964) 778.
17. KULEZNEV, V.N., KONYUKH, I.V. and VINOGRADOV, G.V. *Kolloid Zh.*, **27**(4) (1965) 540.
18. NINOMIJA, K. and FERRY, J. *J. Colloidal Sci.*, **18** (1963) 421.
19. UEMURA, S. and TAKAYANAGI, M. *J. Appl. Polym. Sci.*, **10** (1966) 113.
20. ULYANOV, L.P., YANOVSKY, YU.G. and NEIMARK, V.M. *VINITI*, 1972–78, Moscow, 1978.
21. KULEZNEV, V.N. and ANDREEVA, V.M. *Vysokomol. Soed.*, **4** (1963) 1951.
22. TAGER, A.A. *Physical Chemistry of Polymers*, trans. D. SOBOLEV and N. BOBROV, Mir Publishers, Moscow, 1978.
23. ESKIN, V.E. and NESTEROV, A.E. *Vysokomol. Soed.*, **8A** (1966) 1045; **9B** (1967) 192.
24. LIROVA, B.I., SMOLENSKY, A.L., SAVCHENKO, T.L., *et al. Vysokomol. Soed.*, **14B** (1972) 265.
25. SEMENCHENKO, V.K. *Zh. Fiz. Khimii*, **21** (1947) 1461; *DAN SSSR*, **73** (1950) 331; **80** (1951) 903.
26. TAGER, A.A., DREVAL, V.E. and KHABAROVA, K.G. *Vysokomol. Soed.*, **6** (1964) 1593.
27. BAKEEV, N.F. and LAKOVA, I.S. *Vysokomol. Soed.*, **14A**(5) (1972) 2443.
28. KULEZNEV, V.N., KANDYRIN, L.B. and KLYKOVA, V.F. *Kolloid. Zh.*, **34**(2) (1972) 231.
29. KULEZNEV, V.N. In: *Mnogokomponentnye Polimernye Sistemy* (Multicomponent Polymer Systems), Ed. R. Gold, Khimiya, Moscow, 1974, p. 10.
30. KULEZNEV, V.N., KLYKOVA, V.D., CHERNIN, E.I., *et al. Kolloid. Zh.*, **37** (1975).
31. VOLMER, M. *Zh. Phys. Chem.*, **206** (1957) 181; **207** (1957) 307.
32. KOGANOVA, A.A., FEDOSEEVA, N.P., KUCHUMOVA, V.M., *et al. Kolloid. Zh.*, **39** (1977) 1199.
33. KULEZNEV, V.N., KROKHINA, L.S. and DOGADKIN, B.A. *Kolloid. Zh.*, **31** (1969) 853.
34. ABATUROVA, M.A., VLODAVETA, I.N. and REHBINDER, P.A. *Kolloid. Zh.*, **34** (1972) 315.

35. ULYANOV, L.P., YANOVSKY, YU.G., NEIMARK, V.M., *et al. Zavodskaya Labor.*, **39**(11) (1973) 1402.
36. RUSANOV, A.I., SHCHUKIN, E.D. and REHBINDER, P.A. *Kolloid. Zh.*, **30**(4) (1968) 573.
37. JOANNY, J.F. and LEIBI, L. *J. Phys.*, **39**(5) (1978) 951.
38. FRENKEL, YA.I. *Kineticheskaya Teoriya Zhidkostei* (The Kinetic Theory of Liquids), Nauka, Leningrad, 1975.
39. SEMENCHENKO, V.K. *Kolloid. Zh.*, **24**(3) (1962) 323; **24**(5) (1962) 611.
40. KULEZNEV, V.N., SHVARTS, A.G., KLYKOVA, V.D., *et al. Kolloid. Zh.*, **27** (1965) 211.
41. LIPATOV, YU.S. and VILENSKY, V.A. *Vysokomol. Soed.*, **17A**(9) (1975) 2069.
42. BUDTOV, V.P. *Vysokomol. Soed.*, **25**(3) (1983) 493.
43. CHALYK, A.E. and AVDEEV, N.N. *Vysokomol. Soed.*, **27A**(12) (1985) 2467.
44. MIROSHNIKOV, YU.P. *Mekhanika Komposit. Mater.*, **4** (1984) 104.
45. KRUYT, H. (Ed.). *Colloid Science*, Vol. 1, Holland, Amsterdam, 1952.
46. CHOI, S.J. and SHOUWALTER, W.R. *The Physics of Fluids*, **18**(4) (1975) 420.
47. HELFAND, E. and SAPES, A.M. *J. Chem. Phys.*, **62**(4) (1975) 1327.
48. LIPATOV, YU.S. *Kolloidnaya Khimiya Polimerov* (Colloid Chemistry of Polymers), Naukova Dumka, Kiev, 1984.
49. POKROVSKY, V.N. and YANOVSKY, YU.G. *Rheol. Acta*, **12**(4) (1973) 280.
50. POKROVSKY, V.N., VOLKOV, V.S. and VINOGRADOV, G.V. *Mekhanika Polimerov*, **4** (1977) 186.
51. POKROVSKY, V.N. and VOLKOV, V.S. *Vysokomol. Soed.*, **20A**(12) (1978) 2700.
52. YANOVSKY, YU.G. and VINOGRADOV, G.V. *Vysokomol. Soed.*, **22A**(11) (1980) 2567.
53. VINOGRADOV, G.V., MALKIN, A.YA., YANOVSKY, YU.G., *et al. Eur. Polymer. J.*, **9**(1) (1973) 85.
54. YANOVSKY, YU.G., VINOGRADOV, G.V. and IVANOVA, L.I. *Kauchuk i Rezina*, **3** (1984) 18.
55. PHILIPPOFF, W. *Viskositat der Kolloide*, T. STEINKOPF, DRESDEN, 1942.
56. GRAESSLEY, W.W. *Adv. Polym. Sci.*, **16**(1) (1974) 89.
57. JOHNSON, R.M., SCHRAG, J.L. and FERRY, J.D. *Polym. Jap.*, **1**(6) (1970) 742.
58. KRAUS, G. and GRUVER, J.T. *J. Polym. Sci.*, **3** (1965) 105.
59. AKOVALI, G. *J. Polym. Sci.*, **5**(2) (1967) 875.
60. MASUDA, T., KITAGAWA, K., INOUE, T. and ONOGI, S. *Macromolecules*, **3** (1970) 116.
61. MURAKAMI, K., *et al. Polym. J.*, **2** (1971) 698.
62. PREST, W.M. and PORTER, R.S. *Polym. J.*, **4** (1973) 154.
63. MILLS, N.I. *Eur. Polym. J.*, **5** (1969) 682.
64. ONOGI, S., MASUDA, T., TODA, S., *et al. Polym. J.*, **1** (1970) 542.
65. NINOMIYA, K. *J. Colloid. Sci.*, **17** (1962) 759.
66. BOGUE, D.C. *Polym. J.*, **1** (1970) 542.
67. PREST, W.M. *Polym. J.*, **4** (1973) 163.
68. MANSON, J.A. and SPERLING, L.H. *Polymer Blends and Composites*, Plenum Press, New York, 1976.
69. YANOVSKY, YU.G., VINOGRADOV, G.V. and IVANOVA, L.I. *DAN SSSR*, **282**(5) (1985) 1190.

70. Borisenkova, E.K., Kulichikhin, V.G. and Plate, N.A. *DAN SSSR*, **314**(1) (1990) 193.
71. Borisenkova, E.K., Kulichikhin, V.G. and Plate, N.A. *Rheol. Acta*, **30**(6) (1991) 581.
72. Tsebrenko, M.V. *Ultratonkie Sinteticheskie Volokna* (Ultrathin Synthetic Fibers), Khimiya, Moscow, 1991.
73. Krasnikova, N.P., Dreval, V.E., Kotova, E.V., *et al. Vysokomol. Soed.*, **24A**(7) (1982) 1423.
74. Tsebrenko, M.V., Resanova, N.M. and Vinogradov, G.V. *Polym. Eng. Sci.*, **20**(15) (1980) 1023.
75. Plochocki, A.P. *Trans. Soc. Rheol.*, **20** (1976) 287.
76. Blizard, K.G. and Baird, D.G. *Polym. Eng. Sci.*, **27**(9) (1987) 653.
77. Siegmann, A. Dagan, A. and Kenig, S. *Polymer*, **26**(8) (1985) 1325.
78. Lee, B. *Polym. Eng. Sci.*, **28**(17) (1988) 1107.
79. La Mantia, F.P., Valenza, A., Paci, M., *et al. Rheol. Acta*, **28**(5) (1989) 417.
80. Kulichikhin, V.G., Vasileva, O.V., Litvinov, I.A., *et al. DAN SSSR*, **309**(5) (1989) 1161.
81. Beery, D., Siegman, A. and Kenig, S. *J. Mater. Sci. Lett.*, **7** (1988) 1071.
82. Borisenkova, E.K., Antipov, E.M., Tur, D.R., *et al. Preprints of Proceedings of 32nd Microsymp. on Macromolecules*, Prague, **8**, 1989.

Chapter 4

Rheological and Relaxation Properties of Copolymers

4.1 INTRODUCTION

The tremendous interest exhibited in copolymers by both scientists and experimentalists is explained by the features of their physicomechanical and physicochemical properties. The practical significance of these systems is associated with the hope of obtaining novel materials with 'exclusive' properties in comparison with homopolymers. It can be seen from the other chapters of the present book that this goal can be achieved in different ways, in particular by creating various compositions—blends, filled, combined ones, etc. Analysis of the literature broadly reveals two typical means of 'constructing' such compositions. The first, and apparently the simpler is the mechanical blending of several ingredients which may be either polymers or nonpolymers. The second is 'chemical' blending. The 'chemical' filling technique, which is briefly described in Chapter 5, can be considered to a certain extent as one of its variants. But it will be expedient to consider the formation of copolymers as the 'classical' way of chemical blending.

4.2 GENERAL AND SPECIFIC FEATURES

Statistical or random (alternating), graft, and block copolymers are distinguished. Within the scope of the present chapter, there is no sense in treating even briefly the features of synthesizing such systems. They have been described in sufficient detail in the relevant literature (e.g. Refs 1–3). We shall note below only the most fundamental distinctions of the chemical construction of such systems that we must have in view for interpreting their viscoelastic and relaxation behavior.

As implied by their name, random copolymers have a random distribution of the monomer units along the main chain, whereas alternating copolymers are characterized by their alternation. The viscoelastic and physical properties of such systems are naturally averaged, the best for alternating copolymers. This relates primarily to the glass transition temperature T_g, a number of physicomechanical characteristics, some viscoelastic parameters, etc. Since the sequences of units of one type in these copolymers are small and cannot separate into another phase, these systems are practically single-phase ones.

Graft copolymers can be considered as chemically bound pairs of homopolymers. Their main feature is that the main chains contain polymer A to which a number of sequences B are grafted [2].

In block copolymers, the macromolecules contain chemically different blocks, e.g. A and B, along the main chain bound at their ends. Doube-block (A–B), triple-block (A–B–A), and multiblock $(A–B)^n$ copolymers of this type are distinguished [1–3].

In their thermomechanical and relaxation properties, block and graft copolymers in many ways remind one of incompatible mechanical mixtures. They reveal relaxation or phase transitions typical of each component.

Having related copolymers to chemically blended systems, it is natural to draw a definite parallel between their properties and those of mechanical mixtures. It is just such a parallel, in our opinion, that will help us interpret the observed features of copolymer behavior.

When dealing with the rheological properties of polymer blends, we noted that they are determined to an enormous degree by the compatibility of the components. That blends of polymers with a high molecular weight are practically incompatible can be considered as an established fact. The incompatibility of blend components leads to the appearance of a two-phase morphology. On the other hand, compatible blends should have a single-phase structure. These are two extreme cases.

In view of the unusually high sensitivity of the viscoelastic parameters to the molecular characteristics of the polymer system, it is also natural to employ the rheological approach when studying the properties of copolymers. The techniques traditionally employed for analyzing homopolymers and their blends can also be employed when studying copolymers. However, as will be shown below, it is much more difficult to identify the rheological data obtained for copolymers than for homopolymers. This relates to the greatest extent to block copolymers.

Having in view the chemical 'construction' of copolymers, it is easy to assume that the rheological properties of random copolymers should in many ways be close to those of homopolymers, whereas the properties of block copolymers should sharply differ therefrom. Since block copolymers are featured by several structural forms that are determined by the sequence of the blocks, their rheological behavior will also differ substantially. There is no doubt that the factor of the intensity of deformation is important in the rheological appraisals of block copolymers. Unlike individual polymers, here quite vividly expressed thixotropic effects should be expected.

In a relatively brief chapter, it is virtually impossible to consider the features of behavior of copolymers of various types and various chemical and physical structures, moreso because the main stream of information in this field relates to block copolymers. The increased attention to them is readily explained by the hope of obtaining a new class of materials. Such prospects are quite realistic if we consider the chemical structure of block copolymers, which is exactly what contains the substantial potential possibilities of these systems. When considering the relation between the structure and properties of block copolymers, it is desirable to consider not only the features of their molecular structure but also the morphology of their supermolecular organization.

According to Ref. 4, two-, three-, and multiblock polycondensation block copolymers are divided into one-, two-, three-, and multitype ones. Two-type ones, in turn, may be thermoplastic, elastomers, or elastoplastic, depending on whether the blocks A and B are both rigid or both flexible or flexible and rigid, respectively. Three-block copolymers (ABA) are called thermoplastic elastomers if the content of flexible blocks in them prevails.

Block copolymers consisting of a combination of elastic and rigid components have increased compliance and creep, and stresses relax in them at a high rate. If the matrix is a rigid component, the presence of an elastic block lowers the modulus of elasticity and increases the mechanical losses of the system in comparison with the relevant homopolymer.

In their morphology, block copolymers may be either amorphous or crystallizing. In the former, relaxation (local and segmental) transitions may be observed in the system, while in the latter phase transitions (melting, crystallization, etc.) are also possible. With respect to each component, such relaxation transitions can be called isophase (occurring in a given phase state). In crystallizing block copolymers, a mesophase transition may occur at a definite temperature or within a narrow temperature interval [5].

Analysis, from the literature, of the results of studying block copolymers differing in their composition and structure reveals some quite general trends of the change in their physicomechanical properties. For example, for noncrystallizing block copolymers with rigid and flexible blocks, an increase in the rigid block content raises T_g. On the other hand, an increase in the flexible block content lowers (by the linear law) both T_g and the melting point T_m, the latter changing less, however. Unlike random copolymers in which T_m is stable, in block copolymers of a similar chemical composition T_m may change, i.e. be lowered depending on the duration of holding at this temperature [1]. This effect is associated with the increase in the microphase stratification of the components.

When considering the physicomechanical characteristics of copolymers, one can also discover a definite analogy with the behavior of polymer blends. For instance, for single-phase systems, the temperature dependences of the moduli of elasticity of a system in the glassy state are intermediate with respect to the relevant dependences of the components. For two-phase block copolymers, similar dependences have two temperature regions in which a sharp change in the modulus of elasticity is observed.

The physicomechanical properties of block copolymers are also affected substantially by the block length [6]. In copolymers containing components with sharply differing T_g, an increase in the length of the rigid blocks is attended by an appreciable elevation of T_m. Microdomains can also form inhomogeneous blocks. The most stable microdomains play the role of the physical links of a three-dimensional network, while the block copolymers themselves in this case are typical two-phase systems. The temperature dependence of the mechanical loss factor for them has two peaks. Their relative height is determined by the volume ratio of the phases and the phase morphology [7]. A small blurred intermediate peak may appear in the temperature interval between the regions of unfreezing of the segmental mobility (α-peaks) of the two components. This peak shows the existence in block copolymers of a thermodynamically stable heterogeneous structure formed by the flexible and rigid blocks. The processes occurring at the boundaries of the components play a noticeable role in such systems.

The above cursory analysis discloses the unusually complicated general picture of the relaxation behavior of block copolymers. The same also concerns their viscoelastic properties, in particular in the region of the fluid state. Factors such as the chemical structure, experimental approach, and methodology make it very difficult to register the involved set of laws relating the rheological parameters to the structural features of block copolymers and to generalize them. Success has been achieved here only for the simplest systems.

With a view to the increased attention given to studying block copolymers as a 'new class' of polymeric materials [1–3], we shall illustrate below in greater detail some typical features of their rheological and relaxation properties. They will be exemplified by specific popular and well-known block copolymers. We also presume that separate treatment is also deserved for the properties of compositions which include block copolymers. Such materials can be related to one of the important practical aspects of employing this class of polymers.

4.3 TWO-BLOCK COPOLYMERS

The features of the rheological and relaxation properties of this type of polymer are fully revealed in the elastomeric block copolymers poly(ethylene *co* propylene) (PECP). It must be noted that depending on the content of the polyethylene blocks this type of copolymer may have various structures and properties [1].

Polyethylene has good low-temperature properties, but a low melting point and quite a low modulus of elasticity at room temperature. Polypropylene has a higher melting point and a higher modulus of elasticity, but it is brittle at low temperatures because of its high glass transition temperature. Random copolymers of ethylene and propylene have a lower tensile strength and hardness than the homopolymers. In block copolymers of ethylene and propylene, depending on the content of the relevant blocks, the properties of each of the homopolymers can be combined. From the rheological standpoint, PECP is convenient for study because it allows experiments to be run within a broad temperature range, which is exceedingly important when studying the region of the fluid state.

Let us consider the typical rheological data obtained under conditions of continuous deformation (capillary viscometer, shear plastometer) for a PECP sample containing about 40% of propylene groups with an intrinsic viscosity $[\eta]$ at 408 K in toluene of 1·55 [8–11]. The flow curves of PECP presented in Fig. 4.1 for a broad temperature interval show the

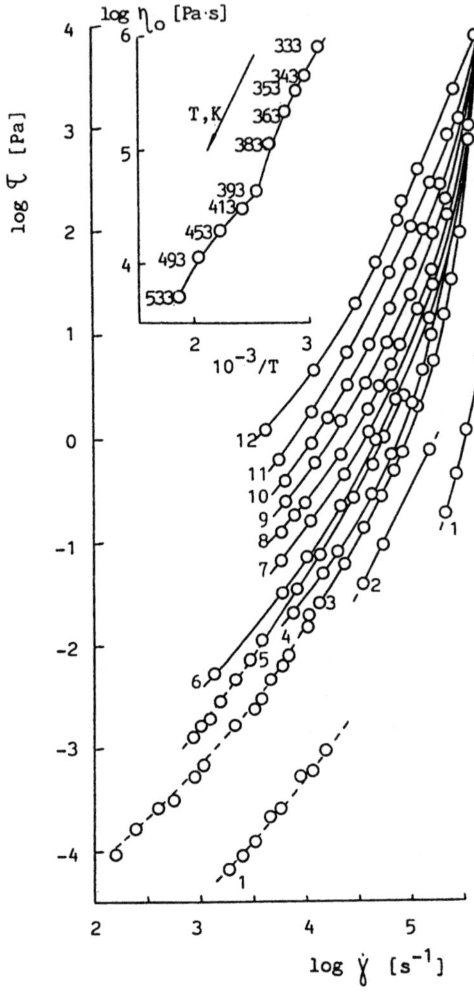

FIG. 4.1. Flow curves, and maximum Newtonian viscosity vs reciprocal temperature, for the block copolymer PECP. The temperatures (K) for curves 1–12 are 298 (curve 1), 313(2), 333(3), 343(4), 353(5), 363(6), 383(7), 393(8), 413(9), 423(10), 493(11) and 533(12).

absence of a monotonic change in the viscosity properties with the temperature. This can be seen especially clearly when considering how the maximum Newtonian viscosity depends on the reciprocal temperature (the upper part of Fig. 4.1). The maximum Newtonian viscosities were determined in Ref. 9 by extrapolation to the zero value of the shear stress (the upper points in Fig. 4.1 on the plots of the rate of strain $\dot{\gamma}$ vs the stress τ show results obtained in a shear plastometer). The activation energy of viscous flow calculated by the Arrhenius equation for temperatures up to 373 K and above 453 K is about 11 kcal mol^{-1} (46 kJ mol^{-1}), whereas in the interval of 383–393 K it rises to 23 kcal mol^{-1} (96·4 kJ mol^{-1}).

The sharp change in the nature of the flow curves in the region of high values of $\dot{\gamma}$ and τ is associated with the appearance of the phenomenon of unstable flow typical of the constricted deformation of a viscoelastic medium in closed channels. It is known to be attended by disturbance of the flow conditions of the system (unstable flow), when the development of elastic deformation causes local breaking away of the material from the capillary walls. The latter is one of the reasons why periodic perturbation of the stream shape appears. It was found [11] that the critical flow parameters corresponding to the onset of unstable flow conditions ($\dot{\gamma}_{cr}$ and τ_{cr}) for PECP vary as a whole with the temperature in the same way as for typical linear stereoregular rubbers, namely, $\dot{\gamma}_{cr}$ grows rather sharply with the temperature, while τ_{cr} is practically constant. However, unlike homopolymers, for which the quantity $\dot{\gamma}_{cr}\eta_0$ is independent of the temperature and is practically constant for a given material, for PECP this rule is not observed [11]. The explanation is the nonmonotonic nature of the change in η_0 within a broad region of temperatures.

At the same time, it should be noted that a criterion of onset of unstable flow conditions such as the Reynolds number of elastic flow Re_{el} [12] has the same value of about 6 for PECP as for homopolymers.

The great change in the viscoelastic properties of PECP registered by the plots of $\dot{\gamma}$ vs τ and η_0 vs T in the regions of 383–393 K and 493–513 K is due to structural transformations—phase transitions connected with melting of the blocks with a structure close to that of low-density polyethylene and polypropylene, respectively, in the macromolecules of the copolymers.

Consideration of the rheological characteristics of PECP shows that beyond the limits of the observed structural transformations, the viscoelastic behavior of this block copolymer differs qualitatively only slightly from that of homopolymers with a linear stereoregular structure [11]. Moreover, it can be seen from Fig. 4.2 that the rheological results for PECP within a definite temperature interval can be presented in a temperature-invariant form as plots of η/η_0 vs $\dot{\gamma}\eta_0$, where η is the apparent viscosity. As for homopolymer melts, for PECP correlation of

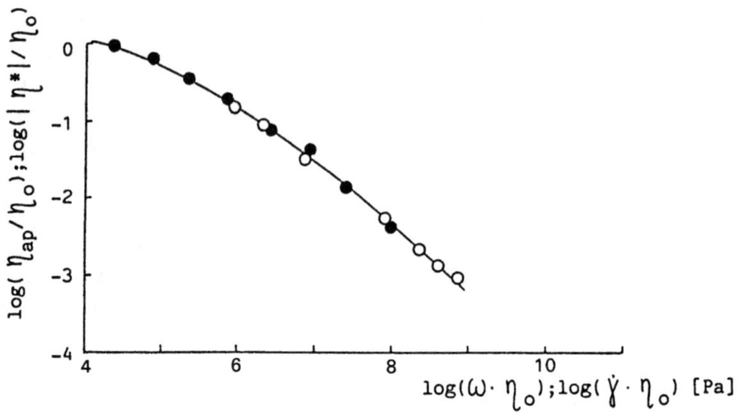

FIG. 4.2. Reduced dependences of the apparent viscosity on the shear rate (●) and of the magnitude of the complex dynamic viscosity on the circular frequency (○) for the block copolymer PECP.

the data obtained in continuous and periodic deformation (low-amplitude periodic deformation) is observed. Curves of $|\eta^*|/\eta_0$ vs $\omega\eta_0$ are also presented in Fig. 4.2 [10].

When speaking of the structural transformations in PECP, it is appropriate to also consider the results of Vinogradov *et al.* [9], who registered these transformations independently by the dynamic and tribometric techniques, and also by the results of differential thermal analysis (Fig. 4.3). The drop in the storage and loss moduli, the peaks on the temperature dependences of the loss tangent tan δ and the extremal change in the force of friction F within the interval of 373–393 K, reliably record the phase transition associated with melting of the polyethylene blocks.

4.4 THREE-BLOCK COPOLYMERS

Butadiene–styrene block copolymers of the SBS type, where S and B are polystyrene and polybutadiene blocks, respectively, are a typical example of three-block copolymers. Owing to the incompatibility of the S and B blocks, such copolymers (thermoplastic elastomers) have a heterophase structure. Here, as indicated by Holden *et al.* [13] and Estes *et al.* [14], the S blocks form domains playing the role of an active filler and, to a

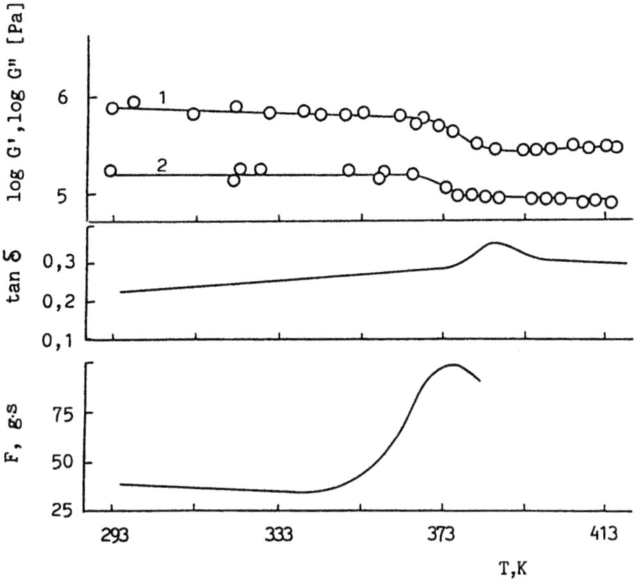

FIG. 4.3. Temperature dependences of the dynamic modulus of storage G' (curve 1), loss modulus G'' (curve 2), loss tangent $\tan\delta$, and force of friction F (a steel hemispherical slide block with radius of 2 mm) for the block copolymer PECP.

certain extent, physical links of the polybutadiene matrix. At normal and low temperatures, the thermoplastic elastomers have properties of compounded rubbers, while above the softening temperature of an S block they acquire fluiditity and can be processed like thermoplastics.

The rheological properties of three-block butadiene–styrene block copolymers of Grade 'Cariflex' with $M = 8 \times 10^4$ and 30% by weight of styrene have been studied in sufficient detail by Vinogradov *et al.* [15, 16]. Let us consider these results briefly in comparison with those for a homopolymer, viz. polybutadiene of the same molecular weight with a narrow MWD.

The flow curves (Fig. 4.4) show that the viscoelastic properties of the SBS-type block copolymers and the homopolymer differ substantially. At stresses τ from 3×10^3 to 3×10^4 Pa, the block copolymer is characterized by a pseudo-Newtonian flow (the slope of the flow curve is close to 45°). In the region of large shear stresses (which are generally the rule when processing the material), the block copolymers, unlike the homopolymer with the narrow MWD, reveal a pronounced non-Newtonian flow. With

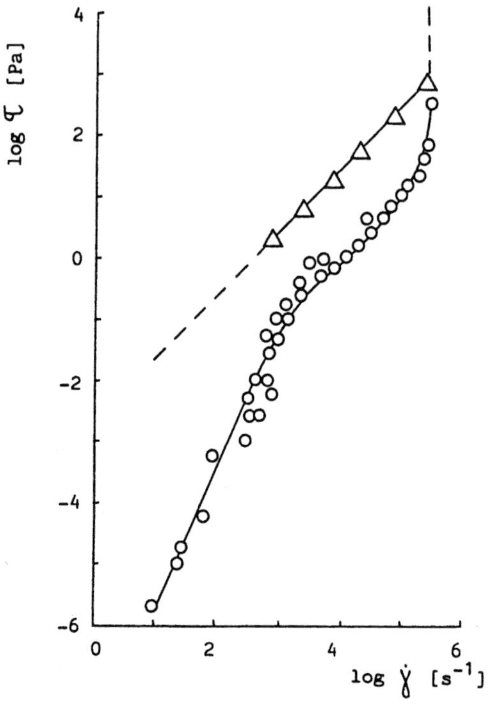

FIG. 4.4. Flow curves of type SBS block copolymer (○) and polybutadiene (Δ) ($M = 8 \times 10^4$) for temperature of 423 K.

an increase in the values of τ as for polybutadiene or other homopolymers, the material begins to slip along the walls of the channel in which it is being deformed (measurements were performed in a capillary viscometer). This spurt effect is a form of unstable flow conditions and is connected with passage of the material into the forced elastic (cold flow) state at high values of $\dot{\gamma}$ and τ [17]. It is important that the value of the critical stress τ_{cr} for both polymers is virtually the same ($\simeq 3{\cdot}5 \times 10^5$ Pa). It is typical that the viscosity of SBS-type block copolymers is scores of times higher than that of polybutadiene at the same molecular mass. Such a behavior of block copolymers at high values of $\dot{\gamma}$ is observed in the temperature interval from 383 K, i.e. a temperature close to the glass transition temperature of the polystyrene block (about 373 K), up to 443 K (Fig. 4.5).

Calculation of the viscous flow activation energy by the Frenkel–Eyring exponential formula shows that it is considerably higher for block

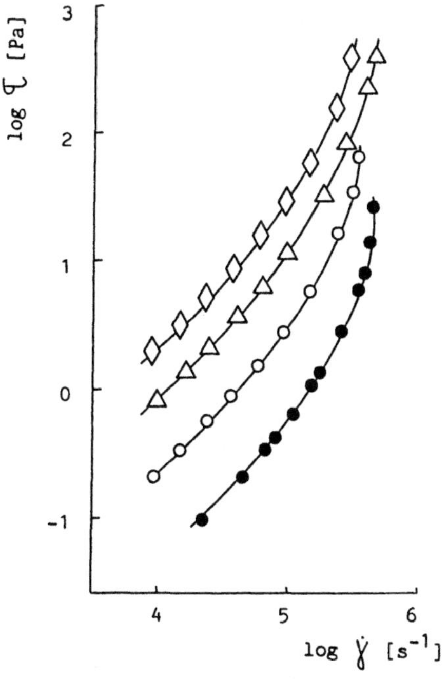

FIG. 4.5. Flow curves of type SBS block copolymer for temperatures of 383 K (●), 403 K (○), 423 K (Δ), and 443 K (◇).

copolymers than for polybutadiene (96·5 and 33·6 kJ mol^{-1}, respectively). This indicates that the viscosity of SBS-type block copolymers depends to a greater extent on the temperature. Finally, it is noteworthy that the flow curves of this polymer are described well by a temperature-invariant characteristic in the coordinates $\log(\eta/\eta_0)$ vs $\log(\dot{\gamma}\eta_0)$ if we assume that the value of η_0 is constant and equals the viscosity of the block copolymer in the region of $\tau = 3 \times 10^3$ to 3×10^4 Pa (Fig. 4.6). However, the temperature-invariant curve for the block copolymer is higher than for homopolymers [18], i.e. an SBS block copolymer in flow reveals a lower viscosity anomaly than homopolymers with a broad MWD.

A sharply pronounced non-Newtonian behavior of SBS block copolymers is also noted in the region of low shear stresses (see Fig. 4.4). The polymer exhibits sharply pronounced thixotropic properties that grow with an increase in the deformation (Fig. 4.7). In the repeated deformation of a material after load removal for three or four hours, its viscoelastic behavior is similar to that of the initial sample (Fig. 4.7). Such

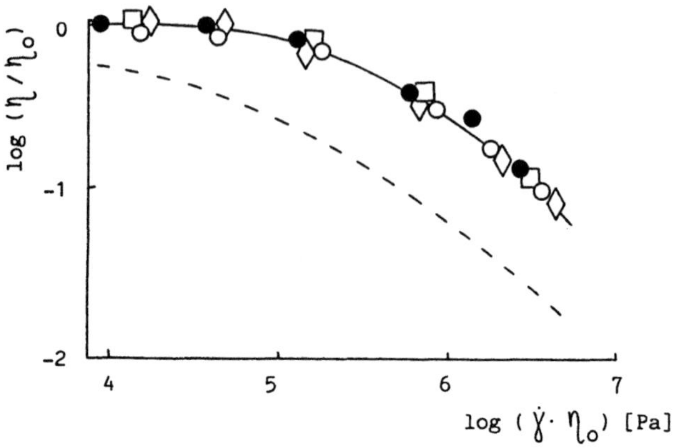

FIG. 4.6. Temperature-invariant characteristic of type SBS block copolymer (——) and homopolymers (– – –) [18] for temperatures of 383 K (●), 403 K(○), 423 K (Δ), and 443 K (◇).

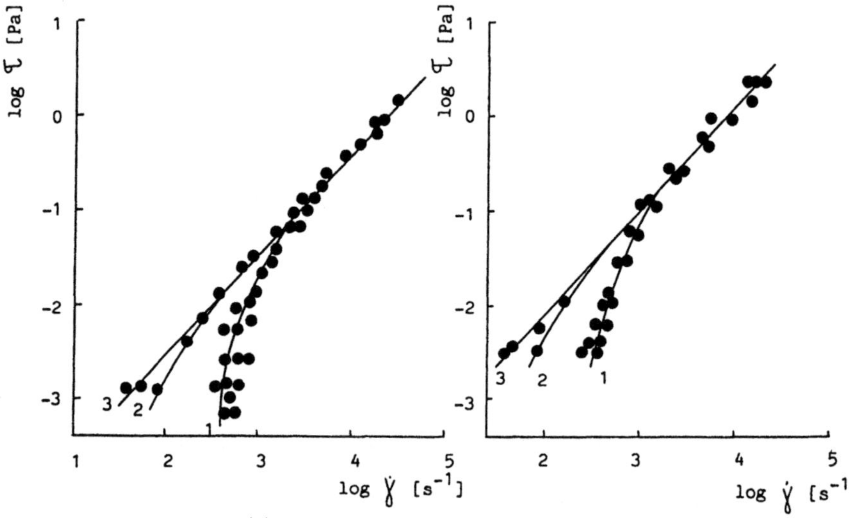

(a) (b)

FIG. 4.7. Flow curves of type SBS block copolymer with consecutive increase of shear rate from 10^{-3} to $10\,s^{-1}$ (curves 1), stepwise diminishing of shear rate (curves 2), and direct passing over from maximum to minimum shear rate (curves 3) at (a) 403 K and (b) 423 K.

behavior is typical of thermoplastic elastomers (see, for example, Ref. 13). The difficulties of control of thixotropy effects appearing when such a polymer is loaded into the working unit of a viscometer or elastoviscometer is the cause of the increased scatter of the experimental results when studying such systems (see also Fig. 4.4).

As a whole, the rheological behavior of block copolymers of such a type is similar to that of structured dispersed systems and filled polymers [19] and is not typical of unfilled homopolymers or statistical copolymers [20].

The unique rheological properties of this type of block copolymers are due to the complicated supermolecular structure of the polymer and its change under the influence of deformation and temperature. The structure of type SBS block copolymers with about 30% by weight of styrene consists of cylindrical polystyrene domains of a colloidal size packed according to a hexagonal system in a polybutadiene matrix [21, 22]. At temperatures above the glass transition temperature of polystyrene, the initial structure of the block copolymer may be destroyed in deformation under the action of the shear stresses [14], the size of the polystyrene domains and their mutual arrangement being determined by the value of τ [23, 24]. With a growth in the shear stress and temperature, destruction of the initial structure increases, and the viscosity of the system diminishes accordingly. After removal of the load, the structure of the block copolymer gradually restores, which is indicated by the results obtained by small-angle X-ray scattering with prolonged annealing [21, 22].

Unlike the viscosity, the rubbery modulus $G_e = \tau/\gamma_e$, where γ_e is the equilibrium rubbery strain, for this block copolymer does not depend on the stress in the region of its low values, and only at sufficiently large values of τ do the values of G_e grow in the same way as for a homopolymer (Fig. 4.8). The temperature only slightly affects the values of G_e for the block copolymer, which is also typical of homopolymers [25]. This allows us to compare the values of G_e measured at room temperatures for type SBS block copolymers and polybutadiene with a narrow MWD. Examination of Fig. 4.8 reveals that in the region of low shear stresses the rubbery modulus of block copolymers is considerably lower than that of polybutadiene, but at high stresses the values of G_e for both polymers coincide. A similar picture is observed when comparing the rheological behavior of polystyrene and block poly(ethylene *co* propylene) [26].

When considering the rheological properties of SBS-type block copolymers, the question arises as to how justified is a comparison of the properties of this copolymer with those of the polybutadiene homopolymer. It should be noted that the reason for this lies in the prevailing contribution of exactly the polybutadiene blocks in the

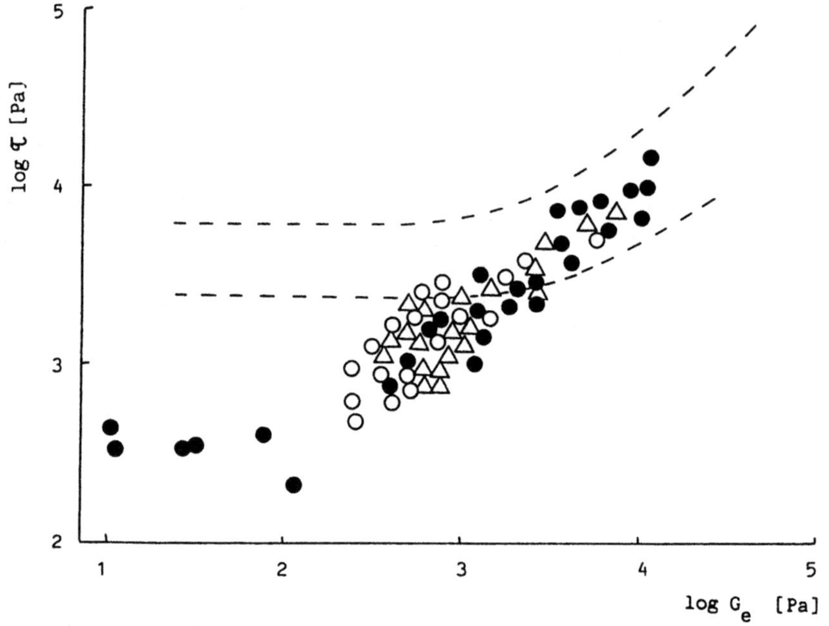

FIG. 4.8. High elastic modulus of type SBS block copolymer (○, Δ, •) and polybutadiene (– – –) [18] on shear stress at 403 K (○), 423 K (Δ) and 443 K (•).

formation of the viscoelastic properties of the copolymer in comparison with the polystyrene blocks. This is explained by the following circumstances. The molecular weight of a polystyrene block of the given copolymer is $1{\cdot}2 \times 10^4$, which is much smaller than its 'critical' molecular weight $M_c = 4 \times 10^4$ [27]. It is known that when $M > M_c$ a fluctuation network of entanglements begins to form in a polymer that is responsible for its exhibiting rubbery properties. The value of M_{cr} for polybutadiene is $(5–6) \times 10^3$ [17], whereas the molecular weight of a polybutadiene block in the copolymer being considered is $5{\cdot}6 \times 10^4$. It is therefore natural that the rubbery properties of the SBS type of copolymer are due primarily to the polybutadiene matrix, and more exactly to the entanglement network formed by the polybutadiene. The lower values of G_e than in polybutadiene for this block copolymer in the region of low values of τ are connected with the lower polybutadiene content in unit copolymer volume in comparison with the homopolymer; the presence of transition layers is connected with an increased mobility at the phase interface [28, 29], and also with the broad MWD typical of industrial products of anionic polymerization.

The pronounced thixotropic properties of the SBS type of block copolymers, which are also typical of other block copolymers, are associated with the features of their morphology, and should be manifest to a much lower degree under the conditions of another kind of rheological studies, namely dynamic experiments. As we have repeatedly indicated on previous pages, here tests can be run under conditions of strictly monitored values of the deformation amplitudes that do not change the morphology of a system under the influence of deformation (nondestructive experiments). It can be seen from Fig. 4.9 that the plots of the dynamic moduli of storage G' and loss G'' versus the frequency of deformation ω within a broad temperature range have an involved nature. For 298 K and 343 K within the frequency range of 10^{-3} to 10^{3}, the curves of G' vs ω have a rubbery plateau, indicating that at

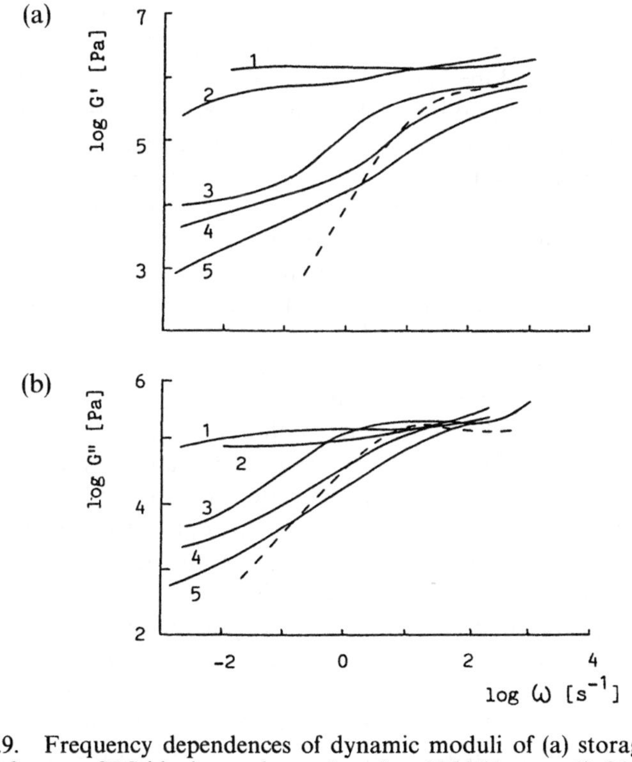

FIG. 4.9. Frequency dependences of dynamic moduli of (a) storage G' and (b) loss G'' for type SBS block copolymer (——) at 298 K (curves 1), 343 K (2), 383 K (3), 403 K (4) and 423 K (5), and for polybutadiene ($M = 8 \times 10^4$) (– – –) at 293 K.

temperatures below T_g of the polystyrene blocks, the material behaves like a rubber. Here the value of the storage modulus on the plateau G'_{pl} is 10^6 Pa, which practically coincides with the value of G'_{pl} for polybutadiene. With elevation of the temperature above 343 K, the rubbery plateau regularly diminishes in size, the values of G'_{pl} decrease, and a descending branch of the modulus G' appears in the region of low frequencies ω. It is due to passing of the system from the rubbery to the fluid state. With elevation of the temperature, these features grow. The dissipative component (loss modulus) varies similarly. The available data indicate the sufficiently high sensitivity of the dynamic characteristics G' and G'' to a change in the structure of a material under the influence of temperature. Lowering of the plateau height on the curves of G' vs ω and the reduction in its length with increasing temperature are equivalent to a drop in the density of the structural organization of the material. They are apparently due to the gradual decomposition of the domain structure of the block copolymer and weakening of the effect of the polystyrene domains playing to some extent the role of a filler.

The increase in the slope of the curves of G' and G'' vs ω in the region of low frequencies with elevation of the temperature connected with transition of the material into the fluid state does not nevertheless reach the level characteristic of homopolymers in the fluid state (it is known [17] that for homopolymers in the region of the fluid state, G' is proportional to ω^2 and G'' to ω). At the lowest frequencies studied by Vinogradov *et al.* [15, 16] for SBS-type block copolymers, $G' \propto \omega^{0.5}$ and $G'' \propto \omega^{0.3}$. This is undoubtedly connected with the retention of the two-phase structure of this polymer in quite a broad temperature interval.

The features of rheological behavior considered in detail above point to the possibility of ambiguous manifestation of anomalous viscoelastic properties of three-block copolymers. Within a broad range of variation of the rates (frequencies) of deformation and temperatures, they can reveal either properties typical of structured, highly concentrated, dispersed systems (with sharply expressed thixotropic effects at low shear rates) or properties close to a homopolymer, in the given case polybutadiene, at high shear rates. With variation of the temperature, a type SBS block copolymer may exhibit polar physicomechanical properties. For example, at temperatures below the glass transition temperature of polystyrene it behaves like rubbers, and above T_g of polystyrene like weakly fluid systems. An increase in the intensity of the deformation effect, in particular the magnitude of the stress, substantially alters the rheological behavior of this polymer.

It is natural to exhibit interest in the relation between the structure of a type SBS block copolymer and its properties, i.e. primarily in the role of the size of the polystyrene and polybutadiene blocks in forming the viscoelastic parameters and relaxation characteristics of a system. This relation to a definite extent can be traced when comparing the properties of samples with various structural characteristics. Let us turn for this purpose to the results of studying samples of type SBS block copolymers with various structural characteristics (Table 4.1) [30, 31].

It is natural that the differences in the structure of these polymers should be clearly manifest on the temperature plots of their relaxation properties, when the main contribution to the general picture of relaxation behavior is introduced by the small-scale mobility of the kinetic elements of the macromolecules. It can be seen from Fig. 4.10 that the structural organization of the block copolymers falls quite sharply in the region of manifestation of the main relaxation transition associated with the glass transition of the polystyrene and polybutadiene blocks. When comparing curves 1–3 and 5 of Fig. 4.10, one must have in view that samples 1 and 5 have close values of the weight-averaged molecular weight $\bar{M}_w$ of the polybutadiene block, while samples 2 and 3 have close values of $\bar{M}_w$ of the polystyrene blocks. All the curves have two main loss peaks (α-peaks) in the glass transition region of the polybutadiene (195–177 K) and polystyrene (373–378 K) blocks. The curves have two important features, namely, (1) a change in the height and width of the loss peak connected with glass transition of the polystyrene blocks depending on their molecular weight, and (2) the appearance of an intermediate loss peak between the regions of the relaxation transition of the polystyrene and polybutadiene blocks.

TABLE 4.1
Structural Characteristics of Type SBS Block Copolymers [31]

SBS sample	*Styrene content (wt %)*	*Microstructure of polybutadiene block*			$\bar{M}_w \times 10^{-5}$		
		cis-1,4	*trans-1,4*	*1,2*	*Polybutadiene block*	*Polystyrene block*	*Entire macromolecule*
1	28	42	46·0	12·0	0·693	0·207	0·90
2	31	38	46·0	16·0	1·576	0·584	2·16
3	46	37	45·5	17·5	0·956	0.694	1.65
4	53	37	45·0	18·0	1·19	1·14	2·33
5	75	45	45·6	19·4	0·65	1·95	2·60

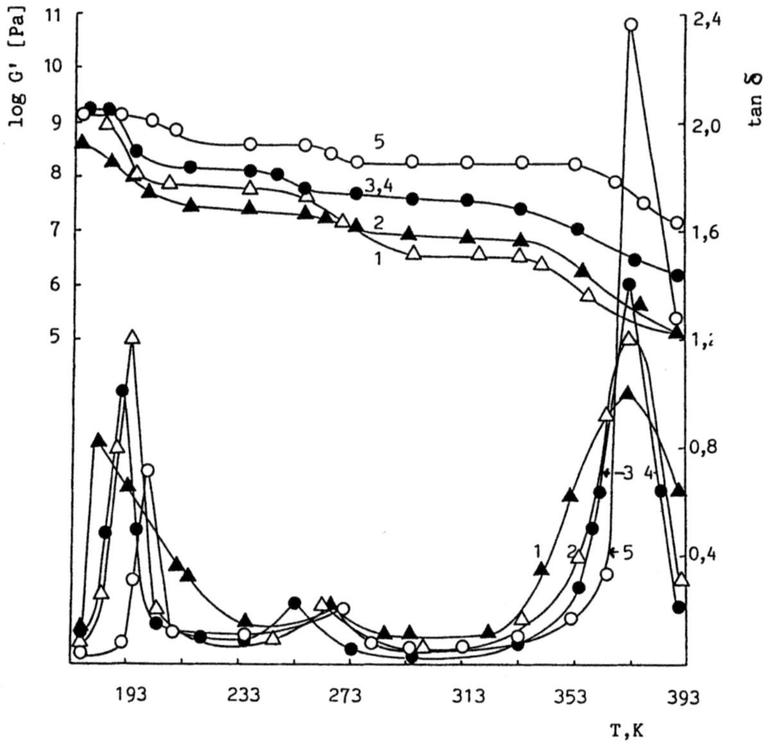

FIG. 4.10. Temperature dependences of modulus of storage G' (1–5) and loss tangent tan δ (1–5) for type SBS block copolymer samples (1–5 (see Table 4.1).

The loss peaks corresponding to the main relaxation transition in the polystyrene blocks diminish in height and become wider with decreasing molecular weight (Fig. 4.11). Close results for other polymers are contained, for instance, in Ref. 32. The above features may be associated either with an increase in the molecular weight of the blocks themselves, or with an increase in the size of the particles (domains) of the dispersed phase of polystyrene in polybutadiene. When the size of the polystyrene domain chemically bound to the polybutadiene is smaller, the specific surface area of the particles, i.e. the area of contact of the polystyrene and polybutadiene, is larger. Another result here is expansion of the region of this relaxation transition.

The appearance of an intermediate loss peak in the temperature region of 258 to 248 K for SBS-type block copolymers (similar peaks were also registered by Marker [33] on differential thermograms in block copo-

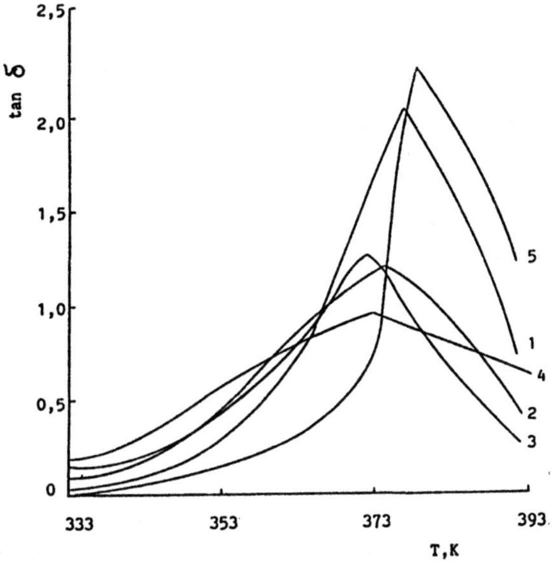

FIG. 4.11. Temperature dependences of loss tangent for type SBS block copolymer samples 1–5 (see Table 4.1).

lymer films obtained from a solution) that does not depend on the ratio of the components can be explained from the concepts of the segmental solubility of polymers at their interface. Such concepts were introduced for polymer blends [34]; they predicted the appearance of the transition layer of a quite considerable size at the interface. The foregoing can also partly explain the expansion of the region of the main relaxation transition (see Fig. 4.11). A similar effect was also observed for polymer blends when measuring the mechanical and dielectric losses [35].

Important information on the behavior of a type SBS block copolymer (and also of any other type) is given by plots of the dynamic viscoelastic characteristics measured in a broad temperature range. This can be seen very clearly in Figs 4.12 and 4.13. A clearly expressed rubbery plateau is observed in the temperature interval between the glass transition temperatures of the polybutadiene and polystyrene blocks (Fig. 4.13). The modulus of storage here remains practically unchanged with the temperature except in the region from 258 to 248 K, where it decreases insignificantly. It is significant that the height of the rubbery plateau determined on the temperature scale (it must be remembered that the storage moduli

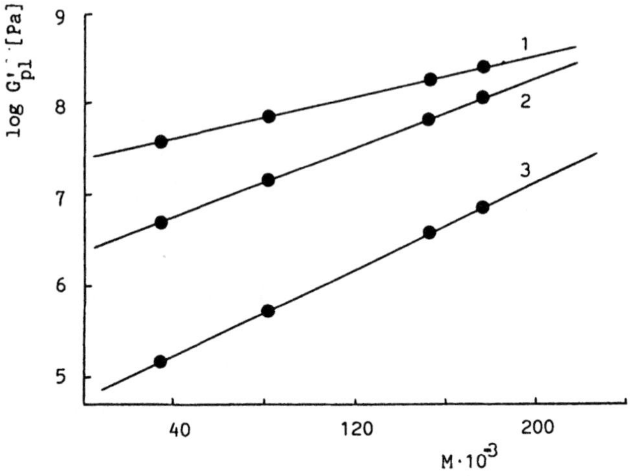

FIG. 4.12. Influence of the molecular weight of PS blocks on the values of the modulus of storage on the plateau G'_{pl} at 233 K (curve 1), 293 K(2), and 393 K (3).

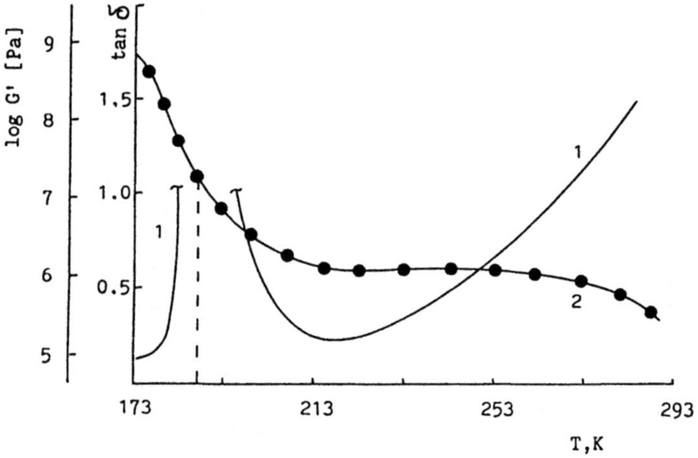

FIG. 4.13. Change in the dynamic characteristics of the loss tangent tan δ (curve 1) and modulus of storage G' (curve 2) with temperature for monodispersed polybutadiene.

plotted in Fig. 4.13 were measured at relatively low circular frequencies [30, 31] depends here on the molecular weight of the polystyrene block. An increase in the molecular weight of the latter makes the system more rigid. The effect is similar to a growth in the concentration of a solid filler.

A comparison of the magnitudes of the moduli of storage for type SBS block copolymers and polybutadiene homopolymers in the region of the rubbery state shows that the plots of G' vs T for the block copolymers are considerably higher (compare Fig. 4.10 (curves 2) and Fig. 4.13; the latter relates to polybutadiene having a narrow MWD with $M = 1{\cdot}5 \times 10^5$). This comparison is quite correct because the microstructures of the polybutadiene blocks of these samples are close.

The above examples convincingly indicate the ambiguity of behavior of type SBS block copolymers depending on the structure of the microblocks, temperature, deformation frequency (rate), and intensity of deformation action. It should be noted that in the region of the fluid state at temperatures considerably exceeding the glass transition temperature of the polystyrene blocks, the decisive contribution to the viscoelastic behavior is made by the polybutadiene blocks, i.e. by the polymer with the higher kinetic flexibility of the macromolecules. But in the rubbery state, at temperatures below T_g of the polystyrene blocks, the latter begin to play the dominating role and in the general case determine the level of the physicomechanical characteristics of the system as a whole.

4.5 FEATURES OF THE RHEOLOGICAL AND RELAXATION BEHAVIOR OF COMPOSITIONS INCLUDING COPOLYMERS

When indicating the important role of copolymers as a new class of polymer systems, we cannot fail to note their significance as materials used as modifying additives to compositions. The following examples illustrate this quite clearly.

As our first example, we shall consider the physicomechanical and viscoelastic properties of compositions based on polypropylene and blocks of poly(ethylene *co* propylene).

Polypropylene (PP), while having many valuable properties, is known to have an inadequate frost resistance, and high glass transition and crystallization temperatures, which underlies its rigidity and brittleness at low temperatures. Since PP is not polar, it blends poorly with polar plasticizers, which are quite popular at present. The plasticization of PP by rubbers (see, for example, Ref. 36), in particular PECP copolymers, was found to be highly effective. In their chemical nature, PP and PECP are close. Unlike other rubbers, PECP has a small number of double bonds and in addition to its high elastic properties is stable to ageing and has a high yield temperature and good frost resistance, as well as good chemical stability and insulating properties. Compositions based on PP and PECP, as should be expected, have better physicomechanical properties than each component separately. However, the optimal properties needed for such compositions can be achieved both by varying their concentration composition and by finding the appropriate molecular weights of the components. This can be illustrated by the following data [37] obtained for compositions of isotactic polypropylene (an atactic content of about 1·2%) and a PECP block copolymer of various molecular weights (see Table 4.2).

The interpretation of both the physicomechanical and viscoelastic behavior of compositions based on polymers is impossible without determining their structural features within a broad temperature range. This can be done, in particular, by analyzing the relaxation and phase transitions in a system. We have already indicated that this is not only a way of appraising the molecular mobility and compatibility of the components of a system, but also to a definite extent a way of analyzing its morphology. Important information can be procured here when considering the temperature dependences of the parameters of internal ($\tan \delta$) and external (e.g. the force F) friction. Yanovsky and Frenkin [38] showed how these quantities are related. Sometimes the characteristics of external friction were measured experimentally within quite a broad temperature interval. This provided independent information on the structural transformations in a material.

A comparison of the temperature dependences of the internal and external friction, viz. the curves of $\tan \delta$ and F vs T for samples of PECP, PP, and blends of PP + PECP, reveals two regions of their extremal change (Fig. 4.14), namely, 343–353 K and 433–443 K. The change in the viscoelastic characteristics of PECP in the region of 363–373 K connected with the melting into the block copolymer of PECP blocks close in

TABLE 4.2
Characteristics of PECP and PP Samples [37]

Item	*PECP-1*	*PECP-2*	*PECP-3*	*PP*
Intrinsic viscosity in decalin at 408 K [η]	1·14	1·55	1·7	5·5
Molecular weight (MW)	$7{\cdot}5 \times 10^4$	$1{\cdot}2 \times 10^5$	$1{\cdot}2 \times 10^5$	$7{\cdot}9 \times 10^5$
Defoe hardness	75	280	460	—
Propylene content, mol %	33–37	33–37	33–37	—

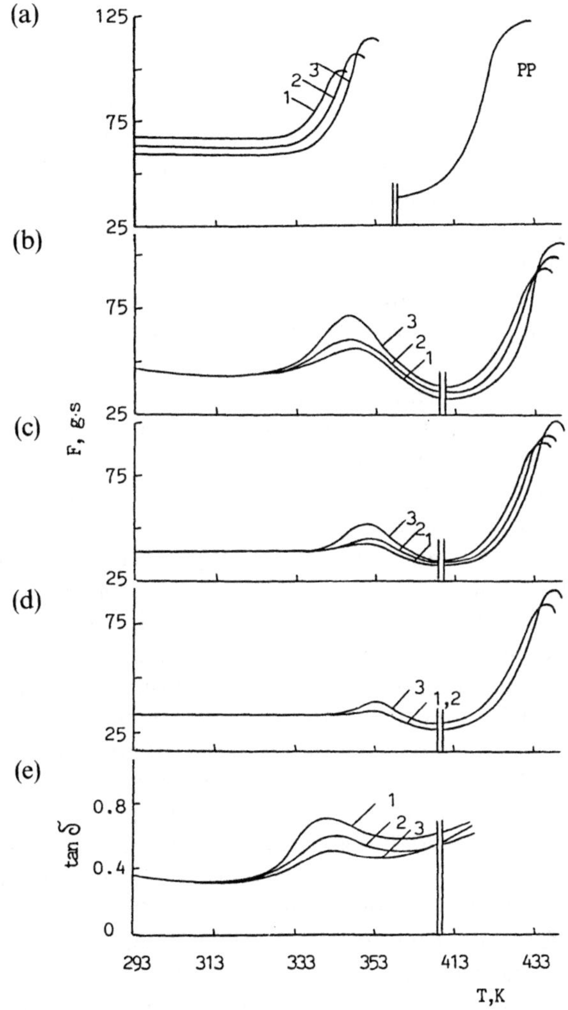

FIG. 4.14. Temperature dependences of force of friction F and loss tangent $\tan \delta$ for (a) PECP and PP, and (b)–(e) blends thereof. Curves 1–3 are for the following samples, respectively: (a) PECP-1, PECP-2, and PECP-3; (b) PP+10% PECP, PP+20% PECP, and PP+30% PECP for PECP-1; (c) the same for PECP-2; (d) the same for PECP-3; and (e) PP+20% PECP-1, PP+20% PECP-2, and PP+20% PECP−3.

structure to low-density polyethylene has already been treated above. In the given case, there is noted a drop in the values of F and shifting of the peak on the curves of F vs T for PECP with an increase in its molecular weight: the former is explained by the increase in the modulus of storage (the Defoe rigidity grows with an increasing molecular weight, see Table 4.2) and the reduction in the contact area of the friction bodies, while the latter is explained by the increase in the van der Waals forces as a result of the higher interaction between the molecules in the samples with the higher molecular weight [39]. A relation between T_m and the molecular weight for some thermoplastics was also observed by Mandelkern *et al.* [40]. The high-temperature peak on the curves of F vs T (Fig. 4.14(a)–(d)) is explained by melting of the PP. The magnitudes of the low- and high-temperature peaks depend on the content of the copolymer in a blend: with an increasing copolymer concentration the low-temperature peak grows and the high-temperature one diminishes. With a growth in the molecular weight of the PECP, the intensity of manifestation of the low-temperature peak decreases.

Consequently, by varying the molecular weight of the block copolymer introduced into compositions with PP, the intensity of the phase transitions can be regulated to a certain extent. From the physical viewpoint this is explained (i) by the difficulty of crystalline block formation in PECP because of the growth of its intrinsic viscosity with an increase in the molecular weight, and (ii) by the influence of the matrix of highly crystalline PP on this process. On the other hand, the plasticizing influence of the PECP in the composition is obvious because the nature of the phase transition process in PP also changes [41].

The properties of materials based on blended compositions depend appreciably on the homogeneity of their morphological structure. Blending of materials with close viscosities in a melt results in better dispersion and more uniform distribution of the components in the mixture. Moreover, the blending of more viscous polymer systems also requires the creation of higher stresses in the mixers, and this, in turn, also facilitates the preparation of homogeneous compositions.

Analysis of the flow curves reveals that the apparent viscosity of the PECP-3 sample (see Table 4.2) is the closest in value to the viscosity of PP, while the composition containing this block copolymer sample has the highest viscosity in the given series (see Fig. 4.15). It was also established that exactly this composition is the most homogeneous [37]. This, in turn, ensures better physicomechanical indices of the material at room temperature, i.e. in the solid state. For example, the relative

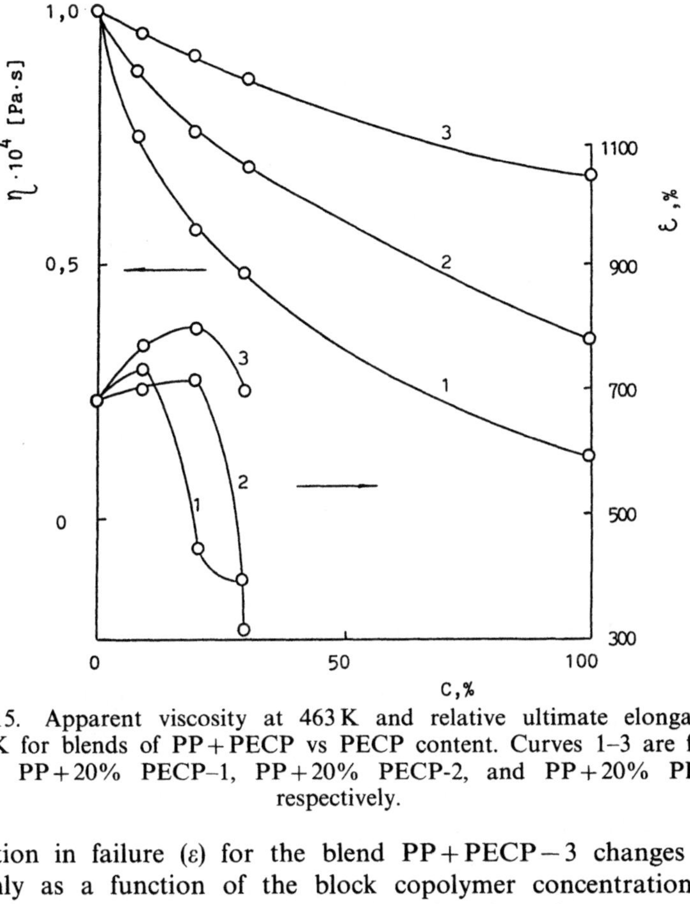

FIG. 4.15. Apparent viscosity at 463 K and relative ultimate elongation ε at 293 K for blends of PP + PECP vs PECP content. Curves 1–3 are for the samples PP + 20% PECP–1, PP + 20% PECP-2, and PP + 20% PECP-3, respectively.

elongation in failure (ε) for the blend PP + PECP − 3 changes more smoothly as a function of the block copolymer concentration. The magnitude of ε is also higher here than for the other compositions (compare curves 1 and 3 in Fig. 4.15). A similar picture is observed qualitatively for the temperature dependences of the ultimate tensile strength. The incorporation of PECP into PP improves the frost resistance of the composition and lowers its brittleness temperature.

Hence, the above example shows convincingly, in our opinion, the advantages of employing compositions including block copolymers. It is also obvious, however, that the preparation of materials with a high level of physicomechanical characteristics depends on many factors, including the molecular weights of the components, their concentrations, and the conditions of blending in a melt.

A second good example of the use of block copolymers is as modifying additives having an alloying effect on a polymer melt and thus improving its viscoelastic properties in processing.

The use of thermodynamically incompatible copolymers underlies one of the methods of physical modification—alloying. It renders a definite influence on the supermolecular organization of a polymer, and changes the nature of the relaxation processes and, therefore, the rheological behavior of the system [42, 43]. Three-block copolymers are quite popular as alloying additives. The important ones include copolymers of butadiene and styrene of the SBS type. The introduction of additives having, besides elastic blocks, sections of various lengths similar in structure to the polymer being modified often affects its properties very significantly.

Blends of homo- and block copolymers occupy a special place among incompatible polymers in their viscoelastic and physicomechanical characteristics because their behavior in many ways is determined by the possibility of partial blending of the homopolymer with the similar block of the copolymer. For example, small amounts of SBS added to polystyrene (PS) have an alloying effect on the latter and improve its mechanical characteristics. On the other hand, a small amount of PS incorporated into the copolymer plays the role of a filler [44].

The influence of additives of type SBS block copolymers on the rheological properties of PS has been considered, in particular, in Refs 45 and 46. It can also be seen quite clearly in the concentration dependences of the dynamic rheological characteristics shown in Fig. 4.16 for a melt of block PS ($M = 2{\cdot}5 \times 10^5$) containing the thermoplastic elastomer SBS with a narrow MWD and having blocks with $M = 2 \times 10^4$, $4{\cdot}3 \times 10^4$ and 2×10^4 respectively. In the region of low frequencies ω, the modulus G' varies with an extremum which corresponds to an approximately 10% content of the block copolymer. This is just what indicates the alloying effect of the thermoplastic elastomer at its concentration (C) up to 10%.

Dreval *et al.* [45] indicate that an extremum on the plot of the modulus of elasticity E vs the concentration C was also observed in experiments involving uniaxial tension. It is important that the changes in the viscoelastic dynamic characteristics of compositions when block copolymers are added are most evident at low deformation frequencies, i.e. in the region of the fluid state (curves 1–3, Fig. 4.16). This is exactly where the highest sensitivity of the rheological characteristics to the molecular and structural features of a polymer system is observed [47].

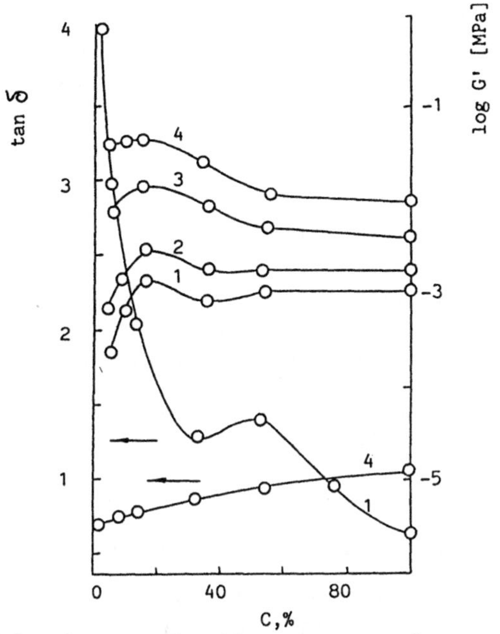

FIG. 4.16. Modulus of storage G' and loss tangent tan δ vs content of type SBS block copolymer in composition with polystyrene at 463 K. Curves 1–4 are for $\log \omega$ (s^{-1}) of $-2{\cdot}4$ (curve 1), $-1{\cdot}0$(2), $-0{\cdot}5$(3) and 0(4), respectively.

Under these conditions, the mobility of large supermolecular formations with prolonged relaxation times is realized. This affects sharply all the relaxation characteristics of a polymer composition. With an increase in the frequency and with the reaching of a region that is transitional to the rubbery state, the noted anomaly of the properties may not be manifest at all (curve 4, Fig. 4.16) because the frequency (duration) of action here is much smaller than the relaxation time of the supermolecular formations, and the experiment is not sensitive to their contribution.

For polymer blends, the nature of the change in the tan δ–composition curves is known (see the preceding sections) to reflect the phase structure of a system. The peak on the curve of mechanical losses (curve 1, Fig. 4.16) in the region of a 50% SBS content indicates exactly the formation of an independent block copolymer in the composition. This was also confirmed by independent structural studies, by electron microscopy [46]. In blends with PS, this block copolymer already forms, at a 5% content, inclusions of a spherical shape that combine into an independent structural skeleton and separate into an independent phase with a particle size considerably exceeding that of the matrix particles. It is exactly the morphological features of melts of PS+SBS blends retaining, to a definite degree, phase heterogeneity, that can explain the anomalous change in the rheological properties of the indicated compositions.

REFERENCES

1. NOSHAY, A. and MCGRATH, J.E. *Block Copolymers—Overview and Critical Survey*, Academic Press, New York, 1977.
2. CERESA, R. *J. Encycl. Polym. Sci. Technol.*, **2** (1964) 485.
3. AGGARWAL, S.L. (Ed.). *Blocks Polymers*, Plenum Press, New York, 1970.
4. VALETSKY, P.M. and STOROZHUK, I.P. *Uspekhi Khimii*, **48**(1) (1979) 75.
5. SEMENCHENKO, V.K. *Izbrannye Glavy Teoreticheskoi Fiziki* (Selected Chapters of Theoretical Physics), Prosveshchenie, Moscow, 1966.
6. MCGRATH, J.E., ROBESON, L.M., MATZNER, M., *et al. J. Polym. Sci.*, **60** (1977) 29.
7. ZELENEV, YU.V., LETUNOVSKY, M.P. and BASCHIROV, A.B. *Acta Polymerica*, **33**(10) (1982) 590.
8. VINOGRADOV, G.V. and IVANOVA, L.I. *Rheol. Acta*, **6**(3) (1967) 209.
9. VINOGRADOV, G.V., YANOVSKY, YU.G., IVANOVA, L.T., *et al. Vysokomol. Soed.*, **10B**(10) (1968) 726.
10. YANOVSKY, YU.G., VINOGRADOV, G.V., MALKIN, A.YA., *et al. Mekhanika Polimerov*, **4** (1969) 698.
11. YANOVSKY, YU.G., IVANOVA, L.I. and FRENKIN, E.I. *Mekhanika Polimerov*, **3** (1970) 530.
12. BAGLEY, E.B. *Trans. Soc. Rheol.*, **5** (1962) 354.
13. HOLDEN, G., BISHOP, E.T. and LEGGE, N.K. *J. Polym. Sci.*, **26** (1969) 37.
14. ESTES, G.M., COOPERS, S.L. and TOBOLSKY, A.V. *J. Macromol. Sci., Rev. Macromol. Chem., C*, **4**(2) (1970) 313.
15. VINOGRADOV, G.V., DREVAL, V.E., YANOVSKY, YU.G., *et al. Kauchuk i Rezina*, **11** (1977) 52.
16. VINOGRADOV, G.V., DREVAL, V.E., MALKIN, A.YA., *et al. Rheol. Acta*, **17** (1978) 258.
17. YANOVSKY, YU.G. *Int. J. Polym. Mater.*, **8** (1980) 257.
18. VINOGRADOV, G.V. and MALKIN, A. *Rheology of Polymers. Viscoelasticity and Flow of Polymers*, trans. A. Beknazarov, Mir Publishers, Moscow, 1980.
19. VINOGRADOV, G.V., YANOVSKY, YU.G., BARACHEEVA, V.V., *et al.* In: *Sb. Trudov Mezhd. Konferentsii po Kauchuku i Rezine* (Trans. Int. Conf. on Rubber), Series C. Vol. 1, p. 19, Moscow, 1984.
20. KRAUS, G. and GRUVER, J.T. *J. Appl. Polym. Sci.*, **11** (1967) 2121.
21. KELLER, A., PEDEMONTE, E. and WILLMONTH, F.M. *Kolloid. Zh.*, **238**(1–2) (1970) 385.
22. DLUGOSZ, J., KELLER, A. and PEDEMONTE, E. *Kolloid. Zh.*, **242**(1–2) (1970) 1125.

23. AGGARWAL, S.L. *Polymer*, **17**(11) (1976) 938.
24. BURKOVA, S.G. *Kauchuk i Rezina*, **5** (1973) 26.
25. VINOGRADOV, G.V. *Pure and Appl. Chem.*, **39**(1–2) (1974) 115.
26. COGSWELL, F.N. and HANSON, D.E. *Polymer*, **16** (1975) 936.
27. YANOVSKY, YU.G. and BARANCHEEVA, V.V. In: *Rheologiya Polimernykh i Dispersnykh Sistem i Reofizika* (Rheology of Polymer and Dispersed Systems and Rheophysics), Part 1, Izd. I-t Teplomassoobmena AN BSSR, Minsk, 1975, p. 13.
28. KRAUS, G. and ROLLMAN, K.W. *J. Polym. Sci., Polym. Phys. Ed.*, **14** (1976) 1133.
29. MEIER, D.J. *Polymer Preprints*, **15**, (1974) 171.
30. AKTSIZER, W.S., OSKIN, N.V, VINOGRADOV, G.V., *et al. Plaste und Kautschuk*, **6**(1) (1971) 44.
31. OSKIN, V.N., YANOVSKY, YU.G., VINOGRADOV, G.V., *et al. Vysokomol. Soed.*, **14A**(10) (1972) 2120.
32. BOYER, R.F. (Ed.). *Transitions and Relaxations in Polymers*, John Wiley, New Jersey, 1966.
33. MARKER, B. *Polymer Preprints*, **10** (1969) 524.
34. KULEZNEV, V.N., DOGADKIN, B.A. and KLYKOVA, V.D. *Kolloid. Zh.*, **30** (1968) 255.
35. KULEZNEV, V.N. and KLYKOVA, V.D. *Kolloid. Zh.*, **30** (1968) 44.
36. NATTA, D. and MAZANTI, S. *Khimiya i Tekhnologiya Polimerov*, **8** (1964) 113.
37. FRENKIN, E.I., ERMILOVA, G.A., YANOVSKY, YU.G., *et al. Plastmassy*, **10** (1970) 32.
38. YANOVSKY, YU.G. and FRENKIN, E.I. *Uspekhi Reologii* (Achievements of Rheology), Khimiya, Moscow, 1970.
39. SYRKIN, YA.K. and DYATKIN, M.E. *Khimicheskaya Svyaz' i Stroenie Molekul* (Chemical Bonds and Structure of Molecules), Khimiya, Moscow, 1966.
40. MANDELKERN, L., FATOU, I.G., DENISON, R., *et al. J. Polym. Sci.*, **B-3** (1965) 803.
41. TINIUS, K. *Plastifikatory* (Plasticizers), Khimiya, Moscow, 1964.
42. SOGOLOVA, T.I., AKUTIN, M.S., TSVANKIN, D.A., *et al. Vysokomol. Soed.*, **17A**(11) (1975) 2505.
43. SVIRIDOVA, E.A., SLONIMSKY, G.L., AKUTIN, M.S., *et al. Vysokomol. Soed.*, **25A**(12) (1983) 2715.
44. AKUTIN, M.S., TKACHEVA, V.S., ANDRIANOV, B.V., *et al. Vysokomol. Soed.*, **12B**(1) (1970) 60.
45. DREVAL, V.E., KASSA, A., BORISENKOVA, E.K., *et al. Vysokomol. Soed.*, **25A**(1) (1983) 156.
46. YANOVSKY, YU.G., KERBER, M.L., VINOGRADOV, G.V., *et al. Vysokomol. Soed.*, **26A**(5) (1984) 1101.
47. YANOVSKY, YU.G., VINOGRADOV, G.V. and IVANOVA, L.I. *Vysokomol. Soed.*, **24A**(5) (1982) 105.

Chapter 5

Rheological and Relaxation Properties of Filled Polymers

5.1 INTRODUCTION

The study of viscoelastic properties of filled polymers is of important scientific and practical significance. We can argue here about the priority of the theoretical or applied approach to the problem, but one thing is certain: the filling of the polymer leads to practical purposes, extending the possibilities of its applications. Some of the questions concerning the creation of new filled polymeric compositions could probably be solved within the framework of applied investigation only. Nevertheless, it is quite obvious that a sufficiently full measure of generalization can be achieved only by using a set of data obtained by physicomechanical, thermodynamic, thermophysical, structural and other methods.

The theoretical description of rheological properties of filled polymers in the broad range of varying velocities (frequencies), deformation and temperature conditions requires a considerably larger number of parameters than in the case of unfilled polymers. The introduction of solid filler particles into the polymer dramatically varies the regularity of its flow. This is primarily due to the appearance of the yield point; the flow is realized only above some definite shear stress value. The existence of the yield point on the whole exerts an influence on the viscoelastic behavior of the system [1–3].

As far as the physicochemical properties of filled polymers are concerned, there is an important problem of reinforcement associated with improving a number of characteristics: modulus, mechanical losses, ultimate strength during extension, resistance to tear and fatigue endurance, hardness, etc. Such an approach should render it possible to raise the service life of the articles. It should be noted that the term 'reinforcement' was first introduced in connection with improving the rubber properties. At the present time it is widely believed that the intensification effect depends considerably on the interphase interaction at the surface boundary of the filler–polymer matrix [4]. We cannot object to this fact. However, it is obvious that such an interphase interaction should influence the complex of the rheological characteristics of the system as well. Moreover, we may suppose that there must exist a definite interrelation between the viscoelastic characteristics of the filled polymer melt and certain parameters characterizing the physicomechanical and strain–strength properties of this material in a solid state. This is indicated by a whole series of data, in particular by those that follow.

With regard to viscoelastic and physicomechanical properties of the filled polymers, we should also take into consideration problems concerning the evaluation of their relaxation characteristics. Within the framework of linear description, of prime importance is the analysis of the relaxation transition conditions that depend on the varying temperature and velocity (frequency) of deformation and that characterize both the boundary of the physical state of the filled system and the fairly fine structural changes (at both molecular and supermolecular level).

The first serious research into the problem of filled polymer behavior, particularly of rubbers, was started in the last century. Although this has been further developed in our time, a satisfactory description of the regularities under observation and their explanation from a general position are far from complete.

Extensive publications have appeared during the last 7–10 years on the rheology of melts of filled thermoplastic materials [5–20]. Moreover, a number of books and monographs contain sections devoted to this problem [21–24]. On the other hand, we must note the considerable number of works on the rheological properties of filled elastomers [25–40]. Quite complete reviews on the problem have been published in recent years [41–43].

Analysis of the works cited above on the rheology of melts of filled thermoplastics and others reveals that many of them deal with the behavior of systems with a relatively low volume content of the filler. Fewer publications discuss the rheological behavior of polymer melts with quite a high filling volume (over 40%). A detailed review and analysis of the rheological behavior of highly filled polymer melts was given in Ref. 41.

The interest aroused by highly filled polymer systems in recent years can be explained, in particular, by the widespread introduction of composites including a polymer matrix or using a polymer as a binder.

The joint use of dispersed and reinforcing fillers combined with a thermoplastic matrix has lately become quite popular in the processing of plastics. The rheological properties of composite melts based on polyolefins containing a coarsely dispersed filler (chalk, talc, glass fibers, kaolin, etc.), their mechanical strength, stability of the size of articles therefrom, and a number of other indices, depend on the size and shape of the filler particles, the concentration, and adhesive interaction at the polymer–filler interface. Moreover, it is possible to monitor the properties of a composition material using small amounts of additives, by alloying.

Almost all rubber articles also contain fillers, among which carbon black is the most popular [25]. The main reason is that it is an excellent reinforcing agent [26]. When processes into articles, compounds based on elastomers containing carbon black acquire their shape in a process of extrusion, compression or blow molding, calendering, etc. [27]. This is why their rheological properties have a fundamental significance.

The compounds sometimes exhibit a rheological behavior similar to the elastomers they are based on, but very frequently this behavior differs appreciably. The primary reason is that the compounds are thixotropic systems. The thixotropy is due to the formation of an internal structure because of the presence of the highly active carbon black filler. The features of this structure and its influence on the viscoelastic behavior in compound deformation depend on the type of elastomer, type of carbon black, its concentration in the compound, quality of mixing, etc.

The influence of the physicochemical nature of an elastomer on the rheological behavior of its filled components was studied in sufficient detail in Refs 28 and 29. The influence of the physicochemical nature of carbon black on the rheological behavior of compounds was considered in Refs 25, 32 and 33. The interaction between the elastomer matrix and the filler was analyzed in Refs 30, 31, 34 and 35. The rheological behavior of such systems is affected substantially by the existence of a yield point [37, 38].

In summarizing the results of numerous studies of compounds based on elastomers containing carbon black, we can note that their rheological behavior is quite complicated and lends itself to generalization with difficulty. The reason is that one must consider several factors acting simultaneously and associated with the features of (i) the elastomer matrix, (ii) the structure and structural formation of the dispersed carbon black, and (iii) the interaction of the elastomer matrix and filler. One must also consider that each of these factors depends in a different way on the temperature, kind of deformation, stress magnitude, etc.

If we survey the rheological data on filled systems, we see that the majority of works refer to a narrow field of investigation. These works do not allow the establishment of a number of necessary general laws of viscoelastic behavior of such materials. It becomes obvious that a satisfactory degree of generalization of the data available can be most easily attained via the study of properties of model filled systems. In this study, special attention should be devoted to the choosing of appropriate polymer materials. As for the filler, the following peculiarities of its structuring capability should be taken into consideration: the possibility of a chemical interaction with the polymer matrix, specific surface magnitude, particle size and shape, etc. The method chosen for investigation should allow us to make a simultaneous evaluation of both the viscous and elastic components of the system. Taking into account the high sensitivity of the filling systems to the strain magnitudes, the investigations should be carried out under their stringent control. The question concerning the yield point of the filling system and its dependence on temperature is very important. The literature data available are extremely contradictory. The reasons for the divergence of the yield point values estimated by different rheological methods are unclear, for example, under the conditions of shear strain and uniaxial extension. Finally, it would be useful to consider the possibility of systematizing the viscoelastic characteristics of the filled polymeric systems. Such a systematization should be carried out by constructing generalized temperature–concentration dependences.

5.2 PROPERTIES OF A POLYMER IN A FILLED SYSTEM

A very simple theoretical model of a filled polymer is structureless medium 1 (polymer) containing isolated particles of dispersed phase 2 (filler), ideal adhesion contact existing between phases 1 and 2. The concept of intrinsic structures being absent in the two phases is a simplification because unfilled flexible-chain polymers have a 'natural' structural microheterogeneity due to the coexistence of crystalline and amorphous regions (for crystallizing polymers) or to the presence of 'frozen' density fluctuations (for noncrystallizing glassy polymers). The need to consider the morphological features of the unfilled polymers when analyzing their properties is undoubted. However, as applied to filled systems, the postulate of a structureless filler and polymer actually requires the absence of microdefects in both phases (e.g. cracks along the spherulite boundaries in polymers or voids between the filler particles in loosely packed agglomerates). The second postulate on the ideal adhesive contact, from the viewpoint of mechanics, follows from the condition of retaining the continuity of a filled system (including continuity at the polymer–filler phase interface) when external fields act thereupon.

In practice, realization of an optimal adhesive contact between two phases is possible provided that the energy of interaction at the phase interface is higher (or at least not lower) than the energy of cohesion of one of the phases. This condition is satisfied by the combination of a 'low-energy' continuous phase 1 (polymer) and 'high-energy' dispersed phase 2 (minerals, metals, etc.). It is quite probable that the boundary layer of phase 1 formed under conditions of the action of 'excess' surface forces of high-energy phase 2 will differ in a structural respect from 'parent' phase 1 under the action of only its own forces of cohesion.

The study of filled polymers by direct structural methods is associated with obvious methodical difficulties. Consequently, indirect data are customarily used to determine the structural state of the polymer phase in a filled system. For instance, for polystyrene (PS), poly(methyl methacrylate) (PMMA), and other noncrystallizing polymers containing inorganic fillers (glass powder, glass fibers, aerosil, etc.), there was revealed a monotonic increase in the glass transition temperature T_g, and also an increase in the sorption capacity in the region of $T < T_g$ with an increase in the filler content [44, 45]. Sometimes, the introduction of fillers differing in nature and dispersion (aerosil, metals, oxides) was attended by a drop in the 'partial' (reduced to the pure polymer) height of the heat capacity jump ΔC_p at T_g.

The set of enumerated experimental data is explained qualitatively within the framework of the following concepts [44]. The adsorption interaction of the macromolecules in a melt with a high-energy solid filler surface limits their thermal mobility (a drop in ΔC_p) and hinders the formation of a closely packed structure (a growth in the specific volume and compressibility). During the subsequent cooling of the filler and melt to $T < T_g$, a loosely packed glassy structure is registered (an increased sorption capacity).

Hence, the change in the kinetics of structure formation when passing through the region T_g is the main cause of the change in the structure and properties in the solid (glassy) state of a filled polymer in comparison with an unfilled one. This conclusion was confirmed in Refs 46 and 47 by the results of a quantitative analysis of the structure relaxation process in the glass transition interval of PS samples filled with a glass powder and aerosil within the framework of the Moynihan model. The main parameters of the relaxation process are the activation energy ΔE and also two phenomenological parameters, namely, the index of nonexponentiality $0<\beta<1$ whose specific value gives an idea of the width of the structural relaxation time spectrum (with a single relaxation time $\beta=1$, while $\beta=0$ corresponds to an 'infinitely great' set of relaxation times), and the index of nonlinearity of the relaxation process $0<\chi<1$ ($\chi=1$ corresponds to the linear region).

The conclusion on the change in the structure of a polymer melt in the presence of solid mineral fillers was also based on the results of studying filled crystallizing polymer [48, 49]. It was found that restriction of the thermal mobility in a melt of macromolecules of oligoesters, polyethylene, polypropylene, and other crystallizing polymers in the presence of aerosil leads to a change in the energy and kinetic parameters of crystalline phase formation in a supercooled melt which usually lowers the depth of crystallization of a polymer. With the maximum filler concentrations, crystallization even suppresses this process completely. Hence, the properties of filled polymers in the solid state retain a 'memory' of the mechanics and kinetics of structure formation in a melt when passing through the glass transition region (for noncrystallizing polymers) and melting point (for crystallizing polymers).

Our conclusion on the change in the structure of a flexible-chain polymer in the presence of a high-energy solid filler surface requires formalization in a specific model. Here two approaches are possible in principle. By the first one, continuous phase 1 in the space between the particles of phase 2 is under the action of van der Waals forces of attraction that monotonically increase as the particles come closer to one another when their concentration grows. In this case, the structure of an interlayer will be a monotonic function of its thickness. The other approach is based on the idea of the restricted scale of action of the surface forces of high-energy phase 2 as a result of which the change in the structure of continuous phase 1 is localized in the comparatively thin surface layer near the phase interface. In this case, the increase in the filler concentration will be attended by a similar growth in the fraction of the polymer with an altered structure distributed in the form of a layer of constant thickness around the filler particles.

5.3 PROPERTIES OF A FILLER IN A FILLED SYSTEM

As has been mentioned, the type of fillers and their properties play an important role. A question arises as to using the minimally and maximally required parameters for the 'serviceability' characteristics of the filler in the polymeric matrix. Within the framework of the present book, I do not have the opportunity to consider in full detail the existing approaches to estimating the properties and 'quality' of the fillers being used as well as their behavior in bulk from the position of solid-phase disperse systems, i.e., outside the polymeric matrix. The latter has been fairly well described in specialist literature [51, 52]. Here, we shall present certain general considerations that can serve as the basis for choosing the filler.

The dispersed solids introduced into polymers can be divided conditionally into three main groups: (1) fine inert particles; (2) particles increasing the strength (reinforcing fillers) which also includes particles finished with silanes, stearic acid, or polymerization (by synthesis of polymer molecules on the surface of the filler particles), anisometric particles, flaky fillers, etc.; and (3) fibrous fillers—glass, carbon, asbestos, basalt fibers, etc. (with or without a finish).

As has been noted by many authors (see, in particular, Ref. 53), the achievement of the reinforcing effect, which is one of the most important tasks, is observed only when the diameter of the filler particles is fairly small (essentially less than 10 μm). Large particles dispersed in the polymeric matrix have a weak influence on the change in physico-mechanical characteristics of the system or can even decrease a number of them. Here, the decrease of fatigue resistance of rubbers can serve as an example [53, 54]. However, experiment has shown that it is evidently insufficient to characterize the filler only by its particle diameter. This slightly improves when such a parameter as the specific surface value is used. (The specific surface value can be estimated from the values for the adsorption of the chemically inert molecules.) This is, of course, understandable; indeed, the 'interaction' between the filler and the polymeric matrix occurs on the filler surface. This interaction can be restricted by the presence of filler micropores or by the macromolecule size.

Can we consider the specific surface and the diameter sufficient criteria for characterizing the filler? It is known, for example, that such a filler as carbon black (CB) consists not of separate spherical particles, but of aggregates containing more or less bound (as if fused) particles.

The degree of aggregation of the particles depends on the number of particles in the aggregate, whereas the aggregate shape or the anisometry can be accounted for, as was supposed in Ref. 55. However, the final answer to this question is not unambiguous, at least in the case of rheological estimates. This will be illustrated. It should be noted that a great deal depends on the deformation conditions—the breaking or cohesion of the filler structure. In this case, it is necessary to consider the evaluation of the thixotropic characteristics. In this manner, we can follow the connection between the technological peculiarities of the process of refining filled material melt, its rheological characteristics, and the physico-mechanical properties of the end material, i.e. the article.

While discussing the general peculiarities of the behavior of the filled polymer materials, it is also worth dealing with certain aspects concerning the interpretation of the possible mechanism for the filler–polymeric matrix interaction. Undoubtedly, the nature of this interaction should have an essential influence on the complex of the rheological and physicomechanical properties of the filled system. For example, according to Ref. 56 it is necessary for achieving a reinforcement to have a strong bond (polymer–filler) of a chemical and physical nature. On the other hand, from Ref. 57, it follows that one of the previously cited types of bonds need not be present for reinforcement. Nevertheless, it is obvious that the presence of a great number of physical bonds, each of which by itself is considerably weaker than the chemical bonds, is capable of ensuring a strong interaction between the polymer macromolecules and the filler surface. A considerable role is played by wetting the polymer filler, the intensive physical binding occurring at the expense of the van der Waals forces, the dispersion forces, etc. Numerous investigations of the mechanisms for the chemical and physical polymer–filler interactions have shown that these interactions are in many respects determined mainly by the filler, whose surface is energetically rather inhomogeneous. As an illustration, let us consider the data on such a widely-used filler as carbon black in industrial rubbers. CB serves as the wetting filler. It has not only surface chemical arrangements, but also free valencies, graphite lattice defects. However, a small part of the available macromolecules of the CB surface is occupied by the high energy centers. According to Ref. 58, about 5% of the surface of such a carbon black as ordinary soot comprises the sections of which chemical sorbtion is possible, whereas on the remaining 95%, the adsorption occurs at the expense of the van der Waals forces.

5.4 PHYSICOCHEMICAL ASPECTS OF REINFORCEMENT

The chemical aspect of reinforcement consists in the interaction between the macromolecules and the surface functional groupings of the filler, resulting in the retention of the polymer on the filler surface formed by the bonds of one or another type (in the case of CB by the carbon–carbon covalent bonds).

The interaction between the polymeric matrix and the filler can lead to the immobilization of a certain amount of polymer on the surface or within the groups of the filler aggregates. With respect to filled CB rubber utilization, such an approach has led to the concept of bound (unamenable to solvent extraction) and occluded rubbers. The interaction on the filler surface can be due to physical adsorption or chemosorption. The latter is supported in particular by the results presented in Ref. 59 in which it was established that the amount of the bound rubber is independent of its molecular weight. This, in its turn, is evidence for the existence of an interphase interaction, characterized by a higher energy level than that of the physical adsorption. The formation of a bound polymer can, in the long run, lead to a change (restriction) in the macromolecular chain mobility, which, in its turn, influences the glass temperature of the composition [60].

According to Ref. 4, the role of the occluded polymer manifests itself only in the process of mechanical effect on the filled material. When the latter becomes deformed, the occluded polymer up to certain stress values appears protected from the effect that the polymer filler undergoes. Although the occluded polymer is captured by the filler aggregates, the particles of the latter do not, as a rule, have an influence on the macromolecular chain mobility. In the case of small strains, the occluded polymer will fill the voids between the aggregates, which should qualitatively resemble the effect of an apparent increase in the filler concentration. In Ref. 61, this phenomenon is explained from the position of the principle of equivalence of the structure and concentration: small concentrations of a high-structuring filler in the polymer matrix may produce an effect such as high concentrations of a low-structuring filler. In fact, the occluded polymer containing a high-structuring filler should promote an increase in the effective filler concentration. It is quite natural in this case to use the function of an effective filler concentration in the polymer matrix—the product of the volume fraction of the filler into the parameter, depending on its structure. However, such a simple interrelation between the structure parameter and the filler concentration [61] can be, in our opinion, regarded more as an exception than as a rule. This is supported by the investigations carried out in recent years.

It has been established by electron microscopy that reinforcement encourages sophisticated aggregation of the fillers. For example, two different structures are observed in the CB reinforced fillers: the aggregates of the primary particle that are bound extremely strongly to one another, and the secondary agglomerate structures—amalgamations of aggregates that are weakly bound to one another [54]. It is essential that the optimal degree of aggregation should depend on the chemical nature of the polymer. It is known that CB dispersed in the rubber maintains a considerable degree of aggregation relevant to its separate phase [62]. Besides, agglomerate formation conditions also originate in the elastomer. The required amount of CB for one or other effect to take place depends on the surface size and the ability of the bulky particles to aggregate.

5.5 SOME ASPECTS OF THE RHEOLOGICAL APPRAISAL OF THE MIXING QUALITY

The ways of introducing a filler into a polymer matrix and, consequently, the 'mixing quality' are exceedingly important for a filled polymer system. Much research has been devoted to these matters. A quite detailed analysis of the problem and the influence of the quality and way of mixing on the rheological properties of mixed thermoplastic systems is given in Ref. 41. Features of mixing elastomers with carbon black and determination of the quality of compound mixing are treated, in particular, in Ref. 36.

As for compositions based on thermoplastic polymers, the quality of mixing carbon black fillers with an elastomer matrix plays an important role. However, in the latter case, there may appear considerable additional difficulties associated with the microheterogeneity of compounds containing carbon black. This is explained, in particular, by the specificity of the physicochemical nature of carbon black capable of aggregating in the elastomer matrix.

We can thus state that the problem of obtaining a homogeneous composition in mixing and determining the mixing quality is exceedingly important. But the treatment of these matters relates more to the practical tasks of technology, which is beyond the scope of the problems being discussed in the present book. With this in view, we shall limit ourselves below only to general, quite trivial remarks and recommendations that will nevertheless help the reader find his way in this field.

The problem of preparing a high-quality homogeneous mixture of a polymer and a filler involves many factors. The optimal conditions can be achieved primarily by the rational choice of the mixing technique, in which important factors are the speed and duration of mixing, the temperature, and the presence of additives improving the miscibility of the components. The temperature and the additives present have the greatest influence here [63–66]. The mixing temperature is chosen to be close to the temperature of the melt or polymer softening to ensure a sufficiently high level of shear stresses in mixing. The temperature must also be high enough to facilitate the filler wettability.

At present, several mixing techniques are employed: addition of the filler to the polymer matrix [65], addition of the polymer to the filler [64, 67–69], or mixing of part of the filler with the entire polymer. But which of these techniques is best is still not clear. Here systematic additional studies are required.

In summarizing, we can note that to a rough approximation any type of high-shear mixer can be selected for a satisfactory mixing quality [66]. As regards the mixing speed and time, high speeds should be employed to obtain the highest possible shear with the minimum mixing time after reaching the equilibrium value of the torque on the mixer shaft. This minimizes polymer degradation and equipment wear [70].

As regards the control of mixing quality, a torque–time diagram obtained for the mixing process provides quite reliable qualitative information on the level of filler dispersion in the polymer. However, rheological analysis is a much more sensitive technique for determining the degree of internal homogeneity of a composite melt.

When considering the rheological techniques for determining the degree of mixing, we must note that capillary viscometry, while providing definite useful information on the degree of agglomeration of the filler at deformation rates close to the mixing speeds, is nevertheless poorly informative. The maximum information at relatively low deformations is provided by dynamic experiments with low-amplitude deformation. Frequency–temperature relations of dynamic characteristics, such as the components of the complex dynamic modulus or the viscosities obtained for various values of the filler concentration, provide information on the agglomeration and degree of dispersion of the filler in a system. These characteristics and the definite structural parameters calculated from them can be used as parameters of rheological models of composites having a high predictive capacity.

An increase in the filler concentration in a composite substantially affects the rheological behavior of the system. The dependence of the relative viscosity on the volume content of the filler is of an exponential nature. This indicates that for systems with a high filler content, the introduction of small additions of the filler produces an appreciable rheological effect that must be taken into account in mixing. Moreover, the shape of the filler particles, their size and size distribution also noticeably affect the rheological behavior [71]. Consideration of these factors as a whole allows us to pose the problem of obtaining composites with a maximum filler content. And although, in our opinion, the question of the optimal filler content from the standpoint of the physicochemical and deformation–strength properties is more expedient in practice, the question of the maximum amount of filler that may be introduced into a system while retaining a good mixing quaity remains urgent. From this viewpoint, the bimodal distribution is potentially better for introducing large filler amounts than a narrow distribution, though theoretically the maximum possible filler content cannot be achieved by a simple chaotic mixing procedure.

Since the rheological characteristics of a composite depend substantially not only on the physicochemical properties of the matrix (MW, viscosity, chemical structure, etc.), but also on its affinity to the filler, considerable changes in the composite properties can be achieved by altering the affinity of the matrix and filler. But here one must not forget the possibility of migration of such incompatible additives at high values of the shear deformation. This may cause inhomogeneity in mixing. Under these conditions, attention must be given to the optimal choice of the mixing process parameters. The level of matrix–filler compatibility (miscibility) can be determined by the scale of increase in the composite viscosity relative to the polymer matrix. By employing the ratio of the dynamic loss moduli of the composite and matrix, one can quantitatively appraise the polymer immobilized on the filler surface [69].

Treatment of the filler surface may sometimes be important for improving the compatibility of the polymer matrix and filler [72–77]. This procedure can yield various levels of surface modification [63, 71, 72, 78, 79]. Study of the technique of processing a filler surface with various media using various durations of treatment conducted by the authors of those references revealed appreciable differences in the rheological behavior depending on the processing technique—the gradual addition of a surface modifier to a polymer melt or pretreatment of the filler prior to mixing. From 0·6 to 0·8% by weight of the modifying additives is optimal for achieving the effect of surface modification [67, 73, 80].

5.6 MODEL BEHAVIOR OF VISCOELASTIC PROPERTIES OF FILLED POLYMERS

In view of the complicated nature of the rheological dependences of filled systems for establishing the general regularities, of important significance is choosing such objects of investigation when the introduction of the filler will have a fairly strong influence. At the same time, the rheological properties of the matrix should themselves be well known. From this viewpoint, the most convenient have proved to be the linear flexible-chain polymers and, in particular, those with a narrow molecular-mass distribution (MMD). The latter are to a certain extent models, since the viscoelastic effects occurring in the process of their deformation are expressed most sharply, especially in the case of the models of relatively high molecular weight (MW) [81, 82]. It has been established [81–85] that the principal regularities of the changing viscoelastic properties of the linear flexible-chain polymers depend on temperature, velocity and strain frequency, deformation regimes, etc. for those with different values of MW and MMD and different microstructures (in particular for polybutadienes, PB). Taking this into account, consideration of the viscoelastic properties of the linear polymer fillers should be very promising in the context of the problem, first in comparison with the available data on unfilled polymers.

5.6.1 Periodic Low-Amplitude Shear Deformation

Dynamic tests under conditions of forced low-amplitude periodic shear deformation of the sample are known to have a number of advantages. In particular, they can be conducted at very small strain amplitude values, γ_0, whereas the characteristics being determined do not depend on the value of γ_0 (the linear deformation region), while the material structure in a broad range of varying velocities (frequencies) will be practically invariable (nondestructive tests). In this case, the component values for the complex dynamic shear modulus G^* are the storage modulus G' and loss modulus G'' characterizing the elastic and dissipative (viscous) properties of the system. The following data were obtained mainly on a Russian-made DKhP mechanical spectrometer for carrying out dynamic investigations [86]. The boundary of the linear region was determined by special experiments wherein the strain amplitudes were varied. The constancy of the G' and G'' values measured at given ω and T served as the criterion for conducting the experiment in the linear strain region, when the deformation amplitude was increased by several times.

As already mentioned, a definite degree of generalization of viscoelastic behavior of the filled polymers can be attained, in our opinion, only by studying the model samples of the linear filled polymers. This assumption, in particular, is supported by the data obtained in Ref. 2 on filled polybutadienes (PB) and narrow MMD. We consider it of principal importance not only for understanding the mechanism that brings about the change in the viscoelastic properties of the filled polymer melts, but also for conceiving the physical character of the processes occurring. This is concerned with the regularities of formation of a secondary plateau related to the yield point of the system [87], its variation with temperature [88], model representations of the rheological behavior, etc. In view of these considerations, we present below a comprehensive analysis of the results obtained.

It should be noted, first of all, that PB is one of the typical representatives of the linear polymers; the data were obtained on a 1,4-*cis* polybutadiene sample with $M = 1{\cdot}35 \times 10^5$ and a polydispersity coefficient $M_w/M_n = 1{\cdot}05$. In the examples presented the fillers were: acetylene black (AB), a rather widespread type of CB possessing a relatively high structuring ability due to the specific shape of its particles (the ratio of the particle length to its diameter is more than 20); kaolin (K), a less 'active' filler; and aluminum spheres (AL), practically inert. (The specific surface (S) of AB determined from nitrogen adsorption equalled $67\ \mathrm{m^2\,g^{-1}}$, those of K and AL 10 and $0{\cdot}4\ \mathrm{m^2\,g^{-1}}$, respectively.) The K and AL fillers are distinguished from those of AB by particle shape, chemistry and specific surface value.

The PB mixture samples containing from 5 to 65 wt% filler (from here on designated as 5C, 10C, etc.) were prepared by the cold milling/rolling technique.

5.6.1.1 *Influence of Concentration: Critical Concentration Value; Concentration-invariant Curves*

From the dependences of G' and G'' on ω (Fig. 5.1), it is seen that all the typical peculiarities of the viscoelastic behavior of the linear high-molecular-weight polymers are represented in the PB frequencies. At low ω values, where $G'' \propto \omega$ and $G' \propto \omega^2$, the fluid state region (terminal zone)

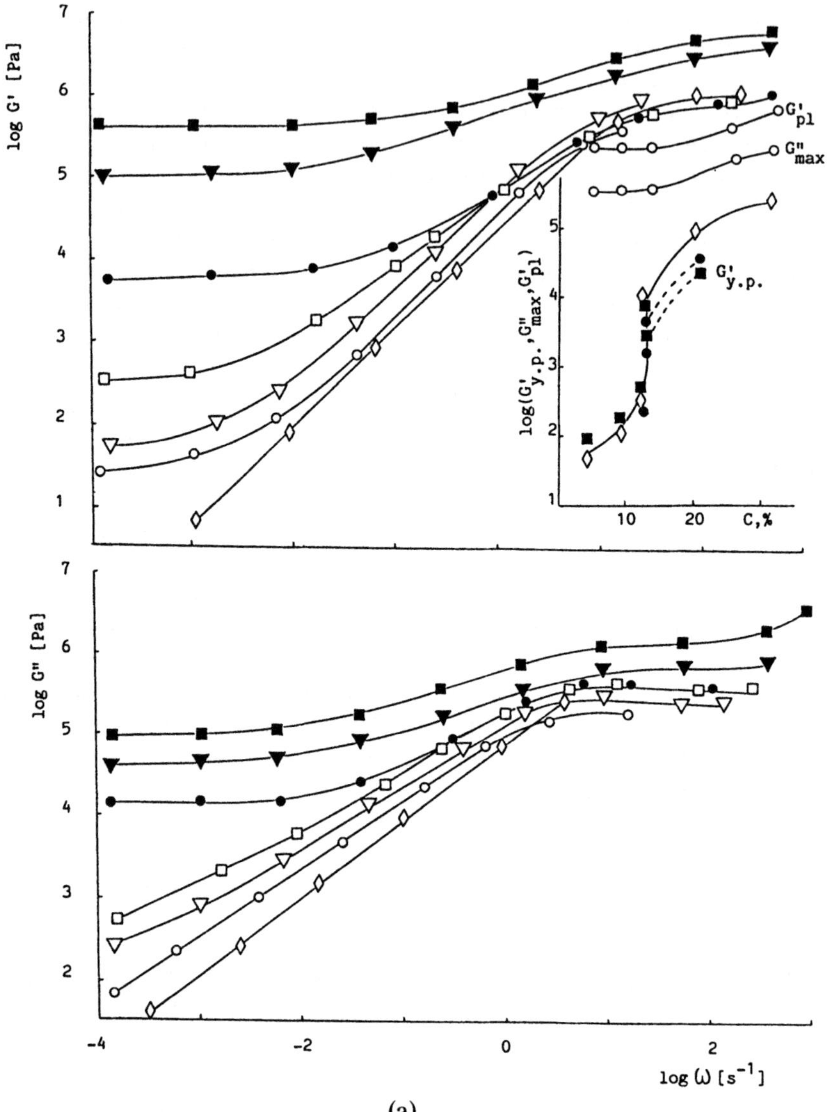

(a)

FIG. 5.1. (a) Storage G' and loss G'' moduli vs circular frequency ω for PB (◇) and PB+AB with filler contents of 5C (○), 10C (▽), 12C (□), 15C (●) 25C (▼) and 35C (■). Inset: $|G^*_{y.p}|$, G''_{max} and G'_{pl} vs AB concentration, C (wt%), for PB ($M = 1{\cdot}35 \times 10^5$)+AB (◇), PB ($M = 10^4$)+AB (●) and silicon oil+AB (■). (b) Kohle–Kohle diagram for PB (◇) and PB+AB with filler contents of 5C (○), 10C (△), 12C (□), 15C (●), 25C (▲) and 35C (■). Temperature 298 K.

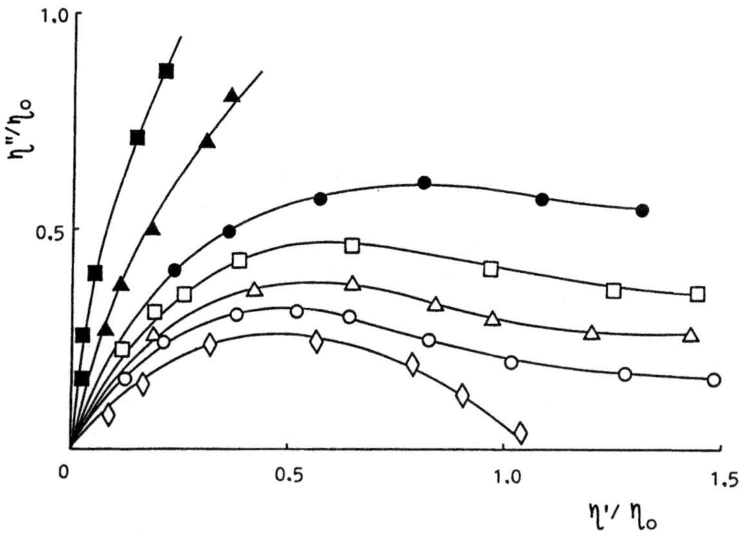

(b)

FIG. 5.1. *Cont.*

is realized. The G''_{max} on the $G''(\omega)$ curve and the appearance of the plateau on $G'(\omega)$ correspond to the achievement of the high-elastic state region, whereas the ascending branches of the $G''(\omega)$ dependence at high frequencies refer to the glass state transition. The addition of AB sharply influences the behavior of the system in the terminal zone on the $G'(\omega)$ dependence. For samples 5C and 10C, the G' moduli practically do not change at low ω values with frequency. There appears a second, low-frequency plateau. A rise in AB content extends this plateau (towards high frequency values) and its height. The appearance of the plateau is undoubtedly due to the presence of a filler in the system and is the result of a formation of definite structural filler particle shapes within it that are united in a steric network. From the physical point of view, the low-frequency plateau is similar to the yield point. According to Refs 1 and 89, we can observe a good correlation between the yield point and the stress corresponding to the low-frequency plateau, which can be related to the generality of their relaxation mechanism. The relaxation of the steric network structure that is formed by solid disperse particles corresponds to extremely low times. This allows us to conditionally refer the low-frequency plateau to the 'yield point' in the frequency range under study. Nevertheless, it should be mentioned that at very low frequencies or shear rates, a further decrease in the characteristics of the system takes place [90] and there is no yield point; therefore in the above-mentioned consideration, we are dealing with an apparently conditional 'yield point'.

The yield point can be characterized by the $G'_{y.p.}$ values that increase with increasing concentration. Judging from the $G''(\omega)$ data, when the concentration of AB reaches ~12 wt%, the network 'nodes' are of a purely elastic nature because the $G''(\omega)$ curves do not have an obviously expressed limit and decline monotonically with decreasing ω. The varying absolute values of G'' in this region witness only a growing dissipation in the system due to the presence of 'free' filler particles that are not amalgamated in aggregates. With increasing concentration of the filler ($C > 15\%$), the fluid state region degenerates and the loss modulus at this point is practically constant—$C''_{y.p.}$. We may suppose that this is due to the formation of the bound polymer having a limited macromolecular chain mobility. The behavior of such a system (at small strains) resembles that of cross-linked rubbers. It should be noted that varying the AB concentration from 5 to 35 wt% leads to an increase in the rheological characteristics of the composition in the terminal zone (by tens of thousands of times), while in the high-elastic state region, the influence of the AB filler appears to be considerably weaker. The values of the storage modulus on the plateau G'_{pl} and the loss modulus at the maximum G''_{max} vary by several times for objects with a high content of AB (>15 wt%). The relation between G'_{pl} and G''_{max} equals ~2·5 and remains practically the same for all filled AB compositions over the whole range of concentrations studied in this work (Fig. 5.1).

Illustrative information on the relaxation processes in filled polymers can also be given by an analysis of their dynamic characteristics involving the plotting of Kohle–Kohle diagrams. Such diagrams were first used when studying dispersions in organic compounds by the dielectric method. Initially, Kohle proposed using these diagrams to describe the relation between the real and imaginary parts of the complex permittivity. The method was later used for studying the mechanical properties of polymers, namely, the relation between the components of the complex dynamic modulus [91], the components of the complex dynamic viscosity of melts [92] and their mixtures [93, 94].

It was noted in Ref. 92, in particular, that the plots of η''/η_0 vs η'/η_0 normalized to the initial viscosity allow two relaxation regions to be singled out for polymers with a linear structure. For systems with a broad MWD and high molecular weight, these regions are represented in the form of a two-hump plot or by asymmetrical arcs. A practically single circular arc can be observed for a polymer with a narrow MWD.

By interpreting the observed picture in terms of relaxation time distribution, we can state that the form of the Kohle–Kohle diagrams reflects the existence of various groups of relaxation times. If these groups are separated precisely, the described relations are a superposition of elementary regions of a diagram. If the relaxation spectrum of the regions is narrow, the relaxation diagrams must differ slightly from two conjugated semicircles. If it is broad, a semicircle transforms into a figure of an irregular shape, and its features enable one to judge whether long or short relaxation times prevail. The latter, for nonpolar polymers with a linear structure, reflect the features of molecular weight distribution. This is just what justifies the conditional use of the name relaxation diagram for plots of η'' vs η'. Qualitative analysis of the Kohle–Kohle relaxation diagram presented in Fig. 5.1(b) for filled PB (see also Fig. 5.1(a)) reveals that the incorporation of a nonrelaxing filler into a relaxing polymer transforms the region of long relaxation times (the right-hand part of the plot in Fig. 5.1(a)). We can thus note that in the given case a filler is a factor retarding the relaxation processes in the region of the fluid state of a polymer. Its addition to a polymer should lower the ability of the system to relax to a certain extent.

The results presented show that the evaluation of the yield point is of exceptional importance for describing the rheological behavior of the composition in the terminal zone. Taking into account the character of the $G'(\omega)$ and $G''(\omega)$ dependences in the region of low frequencies in the broad range of varying AB content, the yield point can be characterized by the values for absolute magnitudes of the complex dynamic modulus:

$$|G^*_{\text{y.p.}}| = \sqrt{[(G'_{\text{y.p.}})^2 + (G''_{\text{y.p.}})^2]}$$

Because $|G^*_{\text{y.p.}}| \cong G'_{\text{y.p.}}$ for compositions with AB concentrations below 15 wt%, the utilization of parameter $|G^*_{\text{y.p.}}|$ renders it possible in a unique way to estimate the 'yield point' on the basis of dynamic measurements. The dependence of $|G^*_{\text{y.p.}}|$ as a function of AB concentration (C) is shown inset in Fig. 5.1(a). At $C > 12\%$, the curve $|G^*_{\text{y.p.}}| = f(C)$ has a bend, which signifies the critical concentration C_{cr} values of the filler in the viscoelastic matrix.

An excess of C_{cr} drastically changes the viscoelastic behavior of the composition (see Fig. 5.1). An essential growth in G' and G'' values at $C > C_{\text{cr}}$ can be explained by the formation of agglomerates of particles [95] (secondary agglomeration), and by the contribution that is being made by the occluded polymer [4]. Thus, the rheological dependences under discussion correspond well to the above-mentioned considerations on the possibility of forming different types of structures in TC [52, 54], and the conclusions in [62] concerning the presence of the indicated structures in the filled polymer matrix.

The set of rheological dependences presented allows us to make some general conclusions concerning the influence of the filler, which has a high-structuring potentiality on the rheological behavior of the polymer system.

Taking into account the rather complicated character of the varying viscoelastic parameters of the filled polymer systems in the terminal zone and the high-elastic state region, we may suggest subdividing them, depending on the filler content, into 'filled' and 'highly-filled'. The concentration boundary between them is determined by the size of the C_{cr} that is a function of individual properties of the filler. In favor of the latter, we may point to the fact that the value $|G^*_{y.p.}|$ at this C_{cr} is practically independent of MW, the polymer matrix viscosity, and its nature. This is confirmed by the data given in Fig. 5.1(a) (inset) that displays the curves $|G^*_{y.p.}|=f(C)$ for compositions on a low-molecular PB base ($M=10^4$; the initial viscosity of this sample is lower by an order of four-tenths than the initial PB polymer matrix involved) and on a silicon oil base (polydimethylsiloxane liquid No. 5). The bend corresponding to the C_{cr} value is distinctly seen on the indicated curves.

At $C \leqslant C_{cr}$, by virtue of the formation of the steric network, the elastic parameters of the system vary drastically. The network strength characterized by the 'yield point' of the system can be estimated by the absolute magnitudes of the complex dynamic modulus $|G^*_{y.p.}|$ measured at low-amplitude deformation frequencies. Apparently, the values for $|G^*_{y.p.}|_{C \leqslant C_{cr}}$ are determined by the individual features of the filler: the density of the structural nodes of the primary aggregates (it is maximal at C_{cr}) and the quality of the individual node (its strength) (see Fig. 5.1). At $C > C_{cr}$, the size of the yield point can be regarded as the sum of two components:

$$|G^*_{y.p.}|_{C>C_{cr}} = |G^*_{y.p.}|_{C=C_{cr}} + A(C) \tag{5.1}$$

where $|G^*_{y.p.}|_{C=C_{cr}}$ denotes the yield point values at critical concentration, and $A(C)$ is the component attributed to the contribution of the agglomerates, the second aggregates. In a rough approximation, $A(C)$ is determined, to a considerable extent, by the MW matrix, its nature, etc. (Fig. 5.1) and the magnitude of γ_0, which will be discussed later.

From Eqn 5.1, it is seen that for high-structuring fillers, accommodated in the viscoelastic medium with a strongly expressed anomaly of properties, the yield point is a complicated function of concentration. Apparently, this should be expected also for the rheologically simpler liquids, in connection with which attention should be given to refining the known Casson equation [96] and its modification for filled polymer solutions, according to Ref. 1, in case the disperse phase with high aggregative capability is used.

Let us turn to the results of the investigation of high-elastic properties of PB filled with kaolin and aluminum spheres. They are presented in Fig 5.2. For PB+K, the elastic properties of the system in the fluid state region change more strongly than the viscosity properties (compare $G'(\omega)$ and $G''(\omega)$ dependences). However, in comparison with PB+AB, the absolute growth of G' is substantially less. For example, with a 35 wt% K content we can observe only a threefold increase in the G' values (as the lowest frequencies realized in this work), whereas in the case of samples with AB values, G' with the same concentration increased 10^4 times. For PB+AL, the modulus G' with the same filler concentrations increased by only 50%. As for the secondary low-frequency plateau on the $G'(\omega)$ dependences, for PB+K and PB+AL, we can say that it tends to appear only in the case of fairly high (>35 wt%) concentrations. Such a plateau for these systems may perhaps manifest itself more sharply with considerably higher filler concentrations than those used in the present work. In comparing the rheological PB+AB, PB+K and PB+AL curves, it would be appropriate to recall the principle of structure and concentration equivalency [61]. Unfortunately, the data under consideration do not allow this principle to be broadly verified. One of the reasons for this is the fact that there arise some difficulties in homogenizing the compositions during the mixing process when large quantities of K and AL fillers are added to PB.

The data in Figs 5.1 and 5.2 allow us to consider the possibility of plotting generalized concentration-invariant dependences of dynamic characteristics of the linear filled polymers.

In the case of the PB+AB system, we should take account of the viscoelastic behavior of the 'filled' and 'highly-filled' samples. At $C \leqslant C_{cr}$, the concentration-invariant dependences of the dynamic characteristics can be plotted by shifting the $G'(\omega)$ and $G''(\omega)$ curves only along the frequency axis up to the alignment which corresponds with the similar dependences for the polymer matrix in the fluid state region, if we take into account the essential difference in the properties of the compositions in the terminal zone and the weak change in the values for G'_{pl} and $G''_{max}=f(C)$. The size of this shift will be determined by the a_{c1}, concentration actuation coefficients. Thus, the polymer matrix characteristics for such an actuation are chosen as the reference point.

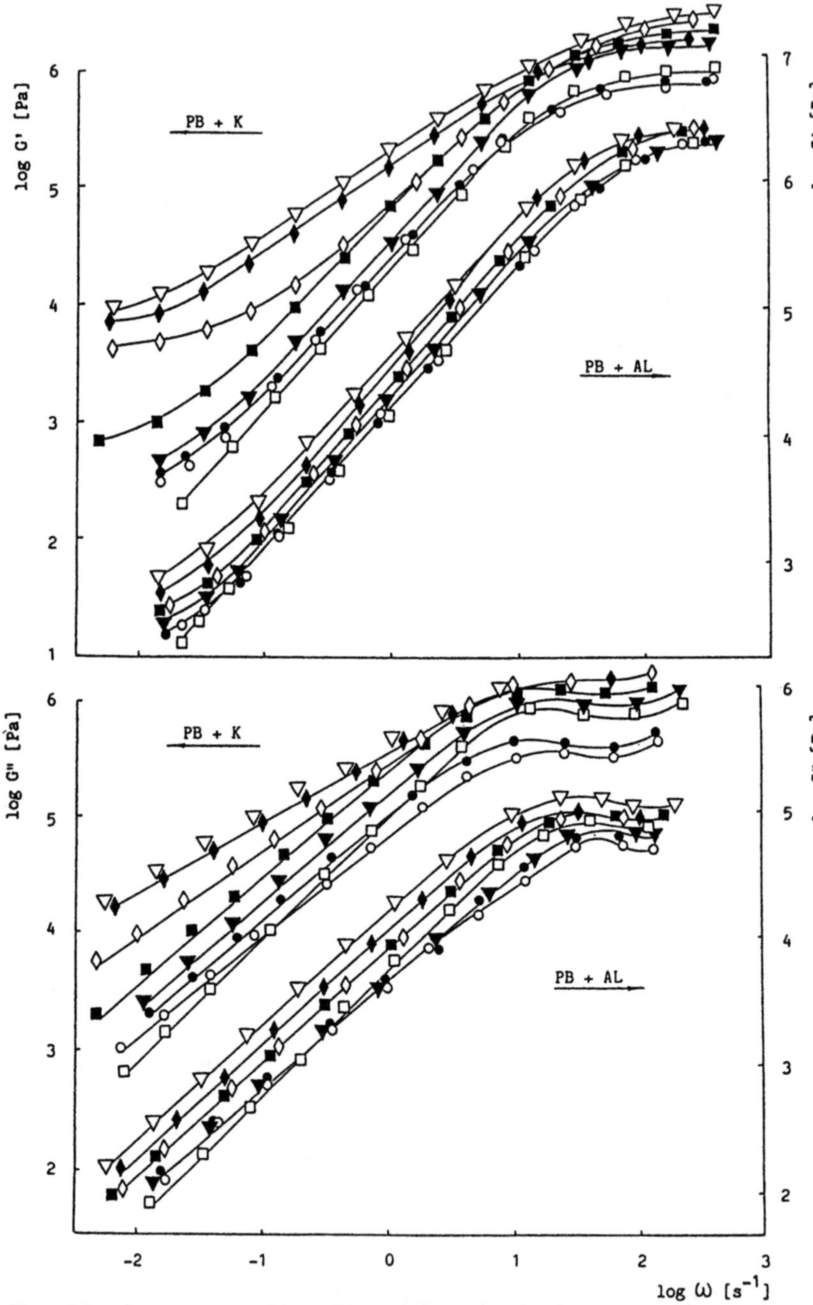

FIG. 5.2. Storage G' and loss G'' moduli vs circular frequency ω for PB (□) and for PB + K and PB + AL with filler contents of 5C (○), 15C (●), 25C (▼), 35C (■), 45C (◆), 55C (◇) and 65C (▽). Temperature 298 K.

For 'highly filled' compositions ($C > 12\%$), the G'_{pl} and G'_{max} values vary several times (Fig. 5.1). Therefore, in order to plot the concentration-invariant dependences, it is necessary to use two concentration reduction coefficients—one along the frequency axis, a_{c1}, and the other along the modulus axis, a_{c2}. In the latter case, it is convenient to use the $G'_{y.p.}$ and G''_{max} values, respectively, as the reference points for the polymer matrix. Since for samples 5C–35C, the G'_{pl} and G''_{max} relationship is practically constant, the a_{c2} coefficient can be determined by both the $G'(\omega)$ and $G''(\omega)$ curves. The concentration invariants obtained by the previously discussed method for the storage modulus $(G'_{red})_c = G'a_{c2}$ and the loss modulus $(G''_{red})_c = G''a_{c2}$ reduced in terms of concentrations, plotted against the reduced frequency ωa_{c1}, are given in Fig. 5.3 for samples 5C–35C. The concentration reduction coefficients a_{c1} and a_{c2} as functions of C are presented in Fig. 5.4. It is seen that such a method renders it possible to obtain reduced dynamic characteristics in the region of fluid and high-elastic states. The method for more effective $G''(\omega)$ dependency reduction is considered.

The concentration-invariant curves have a number of specific features. First, typical 'tails' in the terminal zone, distinctly expressed on the $(G'_{red})_c = f(\omega a_{c1})$ curve, reflect the existence of the yield point. Second, with rising concentrations, the AB maxima on the $(G''_{red})_c = f(\omega a_{c1})$ curve shift toward the large ωa_{c1} values. The size of this shift $\Delta\omega_{max}$ as a function of C is also given in Fig. 5.4, the matrix $\Delta\omega_{max}$ being chosen as the reference point. On the basis of the dependence $\Delta\omega_{max} = f(C)$, we arrive at the following conclusion. In the case of the filled AB samples, the terminal zone taken from the dynamic experiment data is extended in comparison with the pure PB. As a result, we may suppose that the stable laminar flow regime observed during the steady state PB deformation in the channels [81] will be realized in a broader range of $\dot{\gamma}$ shear rates for PB+AB composition. In fact, from Ref. 90, it is seen that the critical deformation parameters corresponding to the attainment of unstable flow regimes and, in particular, to the flow separation in the channels, are slightly higher in filled AB polybutadienes than in unfilled ones.

In the case of PB+K and PB+AL compositions, the concentration reduction of the viscoelastic dynamic characteristics, as seen in Fig. 5.2, can be achieved by a shift of the corresponding $G'(\omega)$ and $G''(\omega)$ dependences along the modulus axis. The same curves for the polymer matrix can be chosen as the reference point. Figure 5.5 shows the concentration-invariant curves for the dynamic characteristics for PB+K and PB+AL respectively. Figure 5.6 displays the concentration

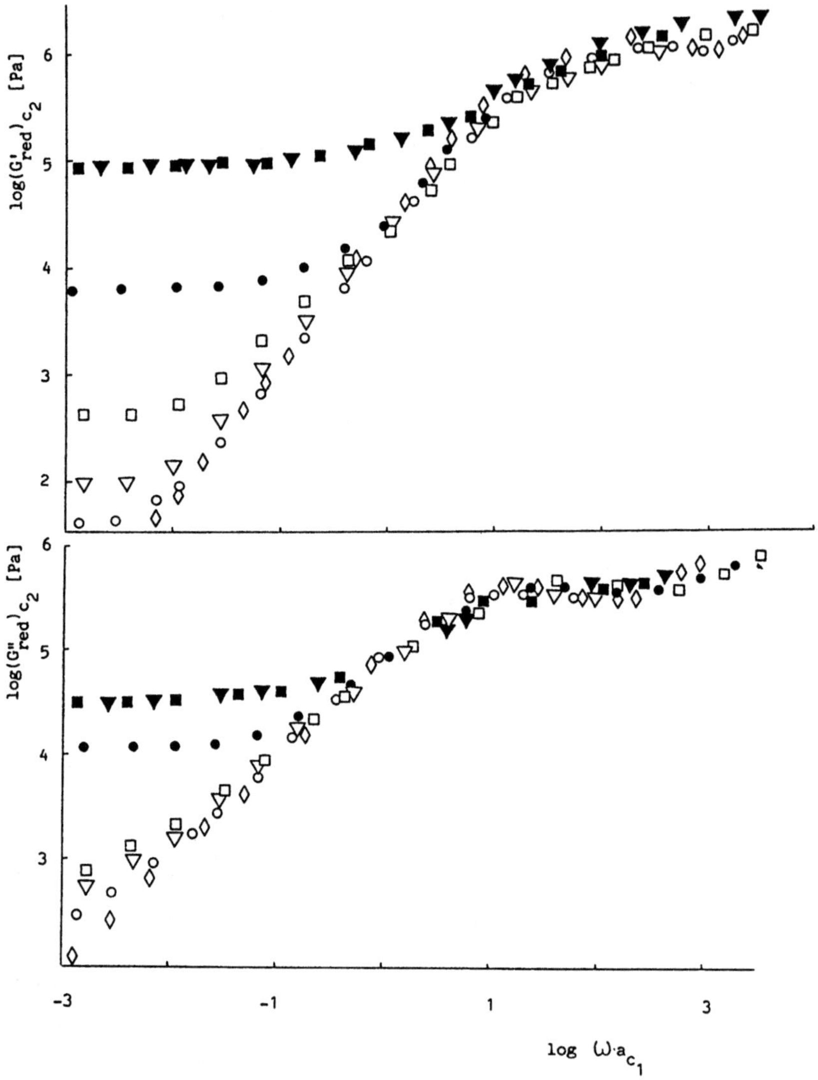

FIG. 5.3. Storage modulus $(G'_{red})_{c2}$ and loss modulus $(G''_{red})_{c2}$ reduced in terms of concentration vs reduced frequency ωa_{c1} for PB+AB at 298 K (see Fig. 5.1(a) for curve symbols).

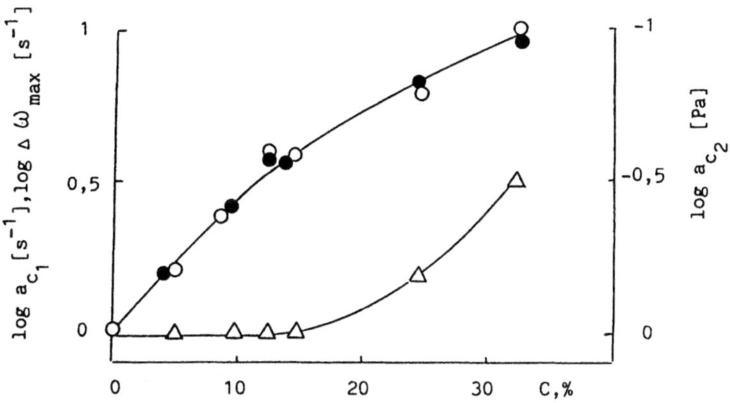

FIG. 5.4. Concentration reduction coefficient for a_{c1} (○) and a_{c2}(△) and $\Delta\omega_{max}$ values (●) vs AB concentration. Temperature 298 K.

reduction coefficients a_{c2} vs the filler concentration. The observed difference in the a_{c2}(C) dependence plotted for the $G'(\omega)$ and $G''(\omega)$ functions should be attributed, at the expense of different sensitivities of the elastic and dissipative components of the complex dynamic modulus, to the structural organizations of the filler in the polymer matrix.

5.6.1.2 *Influence of Temperature: Temperature–Time Reduction*

The influence of temperature on the viscoelastic properties of the filled polymers allows us to conveniently follow the PB+AB system taken as an example. For this purpose, let us analyze the investigation data (see Fig. 5.7) on samples 12C and 35C [97]. They were obtained in a broad temperature range (from 233 to 373 K). The data analysis has shown that there exists a definite difference in the behavior of the 'filled' and 'highly-filled' systems with varying temperature. Nevertheless, we can see that the properties of the matrix itself play a rather important role, and in the long run, a determining role with decreasing temperature. As in the case of unfilled PB [81], we observe a viscosity anomaly intensification, an expansion in the region of transition to a fluid state, and a change in the shape of the $G'(\omega)$ and $G''(\omega)$ curves in the terminal zone. The evaluation of the relative variation in the G' and G'' magnitudes (the comparison is made in the region of the deformation frequencies exceeding the yield point) vs temperature for 5C–35C compositions and the polymer matrix has shown that it is practically the same. The calculation made for the activation energy E_a of the viscous flow for filled samples

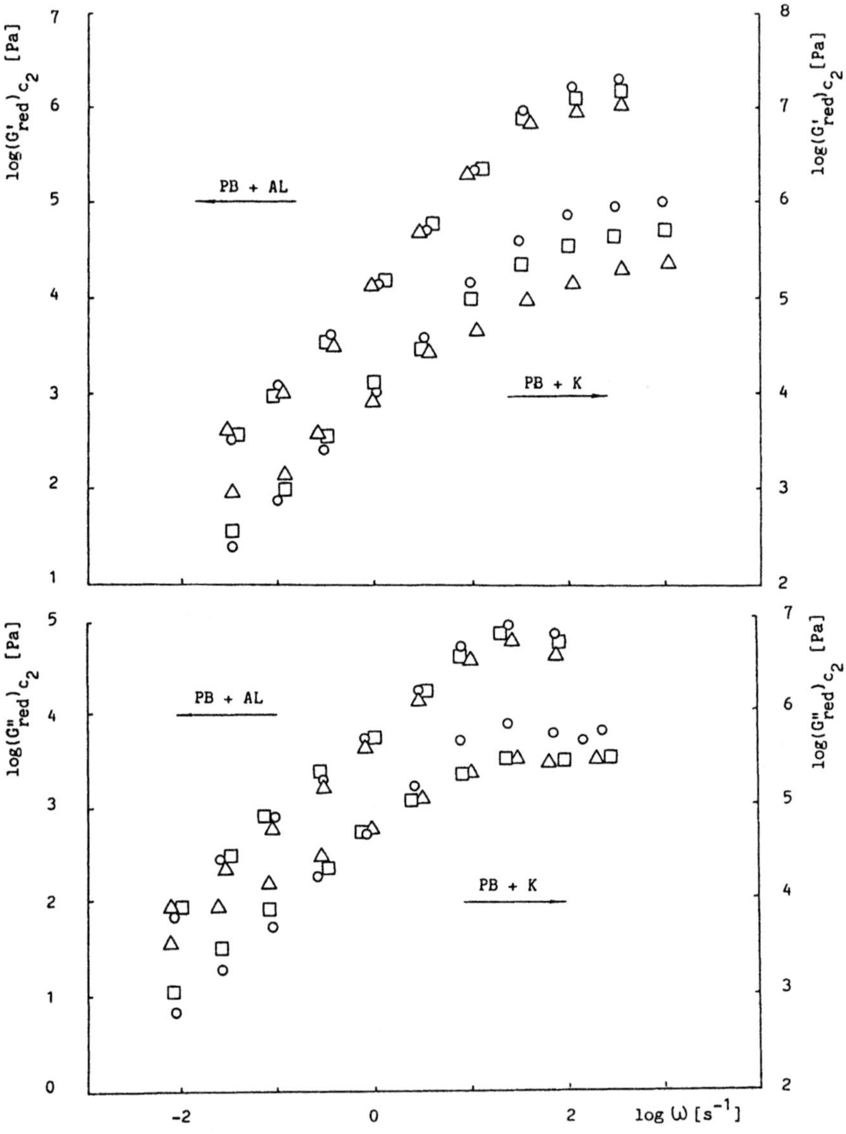

FIG. 5.5. Storage modulus $(G'_{red})_{c2}$ and loss modulus $(G''_{red})_{c2}$ reduced in terms of concentration vs frequency ω for PB+K and PB+AL at 298 K with filler contents of 0, 5C, 15C and 25C (all ○), 35C and 45C (□) and 55C and 65C (△).

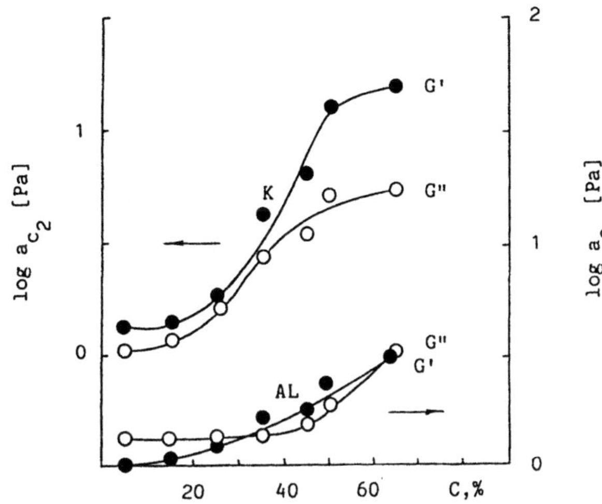

FIG. 5.6. Concentration reduction coefficient for a_{c2} vs filler concentration for kaolin (K) and aluminum spheres (AL) from $G'(\omega)$ dependences (●) and from $G''(\omega)$ dependences (○).

indicates that E_a varies little with increasing AB concentration at room temperature. However, with $T = 253$ K for samples 5C and 10C, the E_a values increase ~2·5 times as compared with those of the PB samples.

The polymer glass transition temperature, T_g, is an exceptionally important parameter, being of practical significance. Figure 5.1(a) shows that the addition of AB fillers leads to a horizontal displacement of the dynamic curves and to a change in their slope in the transition and terminal zone of the region. This is indicative of a certain filler influence on the relaxation properties of the matrix, and in particular, on the possible change in the glass transition temperature of the system that will depend also on the nature of the filler surface [60]. According to the measurement carried out in Ref. 2, the T_g of the PB+AB composition is about 30 K higher than that in the initial PB, and depends weakly on C. The observed change in T_g is considerably greater than that for butadien-styrene rubber [60], which can be accounted for by the formation of a fairly stable bond [54] between AB and PB. It is appropriate here to question the temperature dependence of the yield point. The literature affords little information on this point. For example, Smith studied the filled rubbers with glass balls and noticed a strong dependence of the yield point on temperature [98]. On the basis of the supposition concerning the nature of the structural network of PB+AB for 5C–12C

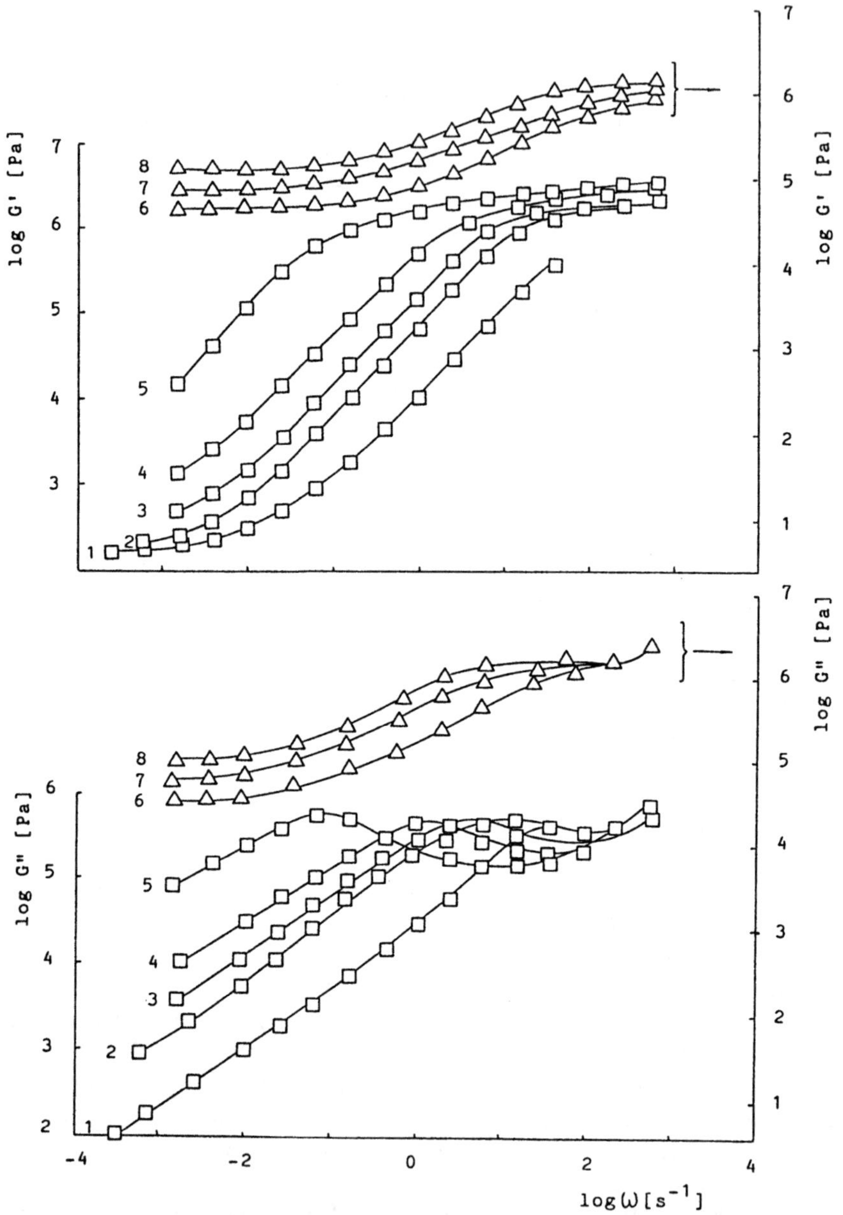

FIG. 5.7. Storage modulus G' and loss modulus G'' vs circular frequency ω for PB+AB samples 12C (□) and 35C (△) at 373 K (curve 1), 298 K (2), 273 K (3), 253 K (4), 233 K (5), 373 K (6), 343 K (7) and 298 K (8).

samples, we may expect a weak dependence of $|G^*_{y.p.}|_{C \leqslant C_{cr}}$ on temperature, which was actually observed for two temperatures (Fig. 5.7, curves 1 and 2). We may say that the change in the polymer matrix viscosity with temperature has practically no influence on the $|G^*_{y.p.}|_{C \leqslant C_{cr}}$ values (see Fig. 5.1). In favor of generality of this conclusion for compositions with a rigid filler structural network, we can also refer to the results of rheological investigations of high-pitch coal tars [89] in which the structural network is formed by pitch and crystalline anthracite particles. According to Ref. 89, $|G^*_{y.p.}|$ does not vary with temperature for these substances.

In the case of 'highly-filled' PB+AB samples, the dependence of $|G^*_{y.p.}|_{C>C_{cr}}$ on temperature is fairly reliably recorded (Fig. 5.7, curves 6–8). The change in the 'yield point' with temperature is similar to the change in the initial polymer matrix viscosity η_0 with temperature (Fig. 5.8).

Thus, we may assume that:

$$|G^*_{y.p.}|^{T}_{C>C_{cr}} = |G^*_{y.p.}|^{T_1}_{C>C_{cr}} \left(\frac{(\eta_0)^T}{(\eta_0)^{T_1}} \right) \tag{5.2}$$

where T and T_1 are the comparable temperatures. Apparently, at fairly high $T = T_g$ values, $|G^*_{y.p.}|^{T_1}_{C>C_{cr}}$ will approach $|G^*_{y.p.}|_{C=C_{cr}}$.

The question arises as to whether the principle of the temperature–time reduction that is so broadly used in the experimental rheological investigations of polymer melts is operative (see Refs 81 and 82). As is known, the principle of construction of the terminal zone of the temperature dependences of the invariant dynamic characteristics provides, in particular, for the $G'(\omega)$ and $G''(\omega)$ curve shifts along the frequency axis by a magnitude of the temperature reduction coefficient obtained for the fluid state region, a_T^{T} [82]. The (G'_{red} and G''_{red})$=f(\omega a_T^{\mathrm{T}})$ dependences reduced in the same manner for PB+AB compositions are given in Fig. 5.9. The reduction temperature T_{red} was 298 K. The $(a_T^{\mathrm{T}})_C = f(T)$ values for compositions and $a_T^{\mathrm{T}} = f(T)$ for matrices (the latter being derived from Ref. 82) are presented in Fig. 5.8 by open and filled symbols, respectively. We can distinctly see that at $T \geqslant 273$ K, the indicated dependences coincide, i.e. in the terminal zone, the temperature dependence of the filled system (PB+AB) is completely determined by the polymer matrix. At the same time, from Fig. 5.9, it follows that temperature reduction for PB+AB, as in the case with PB [82], is not operative in the high-elastic state region.

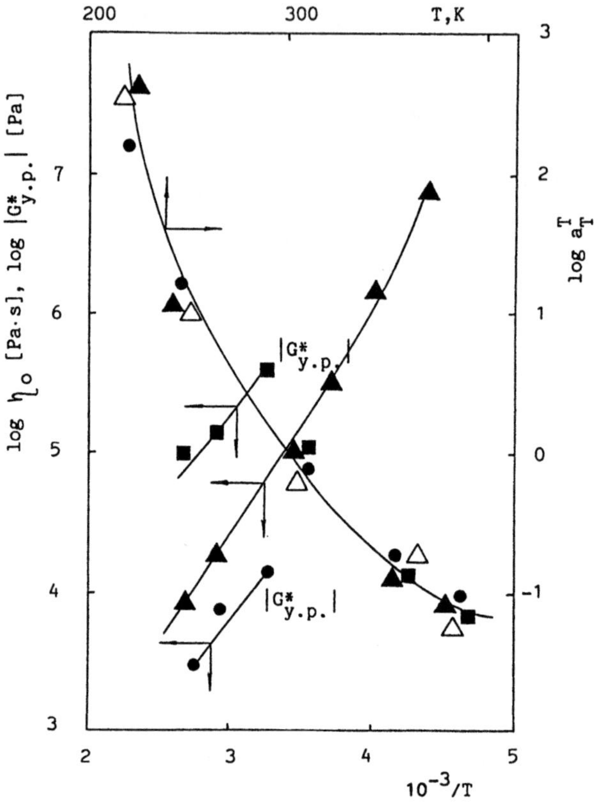

FIG. 5.8. Initial η_0 viscosity $|G^*_{y.p.}|$ values and a_T^T temperature reduction coefficient vs temperature for PB (▲) and PB+AB samples 10C (△), 15C (●) and 35C (■).

5.6.1.3 Temperature–Concentration Dependence

These data allow us to partially touch upon the problems of plotting the temperature–concentration dependence of the dynamic rheological characteristics of the polymer compositions containing a filler having a high structuring capacity. The $(G'_{\text{red}})_{T,C_2}=f(\omega a_{c1}\times a_T^T)$ and $(G''_{\text{red}})_{T,C_2}=f(\omega a_{c1}\times a_T^T)$ curves for PB+AB are illustrated in Fig. 5.10. They were obtained with simultaneous allowance for the concentration and temperature reduction coefficients. The reduction temperature, T_{red}, was chosen to be 298 K. The concentration reduction is realized to the value of $C=0$. The polymer matrix characteristics are taken as the normalizing parameters.

The broad range of circular frequencies covering the fluid and high-

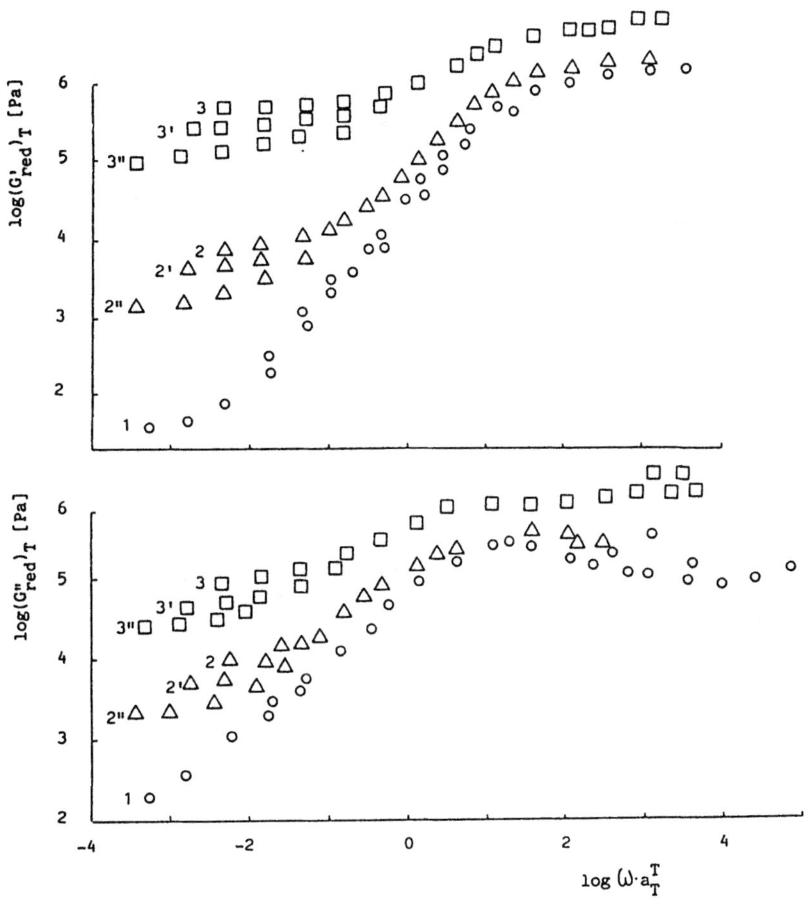

FIG. 5.9. Storage modulus $(G'_{red})_T$ and loss modulus $(G''_{red})_T$ in terms of temperature vs reduced frequency ωa_T^T for PB+AB: 5C (○), 15C (△) and 35C (□) at 233–273 K (curve 1), 298 K (2 and 3), 343 K (2′ and 3′) and 373 K (2″ and 3″).

elastic states for the systems under study does not allow us to achieve a satisfactory generalization for the viscoelastic functions under consideration. In the range of low frequencies, the terminal zone, such a generalization is hindered by the existence of the yield point. In the region of higher frequencies, the high-elastic state, it is hindered by the impossibility of plotting invariant temperature dependences. It is to be noted that the data were obtained in conditions of low-amplitude deformation in which the material structure is not destroyed.

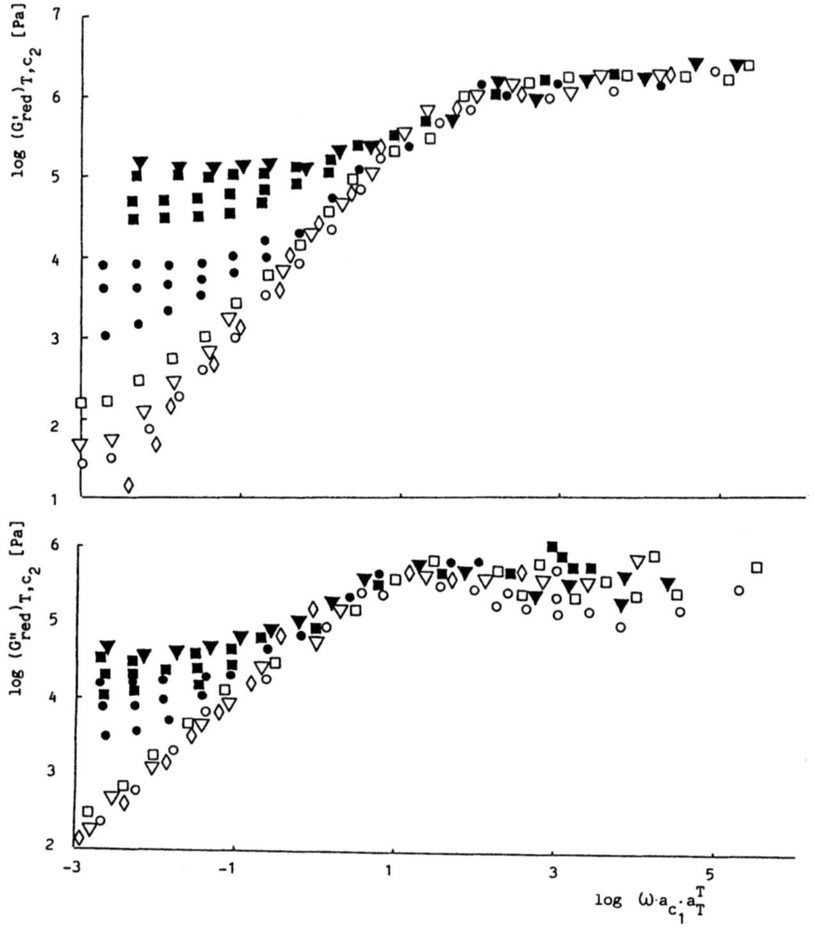

FIG. 5.10. Temperature–concentration dependence of the storage modulus $(G'_{red})_{T,c2}$ and loss modulus $(G''_{red})_{T,c2}$ on the reduced frequency $\omega a_{c1} a_T^T$ for PB + AB. Reduction temperature $T_{red} = 298$ K (see Fig. 5.1(a) for symbols).

5.6.2 Continuous Deformation

5.6.2.1 Some General Knowledge

One of the basic technological requirements when processing melts of filler polymer is their ability to flow. The ideal filler packing shown on the right in Fig. 5.11 (filling over 52%) prevents shear flow, which by definition is isochoric—the volume of the medium, regardless of whether

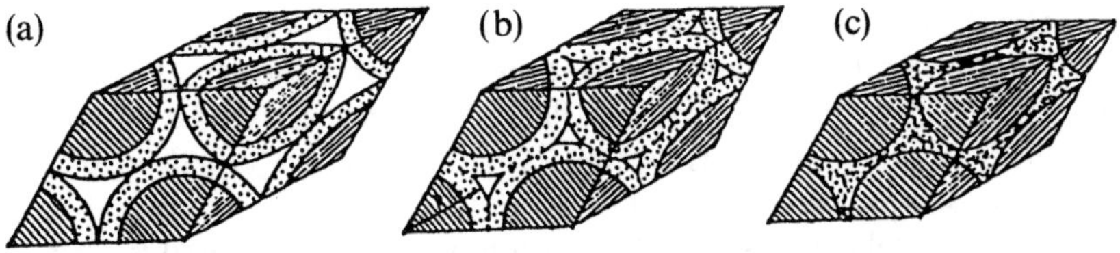

FIG. 5.11. Packing scheme for highly-filled composite. (a) Binder envelopes particles. Porosity $m=26\%$, 'live section' $n=0{\cdot}0921$; length of parallelepiped edge $a=2r$. (b) Formation of closed porosity, $m=3{\cdot}3\%$, $n=0$, $a=1{\cdot}85r$. (c) $m=0$, $n=0$, $a=1{\cdot}81r$.

it is compressible or not, does not change in deformation. For shear to occur in the medium shown in the figure, its volume must grow. This phenomenon, i.e. dilatancy, is exactly what was observed by Reynolds [99]. Filling of the gaps between the solid particles with a viscous liquid does not naturally change matters. Here an experimentalist has to do with a medium that cannot be considered as a liquid; its shear strength will depend on the pressure normal to the shear plane, whose influence is incommensurably greater than that of the hydrostatic pressure on the viscosity and viscoelasticity of high-molecular-weight liquids [100]. Such media are described by models including Saint Venant friction, e.g. models of granular-viscous media [101] and other models well developed in soil mechanics [102].

All the foregoing shows the inexpediency of posing the problem of the maximum replacement of a polymer by a filler in general-purpose polymer compositions. In most cases, one must apparently consider optimal filling from the viewpoints of both the rheological behavior of the composites and the physicomechanical properties of a finished article.

The question of the yield point in shear deformation of filled melts and solutions of polymers is being increasingly discussed and studied. Among the first result to appear was the experimental research described in Ref. 103. Next came a large cycle of studies [104, 105] where it was noted that the introduction of particles over several micrometers (μm) in size into a melt does not produce the observed yield point. In Refs 104 and 105, filling up to 36% by volume was considered, which is higher than what is actually employed in the plastics industry for the processes of extrusion, pressure-die casting, etc. In these publications, the interaction of the particles was assumed to be hydrodynamic. For melts containing particles of submicrometer size (talc, titanium dioxide, carbon black, also possibly chemically deposited chalk), it was assumed that the yield is higher the finer the particles. It was also noted that finishing of the chalk particles, in particular with stearic acid, substantially lowers the interaction between particles, their aggregation, and, consequently, the apparent viscosity. In the cited publications, filled polymer melts and solutions are classified as plastic-viscoelastic media. A rheological equation of state was proposed for such a medium.

Models of a plastic-viscoelastic medium were also proposed in Refs 106 and 107. Those authors presumed a yield point after von Mises in the form of a time function [108, 109] and described the thixotropy of such a medium.

The processing of flow curves for melts of filled and unfilled polymers in Casson coordinates or in these coordinates modified by Onogi and Matsumoto [110] often leads to the yield point. In the modified coordinates, the flow curve of a non-Newtonian matrix becomes a straight line, and usually a filled non-Newtonian matrix is also depicted by a straight line. This allows extrapolation to be performed to a zero shear rate and the yield point to be determined.

The yield point in filled systems can be considered as a conditional one (except for the closest packing mentioned above, when the particles have nowhere to move in simple shear). However, beginning from certain filler concentrations, even ignoring the appearance of 'chain' structures—a space system of particle aggregates—the system of particles has to be 'shifted' in the direction of the deformation rate to realize a flow. Here one usually has to consider the prolonged (in comparison with the duration of an experiment or technological operation) relaxation times. It will be shown below that expansion of the spectrum of viscoelastic relaxation times into the region of prolonged times sometimes yields satisfactory results from experiments describing the viscoelastic characteristics of filled polymer melts.

A considerable number of publications and reviews (see, in particular, Ref. 111) have been devoted to a theoretical description of the flow of liquid media with fillers in the form of solid particles (suspensions).

We can consider that this trend was begun by the works of Einstein on the flow of dilute suspensions with no mutual influence of particle motion [112]. Consideration of the microinertia of suspension particles made it possible [113] to predict the appearance of non-Newtonian properties in dilute suspensions and of the first difference of the normal stresses in shear.

It was shown [108, 114–117] that deviation of the particle shape from spherical may also cause the appearance of non-Newtonian properties in the shear flow of dilute suspensions, namely a dependence of the apparent viscosity on the shear rate, the appearance of a normal stress component in shear, orientation of the particles along the streamlines, etc.

We can continue the list of studies in this direction, but this will introduce nothing fundamentally new into the problem.

Theoretical works treating the flow of concentrated suspensions that take the hydrodynamic interaction of particles into account also include Refs 109, 118 and 119. In Ref. 120, the hydrodynamic interaction of the particles is considered by a power series with respect to the concentrations.

The problem of the flow of spherical particle suspensions with a viscous matrix was dealt with in Ref. 121. A situation very close to real conditions was analyzed when the size of the particles is large enough for their Brownian motion to be disregarded. It was concluded that there is a concentration C_{cr} at which solid particles no longer form separate aggregates, but rather a continuous cluster. A core flow appears in the pipe. A core flow of highly-filled polymers was also observed in Ref. 121, but there was no singularity at $C \sim C_{cr}$. Proceeding from the apparatus of the isotropic theory of flow or percolation (see, for example, Ref. 50 used in Ref. 121), it follows that $C_{cr} \sim 1/5$. With such a working scheme, the viscosity of a suspension depending on the solid phase concentration is described by a relation with a special region near $C = C_{cr}$.

In the case of the fibers it was shown in Refs 122 and 123 for a Newtonian matrix that if the length and concentration of the fibers are within the limits of the interval ensuring conditions of fiber engagement and the condition of a liquid after Doi and Edwards [124, 125], the initial viscosity grows sharply, while in the second (lower) Newtonian region the viscosity of a composite differs only slightly from that of the matrix, namely, oriented accumulations of fibers arrange themselves along the streamlines. If, on the other hand, the fibers are polydispersed, the initial structure thereof is destroyed within quite a broad range of stresses and rates of shear, and there are no sharp transitions to a new highly organized structure.

When a suspension of monodispersed fibers flows in a restricted interval of shear rates, a superanomaly of the viscosity and negative values of the first difference of the normal stresses are observed. With a further increase in the shear rate, flow of the newly appearing highly organized fiber structure occurs already in the form of ellipsoids of revolution [122, 123]. In the process of producing articles reinforced with dispersed fibers, this phenomenon is not desirable—a definite distribution of the fibers along the length must be ensured. This is achieved for glass fibers, in particular, by cutting up the fibers during the flow (processing).

Experimental verification of some model approaches which should predict the behavior of filled polymer systems was observed in Ref. 126. In order to do this, the rheological equations of state known from the literature were analyzed [126]. The integral models were divided into two kinds, one containing a nonlinear relaxation function depending on the second invariant of the tensor of deformation rates, and the other containing an also nonlinear memory function depending on the second invariant of the rates of deformation or deformations. Each kind of model was also divided into groups according to whether the kernel function alters (1) the values of a modulus or the spectrum of relaxation times (without altering the values of the relaxation times), i.e. modulus nonlinearity, (2) the values of the relaxation times (without altering the values of a modulus or spectrum), i.e. relaxation nonlinearity, and (3) both the values of the spectrum or modulus and those of the relaxation times, i.e. general nonlinearity [127]. The last model provided the best approximation of the rheological functions of filled melts.

5.6.2.2 Continuous Shear Deformation

The general laws of flow of flexible-chain polymer melts have been described in sufficient detail in Ref. 128. As regards the model ideas of the flow mechanism, study of the rheological behavior of substances with a narrow MWD was important here (see, in particular, Ref. 129 and Chapter 2).

Study of the flow of filled model systems has been very helpful in establishing some general laws of the rheological behavior of such systems. Let us treat some specific and general results of studies using the example of polybutadienes with a narrow MWD filled with carbon black. For simplicity of analysis, below we discuss the results obtained on the same substances that were described in preceding chapters (PB with $M = 1{\cdot}35 \times 10^5$, $M_w/M_n = 1{\cdot}05$; the filler was acetylene carbon black with a specific surface area of 67 $m^2\,g^{-1}$).

Figure 5.12 presents the apparent shear rate $\dot{\gamma}$ or the volumetric rate of flow per second Q vs the shear stress τ. The value of Q was normalized to the capillary radius ($4Q/\pi R^3$) [130]. An unfilled PB sample (curve 1), like a 10% composition, exhibits a Newtonian nature of flow up to separation stresses τ_{sp}, which the top vertical branches correspond to. For samples with a 15% carbon black content and more, these relations break up into four sections. Section AB is the creep zone (for a system containing 35%

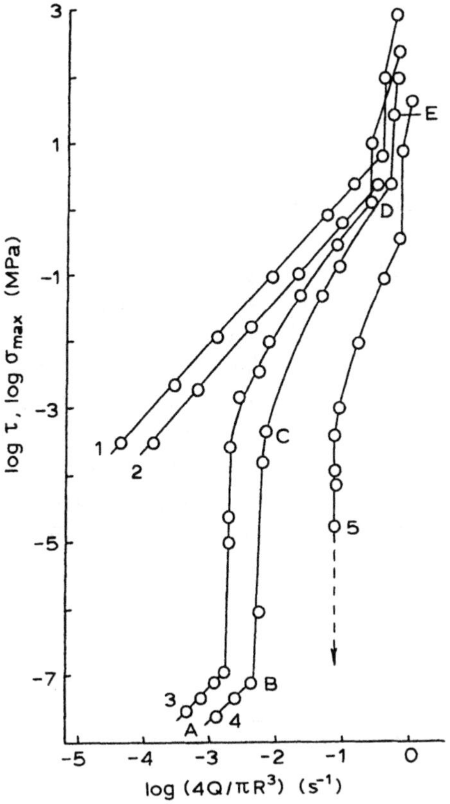

FIG. 5.12. Normalized volumetric rate of flow $4Q/\pi R^3$ vs the shear stress τ (flow curves) for PB (curve 1) and PB+AS with filler contents of 10 (2), 15 (3), 25 (4) and 35 wt% (5). Temperature 298 K. Points A–E are discussed in the text.

of carbon black it could not be realized). On this section, the viscosity of the composition exceeds that of the matrix a million times. Such viscosity values are typical of plastic dispersed systems with an undestroyed three-dimensional carbon black structure [131]. Point B corresponds to the yield point $\tau_{y.p.}$. In the temperature range from 296 to 373 K, $\tau_{y.p.}$ does not depend on the temperature, which is also typical of plastic dispersed systems with a low-molecular matrix [131]. Section BC is the first vertical branch and characterizes sharply pronounced failure of the structure. In this respect, $\tau_{y.p.}$ can be considered as the ultimate strength. The rate of flow on section BC increases by three or four orders of magnitude, and the viscosity diminishes by the same amount. Section CD is the flow zone with a destroyed carbon black structure. In this zone, the viscosity of the filled system exceeds that of the matrix only several times. Here flow by a power law is observed with a coefficient *n* constant for each composition. The equality on section CD of the activation energy of PB and its carbon black mixtures shows that the flow here is chiefly determined by the properties of the matrix. On section DE, the filled systems, like PB, exhibit a separation effect at the stress τ_{sp} changing only slightly with the carbon black concentration.

A glance at Fig. 5.13 presenting the concentration dependences of τ_{sp} and $\tau_{y.p.}$ plotted by the data of Fig. 5.12 reveals that the zone of the fluid state of the filled polymer system is the region limited by the curves of τ_{sp} and $\tau_{y.p.}$ vs C. It diminishes sharply with an increase in the filler concentration. This can explain the considerable difficulties encountered when processing highly filled polymers. As indicated in Ref. 132, the high-elastic modulus G_0 determined by the elastic return to regions AB and BC does not change when passing through the ultimate shear strength; the plot of G_0 vs C has a minimum at an approximately 10%

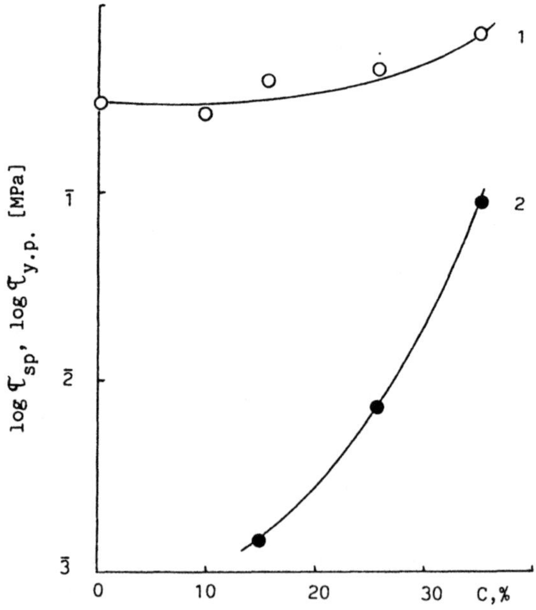

FIG. 5.13. Separation stress τ_{sp} (○) and yield point $\tau_{y.p.}$ (●) in shear vs filler concentration for PB and PB + AS. Temperature 298 K.

content of the carbon black. The lower value of G_0 for the filled composition in comparison with that of the matrix is associated in Ref. 132 with the imperfection of the carbon black skeleton, while the subsequent growth in G_0 is associated with the formation of a continuous carbon black skeleton when the filler concentration grows.

5.6.2.3 On the Initial Viscoelastic Characteristics of Filled Polymers

Interesting results on the concentration dependences of the initial characteristics of filled polymers with a relatively low filler concentration were obtained in Refs 126 and 133.

Proceeding from the correspondence principle of solving problems involving elastic and viscoelastic materials (or from the elastic-viscoelastic analogy) when finding the apparent moduli of a composite consisting of spherical inclusions and a viscoelastic matrix, the author assumed that the ratios of the relaxation moduli, initial viscosity coefficients, and also of the coefficients of normal stresses of the composite and matrix, should equal certain numbers that at a given time or frequency depend only on the filler concentration. The frequency dependences of the complex dynamic shear modulus components may sometimes be made to coincide by shifting along the axis of the logarithm of the moduli. This confirmed to a certain extent the conclusions following from the correspondence principle, but refuted the opinion on the possibility of a 'temperature–concentration' reduction, at any rate for the terminal zone of the spectrum of relaxation times $H=f(\theta)$.

The correspondence principle was found to be correct only when the influence on the modulus or initial viscosity is determined by the hydrodynamic interaction of the particles. Up to a certain critical particle concentration $C \approx C_{cr}$, the concentration dependences of the ratios of the initial rheological parameters of the matrix and composites (the characteristics of the viscoelastic properties at frequencies and rates and stresses tending to zero), viz. the initial viscosity $\eta_0 = \int_{-\infty}^{\infty} H(\theta)\,d\theta$ and the coefficient of normal stresses $A_G^0 = \int_{-\infty}^{\infty} \theta H(\theta)\,d\theta$, as well as the initial high elasticity modulus $G_e^0 = \eta_0^2/2A_G^0$ are shifted only along the axis of the logarithms of the initial characteristics. They are approximated satisfactorily by the formulas of Lee, Mooney, and Thomas, but provided that the apparent concentration of the solid phase is about 2·5 times the nominal one ($C_a = 2{\cdot}5C$). This can be related traditionally to the surface layers of the polymer on the particles. Their thickness is small and depends little on the concentration. A spatial filler structure forms at the critical concentration. Under these conditions, a macromolecule can participate in the formation of the surface layers on several particles at once. The conclusions contained in Refs 126–133 are presented schematically in Fig. 5.14. As was shown when studying a number of composites at a concentration above the critical one, the initial high elasticity modulus G_e^0 may grow with a further increase in the concentration ($C > C_{cr}$). Sometimes near C_{cr} the initial modulus of a composite may be smaller than the initial shear modulus of the matrix (the compliances of two networks—the fluctuation network of the matrix macromolecules and the appearing filler network—are summed) (Fig. 5.14).

The data presented in Fig. 5.15 relating to the kinds of streams flowing out of a capillary illustrate an important result that the rubbery state is suppressed in a polymer filled with carbon black. Figure 5.15(a) is a view of the matrix stream. Distortions of the latter, having typical helical threads (the effect of elastic turbulence), can be seen in the bottom part of the photograph. The middle part shows the separation effect—it is a smooth section whose size corresponds to that of the capillary. The top zone in Fig. 5.15(a) shows the effect of melt destruction. The typical

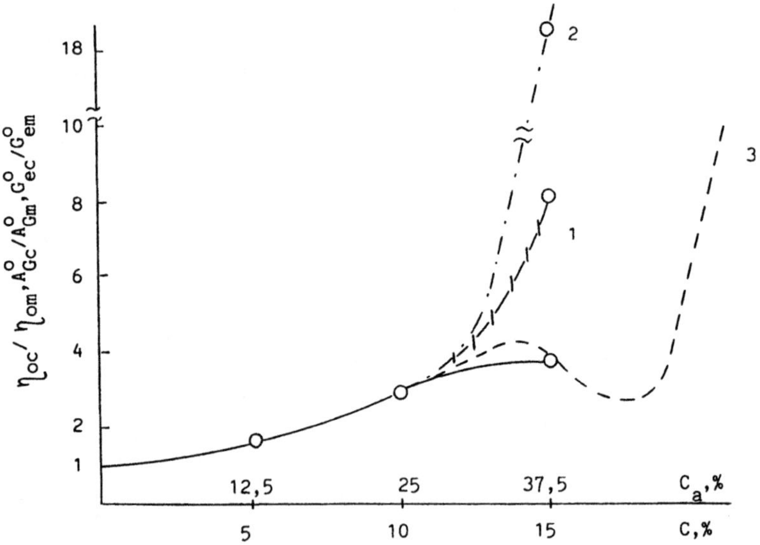

FIG. 5.14. Concentration dependence of ratios of the initial viscoelastic characteristics of a composite and matrix. C_a is the apparent (including the boundary layers) concentration, C is the nominal concentration.

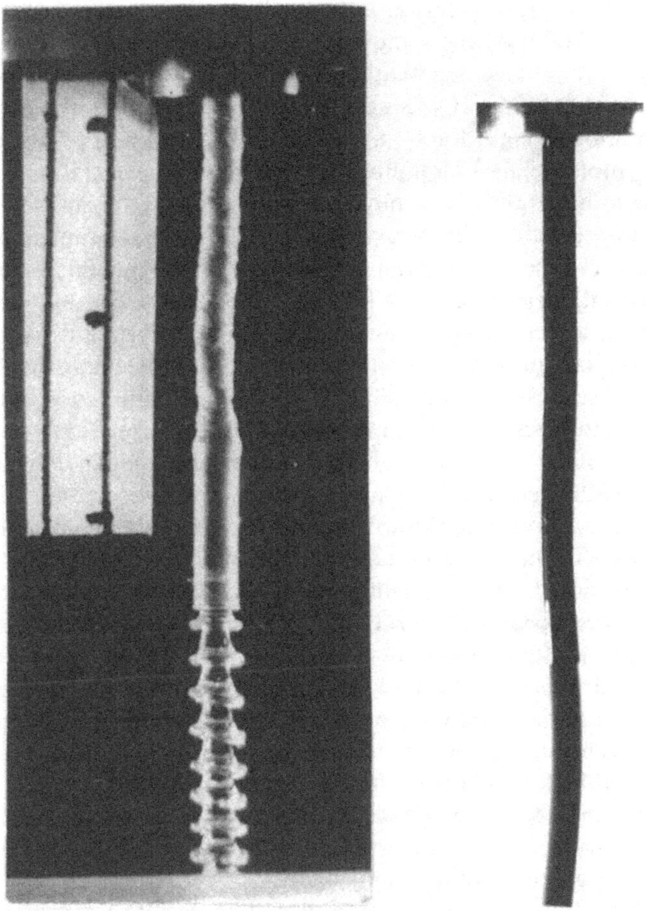

(a) (b)

FIG. 5.15. Appearance of stream of (a) PB and (b) PB + 35 wt% of AS after emergence from capillary under separation conditions.

bottom and top sections are absent for a filled polymer, and the stream is smooth (Fig. 5.15(b)).

The influence of the molecular characteristics of a polymer matrix on the yield point of its filled composites is interesting to consider. It was noted in Ref. 134 that for filled polydispersed polymers with various matrices whose viscosity varied within very broad limits, $\tau_{y.p.}$ depends only slightly on the viscosity of the medium, the matrix, and the average molecular weight. At the same time, a substantial growth in $\tau_{y.p.}$ with an increase in the molecular weight was observed for filled polymers with a narrow MWD [134]. It was also found [134] that only the presence of a low-molecular-weight fraction, and not polydispersion *per se*, causes the formation of a spatial structure with low filler concentrations.

As mentioned above the physicochemical properties of a filler surface should appreciably affect the value of $\tau_{y.p.}$. A considerable contribution here can also be made by the thickness of the polymer matrix layer adsorbed on the surface of the filler particles. By the concept developed in Refs 135 and 136, the shorter molecules are selectively adsorbed from the dispersion medium with an inhomogeneous molecular weight on the surface of the solid phase. They produce a small adsorption layer approximately identical in thickness. This can apparently explain the close values of $\tau_{y.p.}$ for low- and high-molecular-weight polymers with a broad MWD because they are determined by adsorption of the low-molecular fractions and do not depend on the average molecular weight. An increase in the adsorbed layer thickness with an increasing molecular weight lowers the forces of interaction between the filler particles and weakens the structure. On the other hand, a macromolecule, being adsorbed simultaneously on several particles of the solid phase, participates in the formation of a spatial skeleton. This strengthens the structure, and the more it does so the longer is the macromolecule. At low filler concentrations the first factor apparently predominates. With an increase in the filler content, the second process causing $\tau_{y.p.}$ to grow with increasing molecular weight begins to play the major role. Such a relation is not observed for polydispersed samples owing to blocking of the filler surface by low-molecular-weight fractions.

In concluding this small section, it will be useful to consider again the possibility of constructing generalized temperature–time and temperature–concentration dependences of the rheological properties with continuous shear deformation for melts of filled polymers.

In experiments with continuous deformation (flow in capillaries) involving considerable deformations, including those exceeding the 'yield point', the structure of the material is destroyed and the flow of a viscoelastic polymer matrix filled with fragments of the filler structural skeleton occurs in the working gaps. For these reasons, in a definite restricted region of rates of deformations and stresses, but a region that is important from the technological viewpoint, temperature–time or temperature–concentration reduction may be useful. To prove this, we cite

Refs 39 and 40 in which it is noted that for elastomers (rubber compounds) filled with carbon black the principle of temperature–time reduction may be employed only in a limited range of reduction times, reduction temperatures, and filled concentrations.

In view of the above, we can also treat the results given in Ref. 137, where the literature on the flow of polymer oligomer compositions containing a highly dispersed filler was analyzed. An attempt was made in Ref. 137 to obtain a general relation describing the flow of such systems in coordinates of $\log(\dot{\gamma}\eta_0)$ vs $(\tau-\tau_{y.p.})/C$. But we consider that the conclusion [137], that the maximum Newtonian viscosity and filler concentration are the parameters determining the flow of filled systems, is simplified. Apparently it could have been made proceeding simply from general ideas, viz. consideration of the pattern of flow of a Newtonian medium with fragments of a structural skeleton of any filler. As regards the yield point, the above expression contains not the yield point itself, but the difference $\tau-\tau_{y.p.}$, i.e. only the range of shear stresses where there is no longer a yield point, because the structural skeleton of the filler was destroyed under such deformation conditions. In this representation [137] the physicochemical nature of the filler and matrix, the nature of filler–matrix interaction, the individual features of the filler, etc., remain outside the framework of generalization.

5.6.2.4 Uniaxial Extension

When processing polymers from a melt by blow molding, pressure-die casting, or extrusion, the polymer experiences not only shear, but also tensile stresses. This substantially affects the structure (molecular orientation) in the finished articles. Hence, anisotropy of the physicomechanical properties of the latter is determined in many ways by the longitudinal characteristics of the polymer in the melt [138]. This is why a description of the rheological behavior of model filled polymers would be incomplete without considering uniaxial tension.

Polymer melts can be stretched homogeneously scores of times [128]. For a filled composition, homogeneous tension is possible only in a limited region of low deformations [139]. At large deformations, a local contraction or neck appears in a sample. At high deformation rates, the failure of samples appears similar to that of crosslinked elastomers or linear flexible-chain polymers in the cold flow (forced elasticity) state [139]. For polymer melts, homogeneous stretching is attended by passing over to steady conditions, when the stress σ does not depend on the strain (Fig. 5.16, curve 1). For a filled composition (Fig. 5.16, curves 2–13), the

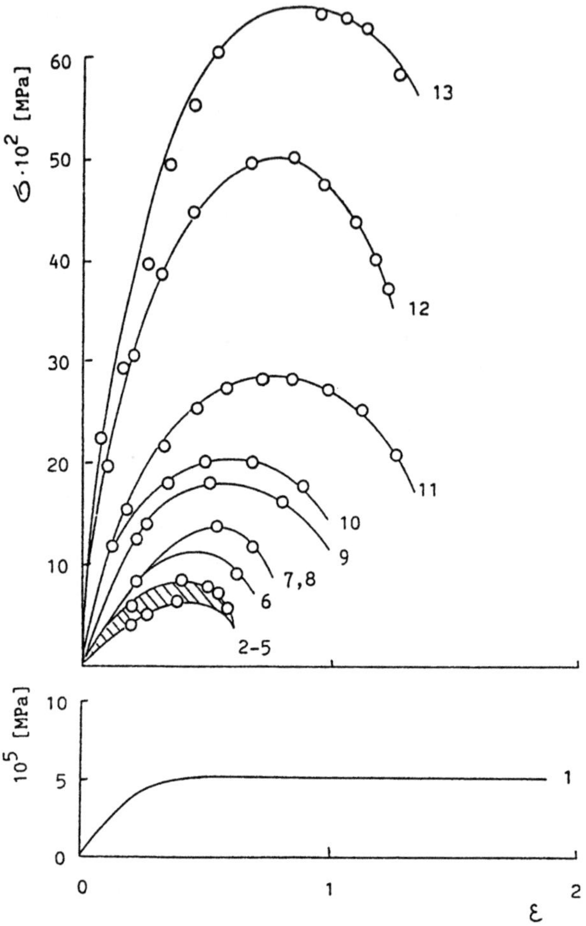

FIG. 5.16. Stress (σ)–strain curve for uniaxial tension of PB ($M = 1{\cdot}35 \times 10^5$, $M_w/M_n = 1{\cdot}05$, $\dot{\varepsilon} = 1 \times 10^{-4}\,s^{-1}$) (1) and PB+35 wt% of AS at $\dot{\varepsilon} = 2 \times 10^{-5}$ (2), 5×10^{-5} (3), 1×10^{-4} (4), 2×10^{-4} (5), 5×10^{-4} (6), 1×10^{-3} (7), 2×10^{-3} (8), 5×10^{-3} (9), 1×10^{-2} (10), 2×10^{-2} (11), 5×10^{-2} (12) and $1 \times 10^{-1}\,s^{-1}$ (13). Temperature 298 K.

formation of a neck corresponds to the appearance of a peak σ_{max} on the stress–strain curve $\sigma = f(\varepsilon)$. In the region of low deformation rates, when $\dot{\varepsilon} \to 0$, σ_{max} corresponds to the yield point $\sigma_{y.p.}$. At higher values of ε, σ_{max} characterizes the ultimate strength of a filled polymer. Figure 5.17 shows that the section corresponding to higher values of $\dot{\varepsilon}$ characterizes hardening of a sample. Figure 5.18 presents an enveloping plot of $\sigma/\dot{\varepsilon}$ vs the

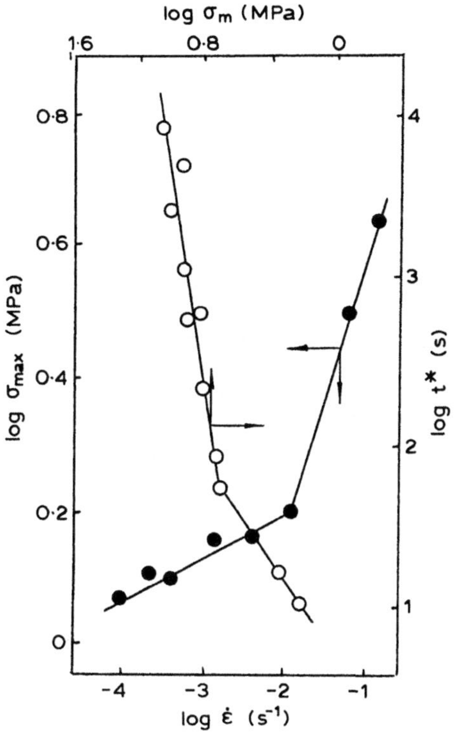

FIG. 5.17. Stress σ_m vs deformation rate in uniaxial tension $\dot{\varepsilon}$ and durability t^* vs σ_m for PB+AS. AS content=35 wt%, temperature 298 K.

time t. It also shows the effect of sample hardening with an increase in the deformation rate [128]. (The curve to the right of σ_{max} is of a conditional nature owing to the inhomogeneity of the sample cross-section.)

The nature of the behavior of a filled composition is due to a set of reasons associated with the deformation rate. In the region of low $\dot{\varepsilon}$ values (10^{-5} to $10^{-4}\,s^{-1}$), the transition from homogeneous to inhomogeneous deformation is due to destruction of the initial structural skeleton of the filler when passing through the yield point ($\sigma_{y.p.}=0{\cdot}13$ MPa) whose magnitude is independent of the deformation rate. In tension, the curve of $\sigma_{y.p.}$ is double that of the yield in shear [128], which is consistent with the von Mises [129] number presuming brittle failure of a structure when passing through the yield point.

In the region of high values of $\dot{\varepsilon}$ (above $10^{-4}\,s^{-1}$), tension and the transition to inhomogeneous deformation should be determined both by

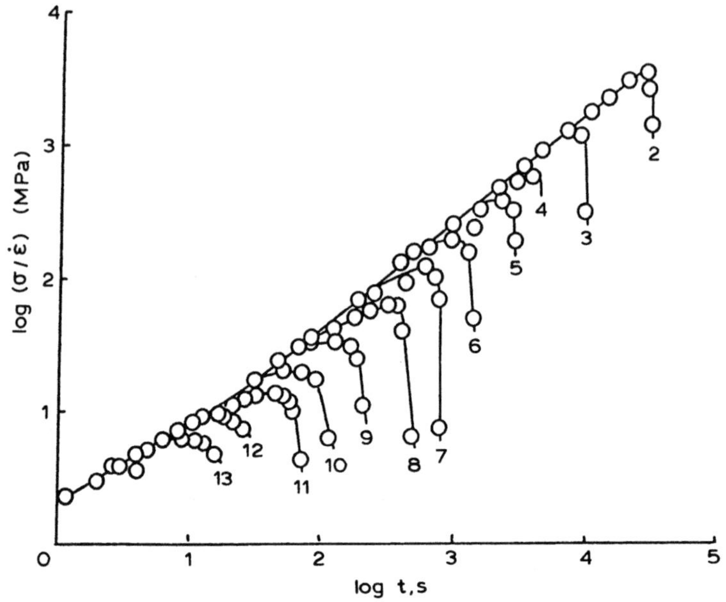

FIG. 5.18. Apparent viscosity in tension ($\sigma/\dot{\varepsilon}$) vs deformation time t at various values of stretching rate $\dot{\varepsilon}$ for PB + AS. AS content = 35 wt%, temperature 298 K. For curve designations, see Fig. 5.16.

the thixotropic failure of the composition structure whose depth grows with an increasing $\dot{\varepsilon}$ and by the development of recoverable strain ε_e playing a stabilizing role and expanding the region of homogeneous deformation. The results obtained in Ref. 138 convincingly prove these assumptions, namely that the elastic strains depend only slightly on the rate of deformation.

The incorporation of an active filler into a polymer also sharply changes the relaxation behavior of the material in tension. This follows from data on the stress relaxation for melts of filled compositions [140]. It was found, in particular, that an increase in the magnitude of the prestraining and the rate of its achievement accelerates the stress relaxation process very greatly. This is associated with higher thixotropic failure of the structure preceding stress relaxation. (It should be noted that a similar effect was predicted theoretically in Ref. 141 for dispersed systems.) But unlike polymer melts, for a filled composition the stresses in relaxation do not drop to zero, but after a definite time reach a quasi-equilibrium value close to that of the yield point. These values depend only slightly on the prestraining rate [140], which is connected with restoration of the structural skeleton of the filler and composition as a whole in the relaxation process.

The strength properties of linear and crosslinked elastomers are closely related to the viscoelastic characteristics [142, 143]. For linear flexible-chain polymers with a narrow MWD that behave within a broad range of stresses and deformation rates like linear viscoelastic bodies, this is manifest in that their durability t^* is a power function of the stress and is proportional to the maximum Newtonian viscosity of the polymer [144, 145]. For a filled composition, $\log t^*$ is a nonlinear function of $\log \sigma$ (Fig. 5.17).

Hence, we can conclude that the features of the change in t^* vs σ for a filled composition are determined by the change in its viscoelastic characteristics with the stress. The change in the latter, in turn, should be determined both by the depth of the thixotropic destruction of the structure and by the manifestation of orientation effects in the process of material deformation.

A criterion of the strength of polymer melts in uniaxial tension was formulated in Refs 128 and 145. The condition of failure has the form

$$\sigma_u/(\varepsilon_e-\varepsilon_e^*)=\text{const.}$$

where σ_u is the ultimate stress, ε_e is the recoverable strain in failure (the Hencky measure), and ε_e^* is the 'critical' recoverable deformation. If crosslinked rubbers may fail at any strains [146], i.e. $\varepsilon_e^*=0$ for them, the fracture of polymer melts is possible only when the recoverable strains realized in practice exceed a certain critical value ε_e^*. As indicated in Refs 128 and 145, ε_e^* does not depend on the loading conditions, temperature, and polymer species, therefore the opinion was advanced on the universal nature of this quantity.

The results of experiments involving the uniaxial tension of filled melts of polymers given in Ref. 147 showed that ε_e^* depends on the volume concentration of the inert filler. It was also shown in Ref. 147 that filled polymer melts at filler concentrations not exceeding a certain 'critical' value behave like viscoelastic liquids while at concentrations exceeding these critical values they may fail at any recoverable strains. This may probably be explained by the formation at the critical concentrations of a sufficiently strong structural skeleton of the filler in the polymer matrix. These results are well consistent with the assumption of the existence of a critical filler concentration in the polymer matrix that sharply alters the viscoelastic behavior of the filled system noted above.

5.6.3 Deformation Testing

As has been mentioned (see Ref. 95), the viscoelastic characteristics of the filled polymers registered in the process of the experiment are, to a great measure, determined by the strain magnitude specified during deformation tests. In the filled samples, the transition to the nonlinear deformation region starts to manifest itself at considerably lower absolute strain magnitudes than in the case of unfilled samples [61]. It has been shown in Ref. 148 that the absolute Young's modulus values for the filled natural rubber decrease with increasing strain above 0·1%, approaching the modulus of the unfilled rubber. This effect is to some extent reversible. It can be attributed to the detachment of the filler particles from the macromolecular rubber and to the destruction of the filler and its surface properties [4]. The stronger the interaction between the filler and the polymer, the higher the number of the primary chemical bonds to be destroyed, and the greater are the strains necessary for the separation [149]. Therefore, in analyzing the temperature–time reduction of the filled polymers, the specific deformation magnitudes, even in their relatively small values, can be employed as essential variables.

Periodic deformation using finite amplitudes is in essence a well-known, generally adopted technique for the mechanical plasticization of polymers [150–158]. This includes rolling, processing in double-worm extruders (the application of periodic shear to a steady flow), and processing of polymers in an extruder with an oscillating mandrel forming a pipe head [156]. An increase in the deformation amplitude is attended by an increase in the loss tangent, and the material liquefies. Nonlinearity of a material is also expressed in appearance in sinusoidal deformations of higher odd harmonics of the stress signal. The amplitude of these higher harmonics passes through a maximum, and then the material becomes linearized [157]. According to Refs 90 and 159, the presence of the filler in the system will behave differently depending on the magnitude of the applied strain.

In the case of extension of filled CB rubbers, the author in Ref. 159 suggests considering three deformation regions: weak (strain up to 5%), medium (15–20%), and strong (>20%) or capable of causing material fracture. At low strains, we observe an overlapping of effects due to rubber occluding and amalgamation of aggregates into agglomerates that will form a continuous network. This is due to the fact that the samples of the filled rubbers with soot will have a greater modulus at small strains [95]. If the strain intensity is increased, the destruction of the agglomerates brings about a substantial and usually irreversible decrease in the modulus. Besides, under such conditions, one can observe a desorption of the polymer molecules or their escape from the filler surface. As a result, the filler separates into an independent phase. A further increase in the strain, when the remaining macromolecular chain cannot withstand the growing strain, will lead to a complete fracture of the sample [4, 160].

The influence of the strain size on the rheological investigation data for filled polymer melts and solutions is considered in Ref. 1.

Let us compare some of the PB+AB rheological investigation results obtained in different conditions of shear stress, both periodic and continuous [81]. It should be noted that the viscoelastic behavior of the filled AB systems during continuous deformation under uniaxial extension conditions and fixed shear flow has been comprehensively discussed in Ref. 90. The 'yield point' measured in periodic $|G^*_{y.p.}|$ and continuous $\tau_{y.p.}$ shear and uniaxial extension $\sigma_{y.p.}$ is illustrated in Fig. 5.19. It is seen

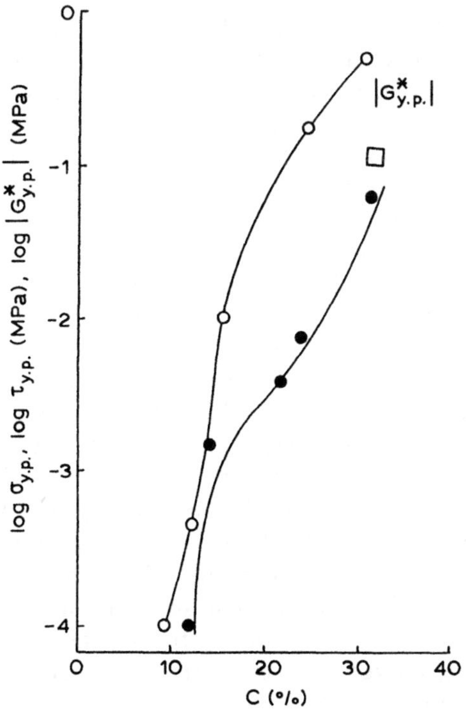

FIG. 5.19. Dependence of 'yield points' according to data on dynamic measurements $|G^*_{y.p.}|$ (○), of continuous shear strain on fixed flow regimes $\tau_{y.p.}$ (●) and in uniaxial extension $\sigma_{y.p.}$ (□). Temperature 298 K.

that the divergence between these data increases with rising filler concentrations and stresses realized in the comparative experiments. The magnitude of $|G^*_{y.p.}|$ for sample 35C exceeds $\tau_{y.p.}$ by an order of one-tenth. On the other hand, the yield point in the uniaxial extension, $\sigma_{y.p.}$, is 1·3 times higher than that of $\tau_{y.p.}$ for the sample (see Fig. 5.19), which is notably lower than the predicted von Mises plasticity theory, but closer to the values obtained for the plastic-disperse greases (see Ref. 161). At the same time, the concentration dependences $G''_{max}(C)$ and $\tau_{cr}(C)$, where τ_{cr} is the critical spurt stress, are practically coincident.

Summing up these results, note that the regularity and the conclusion should have, in our opinion, a general value for filled high-molecular-weight linear flexible-chain polymers at $T > T_g$.

5.7 INFLUENCE OF FILLER 'ACTIVITY'

Let us now consider the possibility of varying within definite boundaries the viscoelastic characteristics of the filled polymer melts, by applying, for example, one or another means of filler 'activity'. Let us now also follow the change in the stress–strength characteristics of the material in the solid state. Their level is known to be an important operational index and determines the serviceability of the articles prepared from those materials. It is obvious that such an approach forms the basis for predicting the material properties starting from the processing stage.

The bond of a filler to the matrix and the role of interface influence deserve major attention. Finishing of a filler with silanes not only increases its tensile strength $1\frac{1}{2}$–2 times in comparison with the matrix and not only diminishes the ultimate elongation, but also lowers the viscosity of the melt and the first difference of the normal stresses with identical shear stresses [162, 163]. This can be exemplified by studies of polystyrene and a copolymer of acrylonitrile containing a variety of finished and unfinished fillers [164]. The finishing was polymerizational, i.e. the particles were surrounded by a polymer shell bound to the surface of the solid particles by covalent bonds. Experiments with aerosil and hollow glass sphere filler showed that such finishing with an adequate thickness of the 'coat' grafted to the particles retains the fluid properties of such a system with larger filler concentrations than when an unfinished powder filler is employed [164–166].

The range of problems can be discussed by using an example of a definite industrial perspective project. For this purpose, let us turn our attention to composition material, produced by the method of chemical (polymerizational) filling wherein highly-filled polymer materials are formed immediately in the process of synthesis by polymerization of the appropriate monomer or monomer mixture on the filler surface. The principles of a novel technology for producing such a composite material are described in Ref. 167. The process of polymerization is initiated by metallocomplex catalysts, free radicals or ions fixed on the filler surface. Inorganic material surface can also accommodate functional groups (e.g. hydroxyl) that can be converted into active polymerization centers by appropriate reactions, and can then initiate the polymerization of practically any monomer.

Let us now consider the results obtained in Ref. 168 for composites based on commercial high-density polyethylene (CHDP) with $M = 1{\cdot}2 \times 10^5$ and melt fluidity index 3·2 g/10 min. Two types of composites, composite 1 and composite 2, have been studied. Composite 1 was obtained based on CHDP and filler concentrate—expanded pearlite (volcanic glass) grafted with linear polyethylene on its surface during polymerization, $M = (1{\cdot}2\text{–}1{\cdot}5) \times 10^6$. The polyethylene–pearlite ratio in this filler is 12:88 by weight. Composite 2 is the same kind of CHDP filled with ordinary expanded pearlite introduced by the usual mechanical mixing technology. In both cases, the pearlite content in the compositions was varied from 10 to 60 wt%. All the samples stabilized 0·3 wt% Irganox 1010.

Let us compare the dynamic rheological characteristics derived at 473 K under low-amplitude deformation, for initial CHDP and composites 1 and 2 (Figs 5.20 and 5.21). The growth of G' and G'' curves with increasing ω for the initial CHDP is related to the transition of the system from a fluid to a high-elasticity state. However, in view of the strong polymer polydispersity, these changes are much weaker when compared with the object model, the narrow MWD polymer, while the transition itself is diffused (compare Figs 5.20 and 5.21). The plateau of the high-elasticity state in the CHDP sample can be seen at $\log \omega > 3{\cdot}0\ \mathrm{s}^{-1}$ frequencies. The addition of the filler results in the appearance of a low-frequency plateau, the yield point. It is essential that with increasing filler content, the height of the plateau grows as well as its extent toward higher frequencies. The fluid state region degenerates and at filler >30 wt% it practically does not appear (as if the filler suppresses the individual properties of the matrix). It is seen that the addition of a substantial amount of filler leads to a decrease in the storage modulus (curve 6, Fig. 5.20), indicating an extreme change in the properties of these

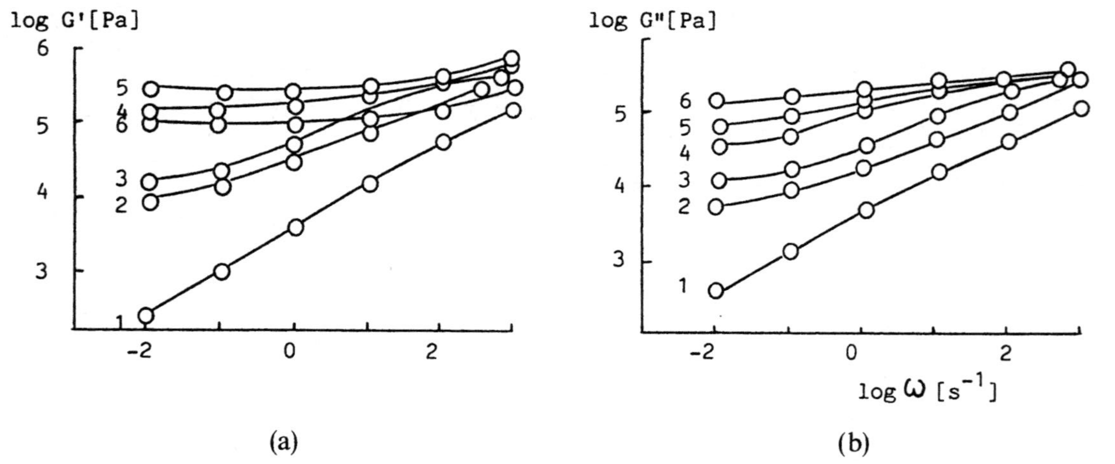

FIG. 5.20. (a) Storage modulus G' and (b) loss modulus G'' vs circular frequency ω for composite 1 at 473 K. Curves 1–6 for filler content correspond to 0, 10, 20, 30, 40 and 50 wt% respectively.

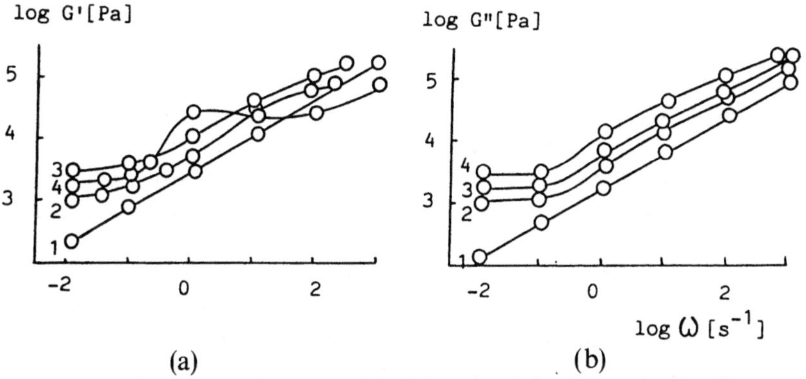

FIG. 5.21. (a) Storage modulus G' and (b) loss modulus G'' vs circular frequency ω for composite 2 at 473 K. Curves 1–4 for filler content correspond to 0, 30, 40 and 50 wt% respectively.

filled systems. In the case of composite 2 (Fig. 5.21) the yield point also manifests itself quite distinctly. However, the absolute rheological parameter values are considerably lower.

When considering the data on the dynamic rheological analysis for filled polymers prepared under low-amplitude periodic shear deformation conditions, the author did not pay special attention to the results obtained under continuous deformation (flow in the channels). Some of the reasons and considerations concerning this have already been exposed. While not deprecating the usefulness of these investigations, especially for processing technology of those materials, it is expedient, nevertheless, to mention one more sufficiently essential limitation of such approaches. The experiments carried out under conditions of continuous deformation on capillary rheometers, particularly under conditions of constant flow rates, allow us to embrace, as a rule, the relatively narrow range of deformation rates [169]. For highly-filled samples, they do not embrace the low rates and deformation stresses. The yield point of the filled compositions, and the high rates and deformation stresses at which the conditions of sluggish or unsteady-state flow accompany the melt destruction, could be evaluated.

5.8 DYNAMIC RHEOLOGICAL CHARACTERIZATION

One way to achieve a qualitative or semi-qualitative evaluation of the rheological dependences under continuous deformation conditions is to employ the approach based on the analysis of the dynamic rheological characteristics [170, 171]. For individual polymers, melts and solutions, the $|G^*|=f(\omega)$ dependences correspond quantitatively to the polymer flow curves, i.e., to $\tau=f(\dot{\gamma})$ dependences, where τ and $\dot{\gamma}$ are the stress and shear rates, respectively, under conditions of equivalency ω and $\dot{\gamma}$. In the case of systems possessing an internal steric structure (the filled polymer systems can be referred to by their number), such a correlation is not always fulfilled quantitatively. In fact, it has been shown that the experiments carried out for continuous deformation lead in certain cases to the destruction of the structural network of the object under investigation. Interesting data were obtained in Ref. 172. Taking into account the considerations presented, as well as the fact that for CHDP samples with 40 and 50 wt% filler content the fluidity is suppressed, experiments under fixed flow conditions cannot be conducted. It may be reasonable to plot $|G^*|=f(\omega)$ dependences, taking them to be qualitatively similar to the flow curves. Such results for composites 1 and 2 are illustrated in Figs 5.22 and 5.23. The $|G^*|\propto\tau$ values and the $|G^*|$ values at minimal frequencies in the experiment, are close to the yield point values, $\tau_{y.p.}$.

The comparison of the G', G'' and $|G^*|$ values for composites 1 and 2 corresponding to the lowest frequencies in the experiment (Figs 5.20 and

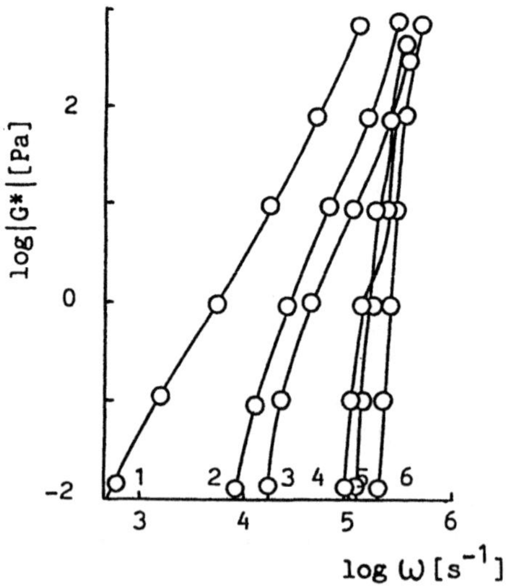

FIG. 5.22. Absolute value for complex dynamic modulus $|G^*|$ vs circular frequency ω for composite 1 at 473 K (see Fig. 5.20 for symbols).

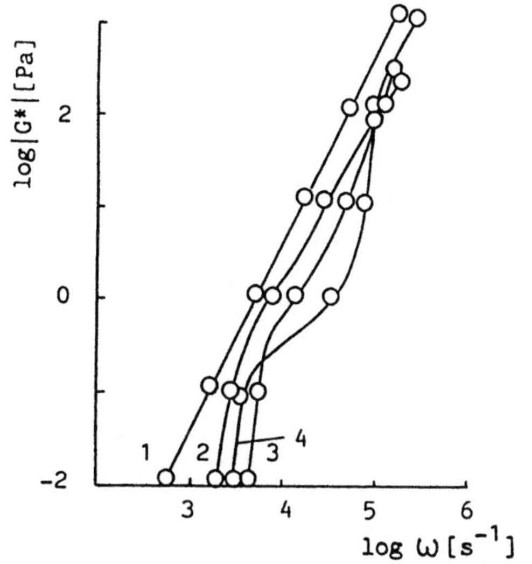

FIG. 5.23. Absolute value for complex dynamic modulus $|G^*|$ vs circular frequency ω for composite 2 at 473 K (see Fig. 5.20 for symbols).

5.21), i.e. the ultimate least parameters of $G'_{y.p.}$, $G''_{y.p.}$ and $|G^*_{y.p.}|$, has shown (Fig. 5.24) that:

(1) For the series of samples in the filler, when plotted against the concentration function C, these parameters have a maximum at $C = 40$ wt%.
(2) $G'_{y.p.}$ and $|G^*_{y.p.}|$ practically coincide, allowing us to use only one dependence for similar evaluations, either $G'_{y.p.} = f(C)$ or $|G^*_{y.p.}| = f(C)$.
(3) The distinction in the $G'_{y.p.} = f(C)$ curves for composites 1 and 2 allows us to evaluate the 'activity' of the filler, which is hundreds of times higher for composite 1 than for the mechanical mixture.

Obviously, the observed distinctions are due to strong interaction between the filler and the polymer matrix in composite 1. Investigations carried out under different temperature regimes (453–493 K) have shown that at high filler concentrations (>30 wt%) the $G'_{y.p.}$ and $G''_{y.p.}$ values do not vary with temperature. This may suggest that not only in the melt, but also in the solid state, composite 1 should possess enhanced physico-mechanical and strength properties in comparison with composite 2. This supposition was confirmed experimentally. (The deformation–strength properties of composites 1 and 2 were tested under conditions of uniaxial

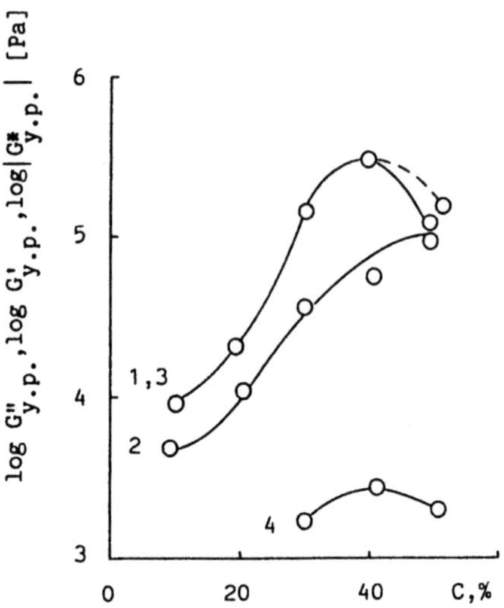

FIG. 5.24. Values for dynamic storage modulus $G'_{y.p.}$ (curves 1 and 4), loss modulus $G''_{y.p.}$ (2) and absolute values for complex dynamic modulus $|G^*_{y.p.}|$ (3) at varied filler concentrations in composites 1 (1–3) and 2 (4).

extension and compression on a multipurpose 'Instron' testing machine, at deformation rates of 0·071 and 0·87 min^{-1} and temperature 295 K.)

5.9 RELATION BETWEEN RHEOLOGICAL AND STRENGTH CHARACTERISTICS

From the analysis of the extension diagram for composites 1 and 2 obtained for different filler concentrations (Fig. 5.25) it is seen that with a rise in C, the elastic modulus E grows, the tensile breaking strength σ^* passes through a maximum, while the relative breaking strain in the case of breakage ε^* decreases. The σ^* and E values for composite 1 (Fig. 5.25(b), curves 1 and 2) are higher than those for composite 2 (curves 3 and 4). This fact supports the regularity expected from the rheological measurements. From the results of the rheological investigation, it follows that the maximal values for the composite strength should be observed if the filler contains 40 wt%. This is also confirmed experimentally. The maximal tensile strength and comparison values for composites 1 and 2 are attained in the 30–40% filler range (see Fig. 5.26). The example presented shows that the strengthening of the filled material is achieved if both

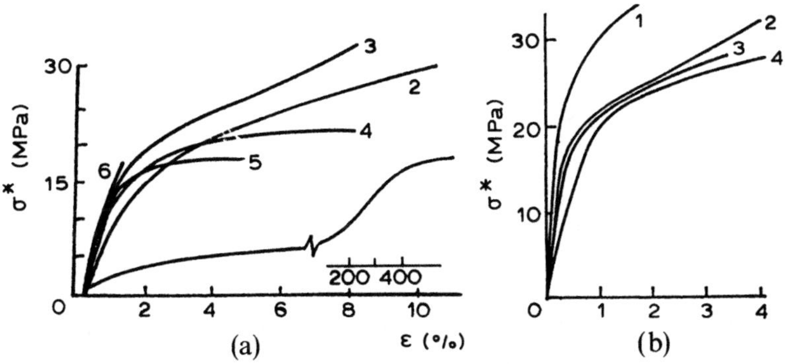

FIG. 5.25. Extension diagram of samples at 298 K. (a) Composite 1 with filler contents of 0 (curve 1), 20 (2), 30 (3), 40 (4), 50 (5) and 60 wt% (6). Deformation rate 0·071 min^{-1}. (b) Composites 1 (curves 1 and 2) and 2 (3 and 4) with filler contents of 20 wt% (2 and 4) and 30 wt% (1 and 3). Deformation rate 0·87 min^{-1}.

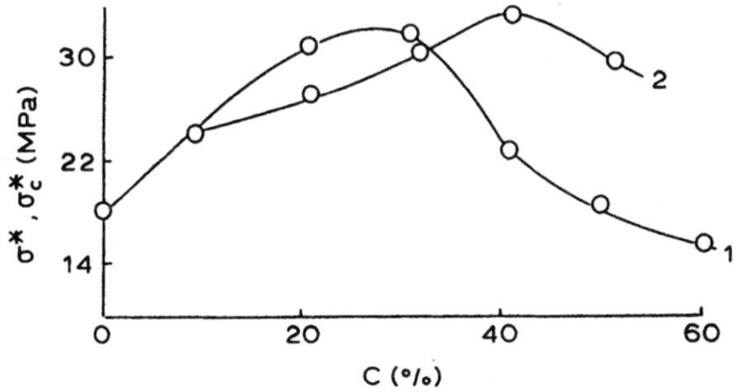

FIG. 5.26. Breaking values for stresses in extension σ^* (curve 1) and compression σ_c^* (2) samples of composite 1 depending on the filler content at 298 K.

types of fillers are used. However, for composite 1 this effect proved to be stronger.

It is essential that certain mechanical properties of solid state material can also be predicted via the analysis of the rheological behavior of the filler polymer melts. For example, the correlation between the size of material strength during compression σ_c^* and yield point $G'_{y.p.}$ determined from the dynamic rheological experiment was found in Ref. 168 (see Fig. 5.27).

The influence of coarsely dispersed fillers and their mixtures on the rheological properties of thermoplastics was analyzed in Ref. 109 for (CHDP) with fillers such as chalk, glass fibers, and combinations thereof in various proportions. The filler content was varied from 10 to 20% by

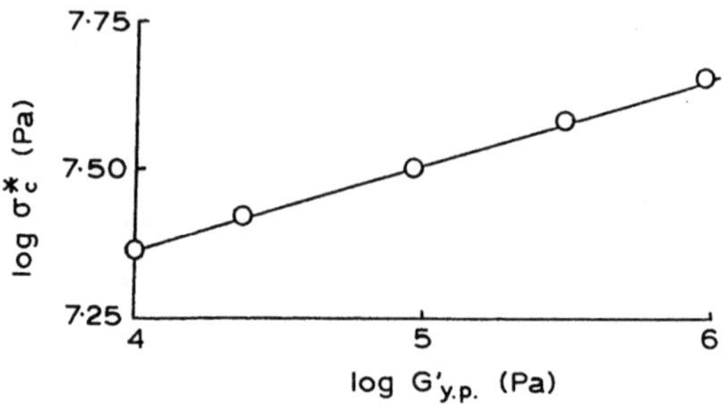

FIG. 5.27. Interaction between strength magnitudes in compression σ_c^* and storage modulus $G'_{y.p.}$ values.

mass (4–9% by volume). The strain and strength characteristics of solid composites were also studied.

It was found that for a number of compositions there is an unambiguous relation according to which an increase in the viscosity and modulus of elasticity of melts of compositions is accompanied by a relevant increase in the strength of the formed material. This was presumed to be due to the similarity of the structure of a filled composite melt and that of the rubbery amorphous phase formed from a melt of the crystalline material. The latter affects its strength characteristics.

We can thus conclude that there is sometimes correlation between the rheological and relaxation properties in a melt and the strain and strength properties of the solid composite. At the same time, it should be noted that the addition of a dispersed filler to a fibrous one lowers the viscosity of the composite melt in shear and uniaxial tension without appreciably affecting its strength [109].

The physical essence of the correlation between the dynamic rheological and strain–strength characteristics of filled polymers (Figs 5.25–5.27) that are of a universal nature can be clearly followed in the study of the viscoelastic parameter changes of the filled polymer system in a broad range of temperatures spanning its different physical states. In the case of filled polyethylene, the subsequent analysis of the change in the indicated characteristics in the following order is of great interest: solid state, phase transition related to the fusion of the polymer matrix; liquid–fluid state. A series of investigations of this kind was carried out in Ref. 172. Essentially, the analysis of the friction characteristics, micro-hardness and structural changes caused by the introduction of the fillers was also carried out in Ref. 172.

Let us now turn our attention to the dynamic test data (Figs 5.28 and

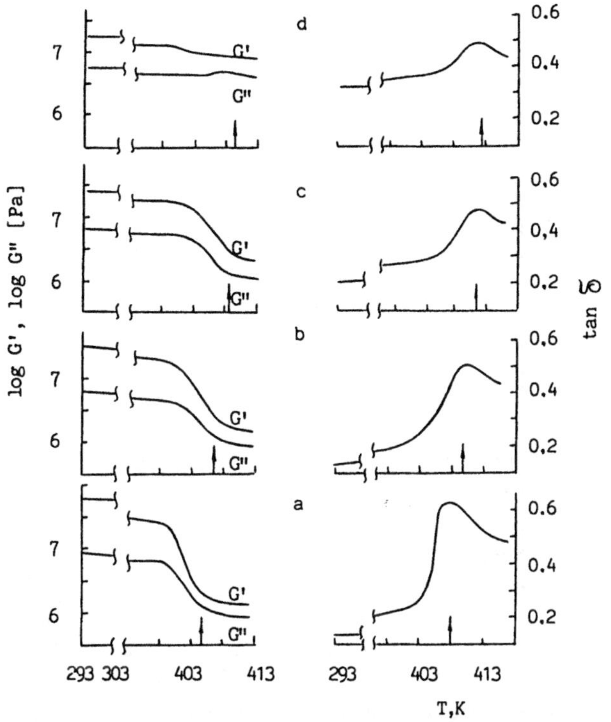

FIG. 5.28. Temperature dependence of dynamic storage modulus G', loss modulus G'', and mechanical loss tangent tan δ, for pure polyethylene and filled 5, 15 and 30 wt% AB, curves a–d, respectively.

5.29) that were obtained on filled AB and AE (aerosil) CHDP samples with melt index 1·2 g/10 min. The specific surface of AE was 300 $m^2\,g^{-1}$. The experiments were carried out under temperature scanning conditions within the range 293–423 K, at a rate of 5–8 K h^{-1}, at a frequency of $\sim 10^4\,s^{-1}$. The arrows in Figs 5.28 and 5.29 indicate the points corresponding to the maxima on the tan δ dependences. These maxima are related to the transition of the material in a fluid state resulting from the fusion of the polyethylene matrix. The melting points of the compositions were close to those recorded by the dynamic technique; the temperatures were determined independently at the same heating rates by the method of differential thermal analysis (DTA); besides, for pure CHDP, they were determined in the polarized light upon the disappearance of the last crystal. However, the scales of variation of G', G'' and tan δ magnitudes (compare the curves in (a)–(d)) are different. For pure CHDP, G' and G''

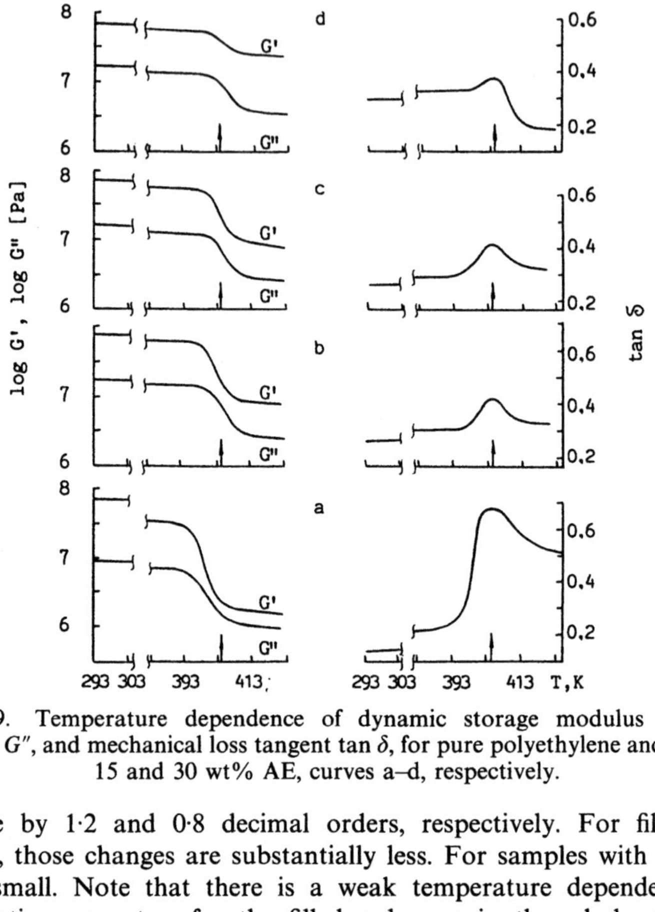

FIG. 5.29. Temperature dependence of dynamic storage modulus G', loss modulus G'', and mechanical loss tangent tan δ, for pure polyethylene and filled 5, 15 and 30 wt% AE, curves a–d, respectively.

decrease by 1·2 and 0·8 decimal orders, respectively. For filled AE samples, those changes are substantially less. For samples with 30 wt% AB is small. Note that there is a weak temperature dependence for viscoelastic parameters for the filled polymers in the whole range of temperatures studied, from 293 to 423 K, and high modulus values for G' and G'' in the melt in comparison with pure CHDP. It is important to underline that judging from the dependence $\tan\delta = f(T)$, the temperature of the phase transition, due to the fusion of the polymer matrix, for the AB samples shifts toward the higher values, while in the AE samples it remains invariable. Moreover, the character of the $G'(T)$, $G''(T)$ and $\tan\delta(T)$ curves (Fig. 5.29d) in the range 393–413 K is qualitatively similar to the same dependences for amorphous materials on their transition from a high elastic to a fluid state. This presupposes that amorphization of the polyethylene matrix possibly takes place in the presence of carbon black.

It is interesting to note that there exists an interrelation between the dynamic characteristics, in particular tan δ, and the frictions, e.g. the frictional force F measured in the broad range of temperatures [172]. The dependence $F(T)$ was obtained on a tribometer in which the contact was accomplished for a steel hemispherical slider (2-mm hemisphere radius) with a flat polymer disc surface under vacuum (10^{-6} torr) at a sliding rate of 0·8 cm s^{-1}, a load of 60 g, and a temperature variation rate of $-1°$ min^{-1}. The variation in the frictional force vs the temperature for the initial polyethylene and filled samples is given in Fig. 5.30. In the 293–393 K range, the frictional force is practically constant, while above 298 K there appear extrema on the $F(T)$ curves. The ascending branches of the curves are due to the variation in the parameters of the crystalline lattice and the mechanical properties of the materials as well as the expansion of the contact area of the polymer sample.

The descending branches of $F(T)$ are due to the development of the polymer fusion process occurring in the surface layer of the sample—to the increase in the melt part and a decrease in its viscosity. The particular similarity of the $F(T)$ and tan $\delta(T)$ appears to be very attractive (compare Figs 5.30 and 5.28). This is due to the displacement of the maximum position and to the change in the slope angle of the ascending sections of the corresponding curves for the CHDP+AB system. The observed character of the $F(T)$ dependence for the CHDP+AB system at $T > 403$ K is accounted for by the large bearing ability of the sample at temperatures substantially exceeding the polyethylene fusion temperature.

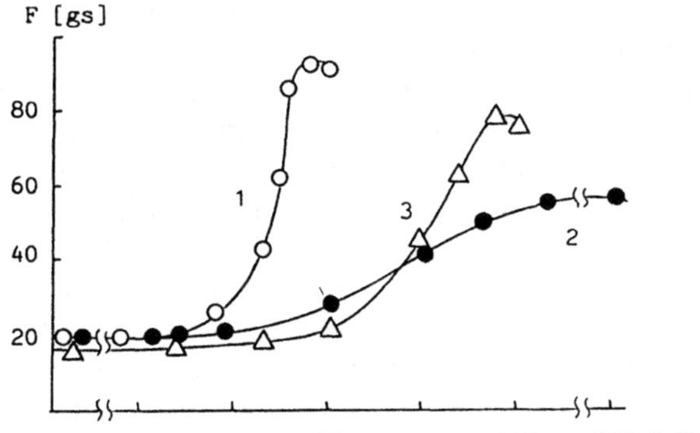

FIG. 5.30. Temperature dependence of frictional force F for pure polyethylene (curve 1) and filled 33 wt% AE (2) and AB (3).

This was also confirmed by the results derived from the hardness measurements (Rockwell–Superficial apparatus; sphere penetrator).

The suppositions concerning the change in the structure of the filled AB polyethylene based on the dynamic data are totally supported by the direct structural investigations. The addition of 30 wt% AE does not change the correlation between the amorphous and crystalline parts of the CHDP X-ray pattern. In the case of the 30 wt% AB, we observe an increase in the amorphous part. The defractometric curves reveal a reflection in the region of a 12·5° deflection angle corresponding to the base planes of the carbon lattice. Electron-microscopic photographs of the objects under description have shown no fibrillous structure in the 30 wt% AB sample, but have revealed a preserving fibrillous structure of the spherolite type in the AE sample.

5.10 RELAXATION BEHAVIOR

In studying the viscoelastic properties of the filled polymers, it is necessary to at least briefly dwell on the characteristic features of their relaxation behavior, particularly on the relaxation transitions due to the specificity of the molecular mobility of different temperature ranges. The presence of a filler substantially changes the relaxation characteristics of the polymer, which in its turn, substantially influences the mechanical properties of the material [173]. The numerous manifestations of the forms of molecular mobility of such systems are accounted for first of all by the peculiarities of their structure. Here we can also refer to a number of works [173, 174] according to which, as a result of irregularity of distribution of the internal stresses in the rubber bulk, part of the length of the molecular chain is capable of detaching itself from carbon black in the process of 'heat motion' that is identified by the relaxation process, the activation energy being of the order of 70 kJ mol^{-1}. In analyzing the literature data, it was observed that the investigation of the relaxation properties of filled polymers has aroused considerable interest among researchers engaged in this area. It should be borne in mind that information in this area is derived from the solid state samples. This is associated both with the ability of the experimental techniques and with the behavior of the systems in glassy and viscoelastic states. The latter is conditioned, in particular, by practical problems.

Investigation of the relaxation processes can be carried out by diverse methods, e.g. for studying the relaxation of stresses and preliminary samples [174] or for analyzing the dynamic properties in non-resonance frequency regimes in the case of low-amplitude deformation. In the first case, the statistic relaxing Young's modulus E can be calculated from the condition: $E(T)=\sigma(t)/(\lambda-1)$, where σ is the stress, t is the time, $\lambda=l_1/l_0$ and l_1 and l_0 are the lengths of the extended and initial samples, respectively. In the second case, the dynamic Young's elasticity modulus E_g and the coefficient of the mechanical loss χ [175] or the G', G'' and $\tan\delta$ magnitudes, respectively, can be determined depending on the extension–compression–deformation conditions of a simple shear.

In this section, it is expedient to concentrate our attention on one example, without dealing with the whole array of relaxation behavior of the filled polymers. From this example, it will be clear what general information can be derived on the basis of study of the relaxation properties. For this purpose, we have taken the data obtained on a divinyl rubber, SKD, at $M=1{\cdot}25\times10^5$ filled with CB–AB of vulcanized sulfur (2·0 wt per 100 wt% rubber) [176]. The crosslinking that occurred during vulcanization fixed, to a certain extent, the structural ordering that existed initially in the system. This allows stress relaxation measurements to be carried out in a relatively broad range of strains and temperatures, within the framework of the viscoelasticity region.

The analysis of the data on stress relaxation at a constant magnitude of the relative extension strain $\varepsilon=(l_1-l_0)/l_0=50\%$ at 150–400 K and the calculation on their basis of the discrete relaxation time spectra, according to Ref. 177, indicates that the $\sigma=f(T)$ curves are described by the sum of exponents; for instance, the sum of six, five, or four exponents by temperatures of 303, 323, or 343 K, respectively. According to this fact, the stress relaxation process can be subdivided into elementary constituents on the basis of the expression $E(T)=E_\infty+\Sigma_i E_i \exp(-t/\theta_i)$, where E_∞ is the conditional equilibrium modulus, E_i is the pre-exponential coefficient, and θ is the discrete relaxation time. The E_∞, E_i and θ_i values are listed in Table 5.1.

The independent dynamic experiments conducted over the same temperature range at different circular frequencies enabled us to calculate the continuous relaxation time spectra $H(\theta)$ by the method of the second Schwazl–Staverman approximation [178] (Fig. 5.31). The comparison of the discrete relaxation times θ_i and $H(\theta)$ has shown that the continuous relaxation spectra have small maxima near discrete times θ_i. This, to a certain measure, indicates the legitimacy of breakdown of the complicated relaxation process into elementary components.

TABLE 5.1
Relaxation Characteristics of 'Slow Physical Relaxation Stage' of SKD Rubbers, Filled 25 wt% AB

T (K)	$E \times 10^{-5}$ $(H\,m^{-2})$	$E_1 \times 10^{-5}$ $(H\,m^{-2})$	θ_1 (s)	$E_2 \times 10^{-5}$ $(H\,m^{-2})$	θ_2 (s)	$E_3 \times 10^{-5}$ $(H\,m^{-2})$	θ_3 (s)	$E_4 \times 10^{-5}$ $(H\,m^{-2})$	θ_4 (s)	$E_5 \times 10^{-5}$ $(H\,m^{-2})$	θ_5 (s)	$E_6 \times 10^{-5}$ $(H\,m^{-2})$	θ_6 (s)
293	31·25	1·32	5·0	1·02	28	1·23	149	0·42	651	1·9	6 203	0·74	$12{\cdot}3 \times 10^4$
303	35·50	1·82	1·82	1·32	15	1·29	107	0·79	585	1·26	2 580	0·70	$32{\cdot}6 \times 10^3$
313	36·62	—	—	1·10	4·6	1·86	23	0·95	155	1·55	1 548	1·20	$12{\cdot}1 \times 10^3$
323	36·90	—	—	1·44	2·3	1·52	35	0·40	195	1·78	911	1·20	6 336
333	35·40	—	—	1·95	4·0	0·96	20	0·98	155	1·44	802	0·91	7 105
343	33·35	—	—	—	—	1·40	5	1·26	54	1·07	401	1·55	3 468

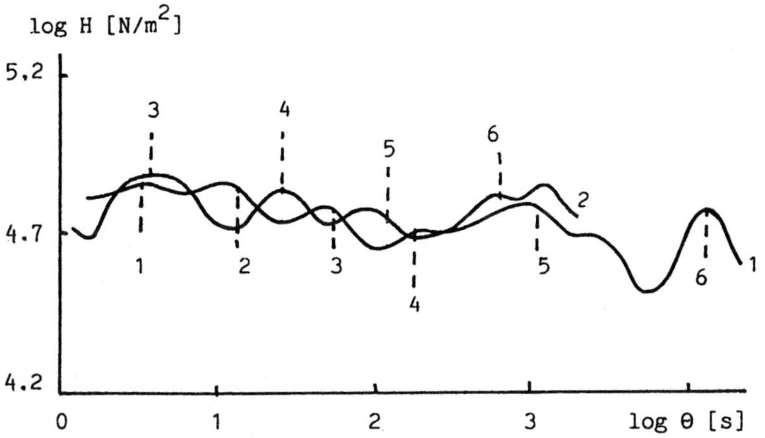

FIG. 5.31. Continuous (solid line) and discrete (dashed line) relaxation time spectra for filled 25 wt% AB SKD rubber at 293 K (curve 1) and 343 K (2).

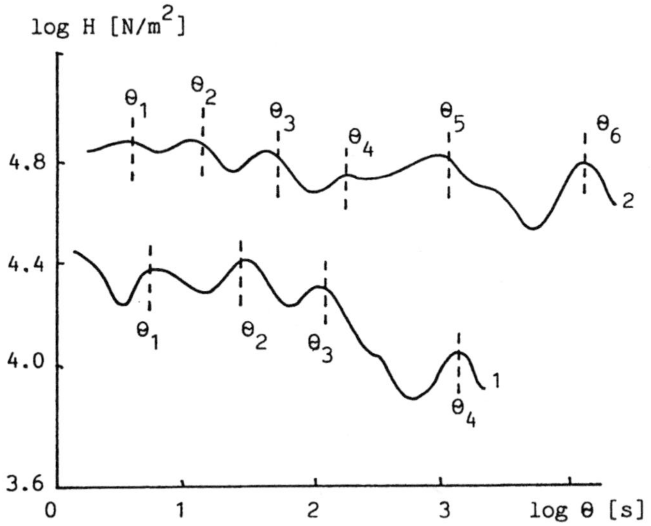

FIG. 5.32. Continuous (solid line) and discrete (dashed line) relaxation time spectra of unfilled (curve 1) and 25 wt% AB (2) SKD rubber at 293 K.

The comparison of the $H(\theta)$ dependences for filled and unfilled samples (Fig. 5.32) leads us to a conclusion on the significant change in the relaxation spectrum of the system due to the influence of the filler. The discrete spectra of the sample differ also substantially in this case.

The nature of each of the registered elementary processes characterized by the θ_i times and maxima on the $H(\theta)$ dependences is found in the $\log \theta_i = f(1/T)$ curves (Fig. 5.33). In fact, the slope angle of $\log \theta_i = f(10^3/KT)$ determines the activation energy U_i of this process depending on the relaxation time θ_i, according to the known correlation $\log \theta_i = \log \theta_{0i} + U_i/2{\cdot}3\,KT$. The fairly good correlation between θ_i and the maxima on the $H(\theta)$ curve involving the evaluation of the U_i magnitude leads us to suppose that each θ_i time corresponds to the process related to the mobility of the structural-inhomogeneous region which occurs at $T > T_g$. The displacement, intensity decrease or disappearance of the maxima indicates a definite reconstitution of the internal structure of the system, an increase in the phase inhomogeneity, etc.

The calculation has shown that the relaxation processes occurring above T_g with $\theta_1 - \theta_6$ times (processes 1–6, respectively) are characterized by different activation energy values. In processes 1, 3, 4 and 5 that are

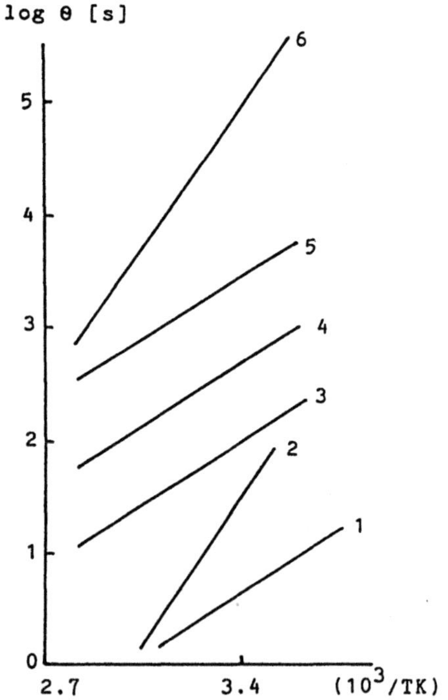

FIG. 5.33. Logarithmic dependence of relaxation times θ_i for processes λ_1 (curve 1), α' (2), λ_2 (3), λ_3 (4), λ_4 (5) and ψ (6) on reverse temperature for filled 25 wt% AB SKD rubber.

registered both in the filled and in the unfilled samples (Fig. 5.32), the U_i values are the same (27 kJ mol^{-1}), indicating the generality of their physical nature due to the mobility of one and the same kinetic forms. Processes 2 and 3 manifest themselves only in the filled samples and have activation energies of 72 and 65 kJ mol^{-1}, respectively. The pre-exponential factors θ_{02} and θ_{06}, respectively, equal $5{\cdot}0 \times 10^{-12}$ and $3{\cdot}2 \times 10^{-7}$ s^{-1}. Since θ_{02} is equivalent to the values obtaining during the calculation of the relation times of the free segments of the SKD molecules [176], it is possible to relate process 2 to the occurrence of the absorbed part on the filler surface (AB) of the rubber layer. The θ_{06} values are considerably higher than those for θ_{02} and are due to the presence of the steric network of the filler.

It is clearly evident that from the physical viewpoint the complete picture of the relaxation transitions observed in the material can be traced by plotting the temperature dependences of dynamic parameters, e.g. E_g and χ. As a rule, the oscillation frequencies in such experiments are chosen to be sufficiently low, thus allowing us to register the mobility of different critical units, also involving those that are large. It is well known that the times for the occurrence of the relaxation processes taking place with the participation of large kinetic units are relatively high. At sufficiently high test frequencies, the duration of the deformation cycle appears to be substantially less than the occurrence period of such processes, by virtue of which the breaking ability of the method decreases.

From the example illustrated in Fig. 5.34 we can see the variety of shapes of the relaxation transitions for filled SKD rubbers appearing as typical maxima on the $E_g(T)$ and $\chi(T)$ curves. Conditionally, they can be divided into α, α^1, λ_i and ϕ maxima. The measurements made at different frequencies (Fig. 5.34) rendered it possible to calculate the activation values for these processes [178]. The α-maxima registered at 160 K correspond to the glass transition temperature of the system, $U_\alpha =$ 37 kJ mol^{-1}. In comparison with the unfilled SKD material, the α-maxima, occurring 3–4 K higher, influence the presence of the network formed by the filler.

The group of maxima registered in the high-elastic state and related to both the crystallization (fusion) of SKD and the occurrence of the processes of a 'slow-physical relaxation stage' [176], α^1, λ_i and ϕ maxima, or the processes 1–6, respectively, have already been considered. At this point we can add that the existence of a network formed as a result of the chemical bonds and filler do not practically influence the 'slow-physical relaxation stage' processes—λ_1, λ_3, λ_4 and λ_5. The same activation

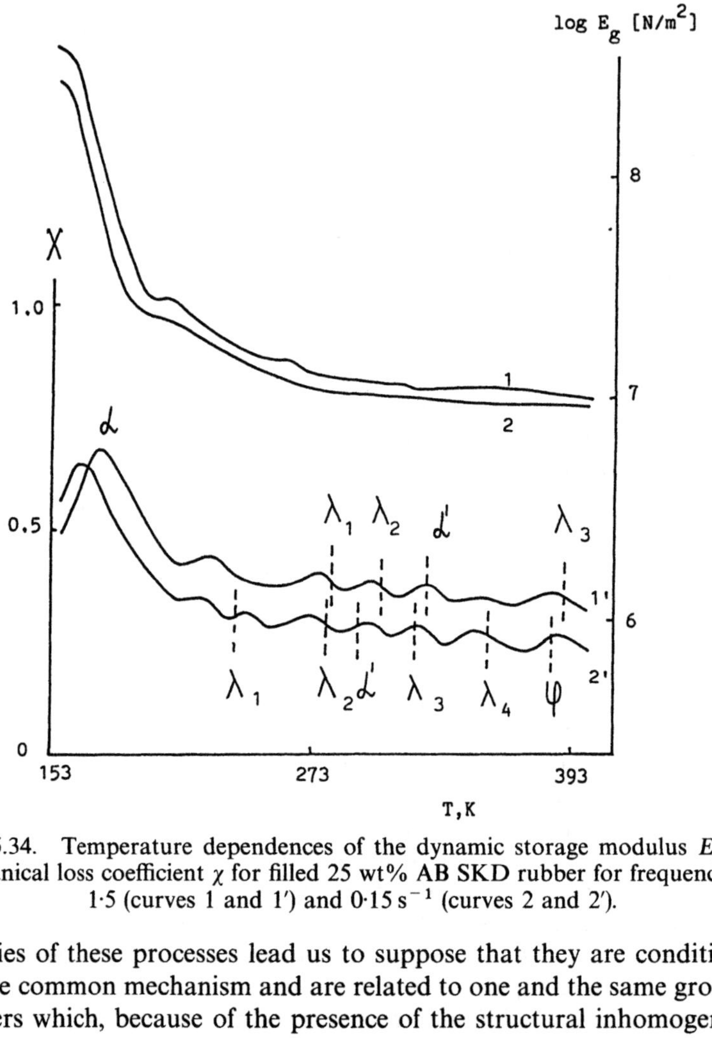

FIG. 5.34. Temperature dependences of the dynamic storage modulus E_g and mechanical loss coefficient χ for filled 25 wt% AB SKD rubber for frequencies of 1·5 (curves 1 and 1′) and 0·15 s^{-1} (curves 2 and 2′).

energies of these processes lead us to suppose that they are conditioned by one common mechanism and are related to one and the same group of relaxers which, because of the presence of the structural inhomogeneity, are distinguishable only by their sizes. The activation energies of these processes are very close to the corresponding value for the viscous filled SKD flow process (23·5 kJ mol^{-1} [174]). This indicates that kinetic forms of the same kind participate in the relaxation processes under consideration. However, the distinction of θ_{0i} values for these processes (Table 5.2) allows us to conclude that the character and mobility mechanism of these segments in the λ_i relaxation processes are different and are, to a great measure, due to the degree of structural inhomogeneity of the filled SKD.

TABLE 5.2

Values for Activation Energy U_i and Pre-exponential Factors θ_{0i} for α^1, λ_i and φ-Relaxations for Unfilled and Filled SKD Rubbers

Unfilled and filled rubbers	*Relaxation processes*	α^1	λ_1	λ_2	λ_3	λ_4	λ_5	φ
0·5 wt% sulfur	θ_{0i}, S	—	$4{\cdot}0 \times 10^{-5}$	$3{\cdot}5 \times 10^{-4}$	$2{\cdot}8 \times 10^{-3}$	$2{\cdot}5 \times 10^{-2}$	$3{\cdot}2 \times 10^{-1}$	—
1·0 wt% sulfur	θ_{0i}, S	—	$2{\cdot}0 \times 10^{-4}$	$1{\cdot}6 \times 10^{-3}$	$2{\cdot}6 \times 10^{-2}$	$1{\cdot}8 \times 10^{-1}$	—	—
2·0 wt% sulfur	θ_{0i}, S	—	$1{\cdot}1 \times 10^{-4}$	$1{\cdot}1 \times 10^{-3}$	$1{\cdot}1 \times 10^{-2}$	$2{\cdot}8 \times 10^{-1}$	—	—
2·0 wt% sulfur and 25 wt% AB	θ_{0i}, S	$5{\cdot}0 \times 10^{-12}$	$4{\cdot}2 \times 10^{-5}$	$8{\cdot}8 \times 10^{-4}$	$5{\cdot}7 \times 10^{-3}$	$4{\cdot}3 \times 10^{-2}$	—	$3{\cdot}2 \times 10^{-7}$
Activation energy U_i	kJ mol^{-1}	72·4	27	27	27	27	27	64·7

The data on weakly linked (0·5 wt% sulfur) unfilled SKD (Table 5.2) also speak in favor of such an interpretation. The data also indicate the existence of five processes, while for the linked (2 wt% sulfur) unfilled SKD sample there are only four processes.

Thus there is evidence that there exists a physical (fluctuational) network alongside the chemical and steric network of macromolecule entanglement nodes distinguished also by their effective size and settled lifetimes (up to destruction at one point and restoration at another).

The comparatively low activation energy of the ϕ process ($\sim$65 kJ mol^{-1}) at relatively large effective filler particle sizes, if they are regarded as separate kinetic forms ($\theta_{0\phi} = 3{\cdot}2 \times 10^{-7}$ s), is accounted for by a subsequent, but not by a simultaneous breakage of the adhesive contacts between the CB particles and rubber molecules. The U_ϕ magnitude corresponds to the value characterizing the strength of separate bonds between the rubber and filler. The rubber macromolecule segments also participate in the heat movement at elevated temperatures in the region of the ϕ-process. A free volume (space) wherein the filler particle diffuses occurs when each adhesive contact between the macromolecular chain segment and filler particles is broken (the filler in this case has linear size of the order of 10^{-8} m). The question concerning the interpretation of the relaxation behavior of filled and unfilled polymers, particularly of the SKD type, is discussed in more detail in Ref. 176. The physical picture under discussion is close to the model scheme of behavior of the filled CB rubber in the course of its deformation, i.e. the formation of filler agglomerates and the occurrence of occluded rubber. This provides evidence for the definite generality and similarity of the physicomechanical behavior of filled polymers in the mechanical and temperature fields.

5.11 CONCLUSION

In conclusion, it is necessary to emphasize the following. The specific features of rheological and relaxation behavior of filled polymers certainly do not encompass the whole spectra of the experimental forms under observation. Nevertheless, the material does not only substantially complement and explain the concepts developed in a number of works of this kind, but also affords a definite generalization, e.g. to single-type (linear flexible-chain) filled polymer systems involving model, physicomechanical behavior of composites in various physical states (fluid–high-elastic–glass transition). The interrelation between the viscoelastic (rheological), frictional, relaxation and strain–strength parameters of the sample that one can attempt to trace is exceptionally important for understanding the general nature of their behavior in changing temperature and mechanical conditions. The similarity of such behavior in the area of linear viscoelasticity, the theory of which is fairly well developed at the present time, and under conditions of large non-linear strains, up to breaking, raises the hope that a uniform theoretical approach will be elaborated that will be able to describe the varied forms of physicomechanical behavior of a composite, starting from the stage of its melt and ending with the production of solid materials [179].

REFERENCES

1. LOBE, V.M. and WHITE, J.L. *Polym. Eng. Sci.*, **19** (1979) 617.
2. YANOVSKY, YU.G., VINOGRADOV, G.V. and BARANCHEEVA, V.V. *Mekhanika Kompozit. Mater.*, no. 6 (1986) 1073.
3. YANOVSKY, YU.G., VINOGRADOV, G.V. and BARANCHEEVA, V.V. Proceedings of the Second Conference of European Rheologists, 1986, *Rheol. Acta*, **26** (1988) 218.
4. DONNET, J.B. and VIDAL, A. *Zh. Vsesoyuzn. Khim. Obshestva im D.I.Mendeleeva*, **31**(1) (1986) 10.
5. HINKELMANN, B. *Rheol. Acta*, **20** (1981) 561.
6. HAN, C.D., VAN DER WEGHE, T., SHETE, P. and HAW, J.R. *Polym. Eng. Sci.*, **21** (1981) 196.
7. NAKATSUKA, T., KAWASAKI, H., ITADANI, K., *et al. J. Appl. Polym. Sci.*, **27** (1982) 259.
8. JUSKEY, V.P. and CHAFFEY, C.E. *Can. J. Chem. Sci.*, **60** (1982) 334.
9. UTRACKI, L.A. and FISA, B. *Polym. Composites*, **3** (1982) 193.
10. HINKELMANN, B. *Rheol. Acta*, **21** (1982) 491.
11. SHENOY, A.V., SAINI, D.R. and NADKARNI, V.M. *Polym. Composites*, **4** (1983) 53.
12. SHENOY, A.V. and SAINI, D.R. *Colloid. Polym. Sci.*, **261** (1983) 846.
13. LEM, K.W. and HAN, C.D. *J. Rheol. Acta*, **27** (1983) 263.
14. LUO, H.L., HAN, C.D. and MIJOVIC, J. *J. Appl. Polym. Sci.*, **28** (1983) 3387.
15. KITANO, T., KATAOKA, T. and NAGATSUKA, Y. *Rheol. Acta*, **23** (1984) 20.
16. KITANO, T., KATAOKA, T. and NAGATSUKA, Y. *Rheol. Acta*, **23** (1984) 408.
17. HINKELMANN, B. and MENNIG, G. *Chem. Eng. Comm.*, **36** (1985) 211.
18. BRETAS, R.E.S. and POWELL, R.L. *Rheol. Acta*, **24** (1985) 69.
19. MITEL, A.T. and KAMAL, M.R. *Polym. Composites*, **7** (1986) 283.
20. SHENOY, A.V. and SAINI, D.R. *J. Reinf. Plastics Comp.*, **5** (1986) 62.
21. NIELSEN, L.E. *Mechanical Properties of Polymers and Composites*, Marcel Dekker, New York, Vol. 2, 1974, Ch. 7, p. 379.
22. HAN, C.D. *Rheology in Polymer Processing*, Academic Press, New York, 1976, Ch. 7, 182.

23. NIELSEN, L.E. *Polymer Rheology*, Marcel Dekker, New York, 1977, Ch. 9, 133.
24. PAUL, D.R. and NEWMANN, S. *Polymer Blends*, Academic Press, New York, Vol. 1, 1978, Ch. 7, 295.
25. DONNE, J.B. and VOET, A. *Carbon Black*, Marcel Dekker, New York, 1976.
26. KIRK–OTHMER. *Encyclopedia of Chemical Technology*, 3rd ed., Elastomers, Synthetic (Nitrile Rubber). Wiley, New York, 1979, 534.
27. NAKAJIMA, N., HARRELL, E.R. and SEIL, D.A. Rubber processing. In: *Encyclopedia of Material Science and Engineering*, Ed. M.B. Bever, Pergamon Press, New York, 1986.
28. NAKAJIMA, N. and HARRELL, E.R. In: *Viscoelastic Characterization of Long Branching and Gel in Elastomers in Encyclopedia of Fluid Mechanics*, Ed. N.P. Cheremisinoff, vol. 9, Gulf Publishing Co., Houston, TX, 1990.
29. NAKAJIMA, N. and COLLINS, E.A. *Rubber Chem. and Technol.*, **48** (1975) 615.
30. DANNENBERG, E.M. *Rubber Chem. and Technol.*, **48** (1975) 410.
31. O'BRIEN, J., CASSHELL, E., WARDEL, G.E. and MCBRIERTY, V.J. *Rubber Chem. and Technol.*, **50** (1977) 747.
32. DANNENBERG, E.M. Carbon black. In: *Encyclopedia of Composite Materials and Components*, Ed. M. Grayson, Wiley, New York, 1983, 230.
33. WHITE, J.L. and CROWDER, J.W. *J. Appl. Polym. Sci.*, **18** (1974) 1013.
34. SERIZAWA, H., ITO, M., KANAMOTO, T., TANAKA, K. and NOMURA, A. *Polymer J.*, **14** (1982) 149; SERIZAWA, H., ITO, M., KANAMOTO, T., TANAKA, K. and NOMURA, A. *Polymer J.*, **15** (1983) 201, 543.
35. NAKAJIMA, N. and SCOBBO, J.J., JR. *Rubber Chem. and Technol.*, **61** (1988) 137.
36. TOKITA, N. and PLISHKIN, I. *Rubber Chem. and Technol.*, **46** (1973) 1166.
37. SONG, H.J., WHITE, J.L., MIN, K., NAKAJIMA, N. and WEISSERT, F.C. *Adv.Polym. Tech.*, **8** (1988) 431.
38. MONTES, S., WHITE, J.L. and NAKAJIMA, N. *J. Non-Newtonian Fluid Mechanics*, **28** (1988) 183.
39. ISONO, Y. and FERRY, J.D. *Rubber Chem. and Technol.*, **57** (1984) 925.
40. ARAI, K. and FERRY, J.D. *Rubber Chem. and Technol.*, **59** (1986) 592.
41. SHENOY, A.V. Rheology of highly filled polymer melt systems. In: *Encyclopedia of Fluid Mechanics*, Ed. N.P. Cheremisinoff, Gulf Publishing Co., Houston, TX, Vol. 7, Ch. 23, 1988.
42. NAKAJIMA, N. and HARREL, E.R. Rheological behaviour of rubber carbon black compounds. In *Encyclopedia of Fluid Mechanics*, Ed. N.P. Cheremisinoff, Gulf Publishing Co., Houston, TX, Vol. 9, Ch. 8, 1990.
43. YANOVSKY, YU.G. and ZAIKOV, G.E. Rheological properties of filled polymers. In *Encyclopedia of Fluid Mechanics*, Ed. N.P. Cheremisinoff, Gulf Publishing Co., Houston, TX, Vol. 9, Ch. 7, 1990.
44. LIPATOV, YU.S. *Fizicheskaya Khimia Napolnennych Polimerov* (Physical Chemistry of Filled Polymers), Khimiya, Moscow, 1977.
45. PRIVALKO, V.P., BESKLUBENKO, YU.D., LIPATOV, YU.S. *et al. Vysokomol. Soed.*, **19A**(8) (1977) 1744.
46. HOWARD, G.J. and SHANKS, R.A. *J. Macromol. Sci.*, **19**(2) (1981) 167.
47. LIPATOV, YU.S., DEMCHENKO, S.S. and PRIVALKO, V.P. *DAN SSSR*, **273**(1) (1983) 128.

48. PRIVALKO, V.P., LIPATOV, YU.S. and DEMCHENKO, S.S. *Vysokomol. Soed.*, **28**(6) (1986) 425.
49. PRIVALKO, V.P. *Fizikochimiya Mnogokomponentnych Polimernych Sistem* (Physico-Chemistry of Polymer Multicomponent Systems), Naukova Dumka, Kiev, 1986.
50. PRIVALKO, V.P., NOVIKOV, V.V. and YANOVSKY, YU.G. *Osnovy Teplofiziki i Reofiziki Polimernych Materialov* (Principles of Thermophysics and Rheophysics of Polymer Materials), Naukova Dumka, Kiev, 1991.
51. REBINDER, P.A. *Selected Works*, Vol. 1, Nauka, Moscow, 1978; Vol. 2, Nauka, Moscow, 1979 (in Russian).
52. URIEV, N.B. *High Concentration Disperse Systems*, Khimiya, Moscow, 1980 (in Russian).
53. KRAUS, G. (Ed.). *Reinforcement of Elastomers*, Interscience, New York, 1965.
54. MANSON, J.A. and SPERLING, L.H. *Polymer Blends and Composites*, Plenum Press, New York, 1976.
55. MEDALIA, A.I. *J. Colloid and Interface Sci.*, **24** (1967) 393.
56. KRAUS, G. *Rubber Chem. and Technol.*, **38** (1965) 1076.
57. KRAUS, G., ROLLMAN, V.W. and GRUVER, I.T. *Macromolecules*, **3** (1970) 92.
58. RIVIN, D., ARON, J. and MEDALIA, A.I. *Rubber Chem. and Technol.*, **41** (1968) 330.
59. SIRCAR, A.K. and VOET, A. *Rubber Chem. and Technol.*, **43** (1973) 973.
60. KRAUS, G. and GRUVER, I.T. *J. Polym. Sci.*, **A-2**(8) (1970) 57.
61. KRAUS, G. *J. Appl. Polym. Sci.*, **15** (1971) 1679.
62. SAMBROOK, R.W. *Rubber Chem. and Technol.*, **44** (1971) 728.
63. BIGG, D.M. *Polym. Eng. Sci.*, **22** (1982) 512.
64. SAINI, D.R., SHENOY, A.V. and NADKARNI, V.M. *Polym. Composites*, **7** (1986) 193.
65. SACKS, M.D., KHADILKAR, C.S., SCHEIFFELE, G.W., SHENOY, A.V., DOW, J.H. and SHEU, R.S. *Advances in Ceramics*, **21** (1987) 495.
66. BIGG, D.B. *Polym. Plast. Technol. Eng.*, **23**(2) (1984) 133.
67. SAINI, D.R., SHENOY, A.V. and NADKARNI, V.M. *Polym. Eng. Sci.*, **25** (1985) 807.
68. SAINI, D.R. and SHENOY, A.V. *Polym. Eng. Sci.*, **26** (1986) 441.
69. SHENOY, A.V. and SAINI, D.R. *Polym. Composites*, **7** (1986) 96.
70. OLMSTEAD, B.A. *SPE-J.*, **26** (1986) 96.
71. BIGG, D.M. Proceedings of the IX International Congress on Rheology in Mexico, *Adv. in Rheology*, **3** (1984) 429.
72. PLUEDDEMANN, E.P. *Silane Coupling Agents*, Plenum Press, New York and London, 1982.
73. MONTE, S.J. and SUGERMAN, G. Kenrich Petrochemicals, Inc., Bayonne, NJ, 1985.
74. COHEN, L.B. *SPE, Plastics Engineering*, 29 November 1983.
75. COHEN, L.B. *SPE ANTEC*, paper 636, New Orleans, April 1984.
76. COHEN, L.B. The Adhesion Society, Savannah, GA, 24 February 1985.
77. COHEN, L.B. The chemistry and reactivity of zircoaluminate coupling agents for filled and reinforced plastics, *SPI RP/C*, Atlanta, GA, 27–31 January 1986.
78. BIGG, D.M. *Polym. Eng. Sci.*, **23** (1983) 206.

79. ALTHOUSE, L.M., BIGG, D.M. and WONG, W.M. *Plastics Compounding*, March/April 1983.
80. SHARMA, Y.N., PATEL, R.D. and DHIMMAR, I.H. *J. Appl. Polym. Sci.*, **27** **7**(1982) 97.
81. VINOGRADOV, G.V., YANOVSKY, YU.G., *et al. Vysokomol. Soed.*, **20A**(11) (1978) 2403.
82. YANOVSKY, YU.G. and VINOGRADOV, G.V. *Vysokomol. Soed.*, **22A**(11) (1980) 2567.
83. YANOVSKY, YU.G., VINOGRADOV, G.V. and IVANOVA, L.I. *Vysokomol. Soed.*, **24A**(5) (1982) 1057.
84. YANOVSKY, YU.G., VINOGRADOV, G.V. and IVANOVA, L.I. *Inzhenerno-Fizicheskii Zh.*, **46**(8) (1984) 974.
85. YANOVSKY, YU.G., VINOGRADOV, G.V. and IVANOVA, L.I. *Doklady Acad. Nauk SSSR*, **282**(5) (1985) 1190.
86. ULIANOV, L.P., YANOVSKY, YU.G., *et al. Zavodskaya Labor.*, **39**(11) (1973) 1402.
87. VINOGRADOV, G.V., *et al. Int. J. Polym. Mater.*, **2** (1972) 1.
88. AHUJA, S.K. In: *Proc. II Int. Congr. Rheology*, Plenum Press, New York, 1980, SD 2.6.3, p. 261.
89. YANOVSKY, YU.G., ZHDANYK, V.K., ZOLOTAREV, A.V. and VINOGRADOV, G.V. *Rheol. Acta*, **27** (1988) 298.
90. VINOGRADOV, G.V., YANOVSKY, YU.G., *et al.* In: *Proc. Int. Rubber Conf.*, Moscow, 1984, C, vol. 1, p. 19.
91. HAVRILIAK, S. and NEGAMI, S. *Polymer*, **8**(4) (1967) 161.
92. MARIN, G. Contribution à l'étude des propriétés viscoélastiques en régime linéaire du polystyrène à l'état fondu. Thèse pour obtenir le grade de docteur des sciences, Académie de Bordeaux, 1977.
93. WISNIEWSKI, C., MARIN, G. and MONGE, PH. *Eur. Polym. J.*, **21** (1985) 479.
94. IAKOBSON, E.E. and FAITELSON, L.A. *Mekhanika Kompozit. Mater.*, **1** (1990) 146.
95. KRAUS, G. *Fortschr. Hochpolymere Forsch.*, **B8** (1971) 156.
96. CASSON, N. In: *Rheology of Disperse Systems*, Ed. C.C. Mill, Pergamon Press, London, 1959, p. 84.
97. YANOVSKY, YU.G., VINOGRADOV, G.V. and BARANCHEEVA, V.V. *Vysokomol. Soed.*, **28A**(5) (1986) 983.
98. SMITH, T.L. *Trans. Soc. Rheol.*, **3** (1959) 113.
99. REYNOLDS, O. *Phil. Mag.*, **20**(5) (1885) 469.
100. COGSWELL, F.N. *Plast. and Polym.*, **41**(151) (1973) 39.
101. KIRBY, I. *Rheol. Acta*, **27**(2) (1988) 326.
102. GOLDSTEIN, M.N. *Mekhanicheskie Svoiistva Gruntov* (Mechanical Properties of Soils), Stroiizdat, Moscow. Vol. 1, 1971, p. 368; vol. 2, 1973, p. 376.
103. VINOGRADOV, G.V., MALKIN, A.YA., PLOTNIKOVA, E.P., *et al. Int. J. Polym. Mater.*, **2** (1972) 184.
104. WHITE, J.L., CZARNECKI, L. and TANAKA, H. *Rubber Chem. and Technol.*, **53** (1980) 823.
105. SUETSUGU, Y. and WHITE, J.L. *J. Appl. Polym. Sci.*, **28**(4) (1983) 1481.
106. GLUSHLOV, I.A., VINOGRADOV, G.V. and ROZHKOV, V.A. *Mekhanika Polimerov*, **5** (1974) 902.

107. SUETSUGU, Y. and WHITE, J.L. *J. Non-Newtonian Fluid Mechanics*, **14** (1984) 121.
108. HINCH, E.J. and LEAL, L.G. *J. Fluid Mech.*, **52** (1972) 683.
109. SIMHA, R.A. *J. Appl. Phys.*, **23** (1952) 1020.
110. ONOGI, S. and MATSUMOTO, T. *Polym. Eng. Rev.*, **1**(1) (1981) 45.
111. HAN, C.D. *Multiphase Flow in Polymer Processing*, Academic Press, New York, 1981.
112. EINSTEIN, A. *Ann. Phys.*, **19** (1906) 289; **34** (1911) 591.
113. LIN, C.J., PERRY, J.H. and SCHOWALTER, W.R. *J. Fluid Mech.*, **44** (1970) 1.
114. JEFFERY, G.B. *Proc. Roy. Soc., Ser. A.*, **102** (1922) 161.
115. GOLDSMITH, H.L. and MASON, S.G. In: *Rheology Theory and Applications*, Ed. F.R. Eirich, Vol. 4, Academic Press, New York, 1967, p. 85.
116. BRENNER, H. *Progress in Heat and Mass Transfer*, Vol. 5, Pergamon Press, New York, 1972.
117. LEAL, L.G. and NIHEN, E.J. *Rheol. Acta*, **12** (1973) 127.
118. MOONEY, M. *J. Colloid Sci.*, **6** (1951) 162.
119. THOMAS, D.G. *J. Colloid Sci.*, **20** (1965) 267.
120. LEE, D.J. *Trans. Soc. Rheol.*, **13** (1969) 273.
121. DE GENNES, P.G. *Le Journal de Physique*, **40** (1979) 783.
122. FAITELSON, L.A. and KOVTIN, V.P. *Mekhanika Polimerov*, (N2) (1975) 326.
123. KOVTIN, V.P. and FAITELSON, L.A. *Mekhanika Kompozit. Mater.*, (N4) (1990) 715.
124. DOI, M. and EDWARDS, S.F. *J. Chem. Soc., Faraday Trans.*, part 2, (5) (1978) 918.
125. DOI, M. and KUZUU, N.I. *J. Polym. Sci., Polym. Phys. Ed.*, **18**(3) (1980) 409.
126. FAITELSON, L.A. and YAKOBSON, E.E. *Mekhanika Kompozit. Mater.* (N2) (1981) 277.
127. FAITELSON, L.A. and YAKOBSON, E.E. *Mekhanika Polimerov*, (N1) (1978) 172.
128. VINOGRADOV, G.V. and MALKIN, A.YA. *Rheology of Polymers*, Mir Publishers, Moscow, 1980.
129. VINOGRADOV, G.V., MALKIN, A.YA., YANOVSKY, YU.G., *et al. J. Polym. Sci., Polym. Phys. Ed.*, **A-2**(10) (1972) 1061.
130. YANOVSKY, YU.G. *L'actualité chimique*, March-April 1991, 109.
131. VINOGRADOV, G.V. and PAVLOV, V.P. *Rheol. Acta*, **1** (1961) 416, 455.
132. VINOGRADOV, G.V., BORISENKOVA, E.K. and ZABUGINA, M.P. *DAN SSSR*, **277**(3) (1984) 614.
133. FAITELSON, L.A. *Int. J. Polym. Mater.*, **8**(2–3) (1980) 207.
134. BORISENKOVA, E.K., SABSAII, O.YU., KURBANALIEV, M.K., *et al.*, *Polymer*, **19** (1978) 1473.
135. LIPATOV, YU.S. *Kolloidnaya Khimiya Polimerov* (Colloid Chemistry of Polymers), Naukova Dumka, Kiev, 1984.
136. VINOGRADOV, G.V., PLOTNIKOVA, E.P., ZABUGINA, M.P., *et al. Vysokomol. Soed.*, **29A**(3) (1987) 211.
137. DREVAL, V.E. and VINOGRADOV, G.V. *Kolloid. Zh.*, **51**(2) (1985) 237.
138. MEISSNER, J. *J. Chem. Eng. Comm.*, **33**(1) (1985) 159.
139. BORISENKOVA, E.K., DREVAL, V.E., VINOGRADOV, G.V., *et al. Polymer*, **23**(1) (1982) 91.

140. Vinogradov, G.V., Dreval, V.E., Borisenkova, E.K., *et al.* *Vysokomol. Soed.*, **23A**(12) (1981) 2627.
141. Bartenev, G.M. and Ermilova, N.V. *Kolloid. Zh.*, **29**(6) (1967) 771.
142. Gul', V.E. *Struktura i Protochnost' Polimerov* (Structure and Strength of Polymers), Khimiya, Moscow, 1978.
143. Bartenev, G.M. *Protchnist i Mekhanizm Razruscheniya Polimerov* (Strength and Fracture of Polymers), Khimiya, Moscow, 1984.
144. Vinogradov, G.V., Malkin, A.Ya., Volosevitch, V.V., *et al.* *J. Polym. Sci., Polym. Phys. Ed.*, **13**(9) (1975) 1721.
145. Vinogradov, G.V., Dreval, V.E. and Yanovsky, Yu.G. *Rheol. Acta*, **24** (1985) 574.
146. Smith, T. *Polym. Sci.*, **A-1** (1963) 3597.
147. Optov, V.A., Borisenkova, E.K. and Sabsaii, O.Yu. *Kolloid. Zh.*, **50** (1988) 1033.
148. Payne, A.R. *Rubber Plastic Age*, **42** (1961) 963.
149. Payne, A.R. *J. Appl. Polym. Sci.*, **3** (1960) 127.
150. Faitelson, L.A. *Izvestiya Akad. Nauk Latvii*, (N3) (1965) 42.
151. Faitelson, L.A. *Mekhanika Polimerov*, (N1) (1969) 182.
152. Faitelson, L.A. *Mekhanika Polimerov*, (N1) (1978) 113.
153. Faitelson, L.A. *Mekhanika Polimerov*, (N7) (1976) 146.
154. Faitelson, L.A. and Alekseenko, A.I. *Mekhanika Polimerov*, (N3) (1970) 558.
155. Faitelson, L.A., Leonov, A.I. and Tshiprin, M.G. *Mekhanika Polimerov*, (N3) (1970) 521.
156. Faitelson, L.A., Tshiprin, M.G. and Briedis, D.B. *Plasticheskie Massy*, (N9) (1970) 38.
157. Tshiprin, M.G. and Faitelson, L.A. *Mekhanika Polimerov*, (N4) (1969) 561.
158. Yakovlev, Yu.P. and Briedis, I.P. *Mekhanika Polimerov*, (N3) (1969) 561.
159. Payne, A.R. *J. Appl. Polym. Sci.*, **6** (1960) 368.
160. Yanovsky, Yu.G., Vinogradov, G.V. and Dreval, V.E. *Trans. CSME*, **8**(2) (1984) 84.
161. Mosiichin, E.P. and Vinogradov, G.V. *Kolloid. Zh.*, **19** (1957) 311.
162. Han, C.D. *Multiphase Flow in Polymer Processing*, Academic Press, New York and London, 1981.
163. Suetsugu, Y. and White, J. *J. Appl. Polym. Sci.*, **28** (1983) 1481.
164. Teteris, G.G. and Briedis, I.P. *Mekhanika Kompozit. Mater.*, (N3) (1989) 514.
165. Teteris, G.G. *Mekhanika Kompozit. Mater.*, (N4) (1990) 715.
166. Briedis, I.P. and Teteris, G.G. *Proc. Vth Nat. Conf. on Mechanics and Technol. Composites*, Varna-Sofia, Acad. Sci. of Bulgaria Publishers, 1988, p. 54.
167. Gevorgyan, A.M., *et al.* *Technology of Chemical Polymers, Proc. Vinogradov, G.V.NITI*, Moscow, 1980, 3233–80.
168. Yanovsky, Yu.G., Vinogradov, G.V., Gevorgyan, A.M., *et al.* *Vysokomol. Soed.*, **28**(11) (1985) 2385.
169. Fridman, M.M., Popov, V.L., *et al.* *Doklady Akad. Nauk SSSR*, **255**(5) (1980) 1185.

170. Cox, W.P. and Merz, E.H. *J. Polym. Sci.*, **8** (1958) 619.
171. Vinogradov, G.V., Yanovsky, Yu.G. and Isaev, A.I. *J. Polym. Sci.*, **8** (1970) 1239.
172. Yanovsky, Yu.G. In: *Theoretical and Instrumental Rheology*, Ed. P. Rebinder). Proc. Inst. Heat Mass Transfer, Minsk, 1970, p. 119.
173. Vinogradov, G.V., Yanovsky, Yu.G. and Frenkin, E.I. *R Rheol. Acta*, **7**(3) (1968) 277.
174. Bartenev, G.M. *Struktura i Relaksatsionnye Svoiistva Elastomerov* (Structure and Relaxation Properties of Elastomers), Khimiya, Moscow, 1979.
175. Aivazov, A.B. and Zelenev, Yu.V. *Zavodskaya Labor.*, **34**(6) (1968) 750.
176. Ligidov, M.H., Cobzar, Yu.N., Yanovsky, Yu.G., *et al. Kompozit. Polimernye Mater.*, **4** (1989) 215.
177. Tobolsky, A.V. *Properties and Structure of Polymers*, Wiley, New York, 1960.
178. Ferry, J.D. *Viscoelastic Properties of Polymers*, 3rd ed., Wiley, New York, 1980.
179. Gul', V.E., Yanovsky, Yu.G., *et al. Doklady Akad. Nauk SSSR*, **294**(4) (1987) 905.

APPENDIX 1
THE PROTOTYPE PROGRAM OF THE BROWNIAN DYNAMIC METHOD

```
#include <alloc.h>
#include <dos.h>
#include <math.h>
#include <time.h>
#include <stdlib.h>
#include <io.h>
#include <stdio.h>
#include <bios.h>
#include <values.h>
#include <conio.h>

#define POINTER far
#define malloc farmalloc
#define calloc farcalloc
#define free farfree
#define YES   1
#define NO    0

float lin_yx( float POINTER *x, float POINTER*y, int T);
void  p_graphic ( float POINTER *r, int l);
void print_matr(float POINTER *x,int n,int m);
void print_e (float POINTER *x, int n, int m);
void stat (float POINTER *x, int n, float POINTER *sred, float POINTER *s2);
float eij(int i,int j,float POINTER *r);
float gauss( float s,float am );
float gauss1( float s,float am );
float gauss2( float s,float am );
float slutch( void );
float slutch1( void );
float slutch2( void );
void  normro(int n,float POINTER *r,float POINTER *p);
int indexrrm( int i,int j, int k, int n );
 void corfun( int n, float POINTER *r, float POINTER *p,
             float POINTER *corrf, float POINTER *ls,
        float POINTER *p1, float POINTER *hcov,
        int POINTER *nvrcorrf, int POINTER *nvrls,
        float POINTER *hls,
        float t, float ht, float tmin,
        int nls, int ncorrf,  int nirrm, int krrm,
        float  POINTER *pcen, float POINTER *pcen1,
        float POINTER *rcen );
int readzap( float *p, long n, int len, int pcenn );
void centrr( int n, float POINTER *r, float POINTER *r1) ;
#define EPSILON  1e-20

FILE POINTER *outfile;
FILE POINTER *fmodes, POINTER *fnul;
FILE  POINTER *frrm;
FILE  POINTER *frrm1;

/* ========================================================== */
void main()  {
FILE POINTER *fp;      /* file of input parameters*/
FILE POINTER *fs;      /* file of plots            */
FILE POINTER *fdata;   /* file of data             */
float  POINTER *r, POINTER *u, POINTER *a, POINTER *p,
       POINTER *p1,
       POINTER *r1, POINTER *u1, POINTER *a1, POINTER *stohas,
       POINTER *corrf, POINTER *ls, POINTER *hcov, POINTER *hls,
```

```
      POINTER *pcen, POINTER *pcen1, POINTER *rcen;
int   POINTER *nvrcorrf,POINTER *nvrls;

float t,ht;        /* t,ht - time and integration interval */
float tmin;        /* tmin - time before calculation      */
float tmaxls; /* tmaxls,hls-the largest arg. and interval of disp.*/
float m,z,b,h,e,pj,dpj_drj2;  /* parameters */
float tmxcor;      /* tmxcor - the largest arg. of corr. function */
float usred;       /* averaged squared velocity U              */
float tmax;        /* calculations do , until  time  t <= tmax  */
float uij,aij,deli2;
float rr,rrr,stoh,skvotk;
int n;             /* ( n + 1 ) - the number of particles        */
int nls;           /* nls - number of disp. intervals            */
int ncorrf;        /* ncorrf - number of points of corr. function  */
int nirrm;          /* nirrm - number points under averaging  (for nrrm) */
int krrm;           /* writing in "rrm" after  "krrm" intervals     */
int kmodel;         /* model choose parameter*/
int ku = 0, kumax;  /* kumax - the volume of data for squared u */
int  i, j,k,ij,ii,iprint;
int n_slides;
float t_slides;
int read_flag = YES, n_read;  /*     for read                */
int  n_max;              /*     for plots of corr. fun-n  .   */
char p_str[80];          /* the line for input of parameters */
float htt;

clrscr();
fp = fopen("parameter.mlk","r");
fs = fopen("slides.mlk","w");
frrm = fopen("filerrm","w");
fclose(frrm);
frrm1 = fopen("filerrm1","w");
fclose(frrm1);

fgets(p_str, 80, fp ); sscanf(p_str, "%d", &n);
fgets(p_str, 80, fp); sscanf(p_str, "%f", &ht);
fgets(p_str, 80, fp); sscanf(p_str, "%f", &tmax);
fgets(p_str, 80, fp); sscanf(p_str, "%d", &ncorrf);
fgets(p_str, 80, fp); sscanf(p_str, "%d", &n_slides);
fgets(p_str, 80, fp); sscanf(p_str, "%f", &tmin);
fgets(p_str, 80, fp); sscanf(p_str, "%f", &tmxcor);
fgets(p_str, 80, fp); sscanf(p_str, "%f", &tmaxls);
fgets(p_str, 80, fp); sscanf(p_str, "%d", &nls);
fgets(p_str, 80, fp); sscanf(p_str, "%f", &skvotk);
fgets(p_str, 80, fp); sscanf(p_str, "%d", &kmodel);
fgets(p_str, 80, fp); sscanf(p_str, "%d", &nirrm);
/*---------------------------------------------------------------*/
iprint=1000;     /* τÑαÑ° iprint Φáú«ó .- »ÑτáΓ∞ */
kumax = iprint;
t_slides = tmax/n_slides;
rr = ( tmax - tmin - tmxcor ) / ht;
krrm = (int)( rr / (float)nirrm );
/*---------------------------------------------------------------*/
r=(float POINTER *)calloc((n+1)*3,sizeof(float));
r1=(float POINTER *)calloc((n+1)*3,sizeof(float));
u=(float POINTER *)calloc((n+1)*3,sizeof(float));
u1=(float POINTER *)calloc((n+1)*3,sizeof(float));
a=(float POINTER *)calloc((n+1)*3,sizeof(float));
```

```
a1=(float POINTER *)calloc((n+1)*3,sizeof(float));
p=(float POINTER *)calloc((n+1)*3+1,sizeof(float));
p1=(float POINTER *)calloc((n-1)*3+1,sizeof(float));
pcen=(float POINTER *)calloc((n+1)*3+1,sizeof(float));
pcen1=(float POINTER *)calloc((n+1)*3+1,sizeof(float));
rcen=(float POINTER *)calloc((n+1)*3+1,sizeof(float));
stohas=(float POINTER *)calloc((n+1)*3,sizeof(float));
corrf=(float POINTER *)calloc((n+1)*(ncorrf+1),sizeof(float));
ls=(float POINTER *)calloc((nls+1),sizeof(float));
hcov=(float POINTER *)calloc((ncorrf+1),sizeof(float));
hls=(float POINTER *)calloc((nls+1),sizeof(float));
nvrcorrf=(int POINTER *)calloc((ncorrf+1),sizeof(int));
nvrls=(int POINTER *)calloc((nls+1),sizeof(int));
// hcov[k] - arg. of corr. functions,   k=0,1,...,ncorrf
// nvr[k] -  time of series number of elements at corr. function culculation

if(nvrls == NULL) { printf("(nvrls)  value is exceeded \n");delay(2000);exit(0)
}

outfile = fopen("results.out","w" );
/* reading of constant */
fgets(p_str, 80, fp); sscanf(p_str, "%f", &h);
fgets(p_str, 80, fp); sscanf(p_str, "%f", &b);
fgets(p_str, 80, fp); sscanf(p_str, "%f", &z);
fgets(p_str, 80, fp); sscanf(p_str, "%f", &m);
fgets(p_str, 80, fp); sscanf(p_str, "%f", &e);
fclose (fp);

/* print of parameters */
fprintf (outfile," parameters of calculations\n");
fprintf (outfile," .the number of particles                 = %d\n", n+1);
fprintf (outfile," .interval of time                        = %f\n", ht);
fprintf (outfile," .largest time                            = %f\n", tmax);
fprintf (outfile," .number of points of corr. function = %d\n", ncorrf);
fprintf (outfile," .the number of slides                    = %d\n", n_slides);
fprintf (outfile," .the deviation of random force           = %f\n", skvotk);
fprintf (outfile," .model chose parameter(0 or 1)           = %d\n", kmodel);
fprintf (outfile," .number of points in series              = %d\n", nirrm);
fprintf (outfile," Constantsof process\n");
fprintf (outfile,"   .constant of  h = %f\n", h);
fprintf (outfile,"   .constant of  b = %f\n", b);
fprintf (outfile,"   .constant of  z = %f\n", z);
fprintf (outfile,"   .constant of  m = %f\n", m);
fprintf (outfile,"   .constant of  e = %f\n\n", e);

/*  for plots  */
fprintf(fs, "%f %f\n", 0.0, t_slides);
fprintf(fs, "%d\n", n_slides);
/* ======================================================== */

    /*  dr[i,k]=u[i,k]*ht;  i=1,2,3;  k=0,1,...,n; \\ r1=r+dr \\ */
    /*  du[i,k]=a[i,k]*ht;  ---"---   -----"-----   \\ u1=u+du \\ */
/*    identification of initial state       */
  fdata = fopen("rua.mlk", "r");
fgets(p_str, 80, fp); sscanf(p_str, "%d", &n_read);
 if (ferror(fdata)!=0)  read_flag =NO;

if (n_read != n)
  read_flag =NO;
if (read_flag == YES)  {
```

```
  for(j=0;j<=n;j++)  {    /* r[]=u[]=a[]=0 */
    ij=3*j;
    fgets(p_str, 80, fp); sscanf(p_str, "%f %f %f", &r[ij], &r[ij+1], &r[ij+2]
    fgets(p_str, 80, fp); sscanf(p_str, "%f %f %f", &u[ij], &u[ij+1], &u[ij+2]
    fgets(p_str, 80, fp); sscanf(p_str, "%f %f %f", &a[ij], &a[ij+1], &a[ij+2]
  }
 if (ferror(fdata)!=0) { read_flag =NO; clearerr(fdata); }
  fclose (fdata);
}
 if (read_flag == NO)  {  /* no data  */
  fclose (fdata);
  for(j=0;j<=n;j++)     /* r[]=u[]=a[]=0 */
    for(i=1;i<=3;i++)   {
      ij=3*j+i-1;
      r[ij]=0.;
      u[ij]=0.;
      a[ij]=0.;
    }
}

ii=0;
t=0;
usred=0;
pj=0;
dpj_drj2=0;
hcov[1] = tmxcor / (float)ncorrf;
i = (int)( hcov[1] / (ht * (float)krrm) + 0.5 );
hcov[1] = ht * (float)krrm * (float)i;
tmax = tmin + ht * (float)krrm * (float)nirrm + hcov[1] * (float)ncorrf;
for( k=0; k <= ncorrf; k++)  hcov[k] = hcov[1] * (float)k;
hls[1] = log10(ht * (float)krrm);
hls[nls] = log10(tmaxls);
rr = ( hls[nls] -hls[1] ) / ( nls - 1);
for( k=2; k <= nls; k++)  hls[k] = hls[1] + rr * ( (float)k - 1 );
for( k=1; k <= nls; k++)  hls[k] = pow(10.,hls[k]);

frrm = fopen("filerrm","a+b");
frrm1 = fopen("filerrm1","a+b");

/* ================================================ */
while (t < tmax)  {            /* while t<tmax */
  /*  averaged squared velocity U   */
  for (i=1; i<=3; i++)
    usred += u[i-1]*u[i-1];
    ku++;
    if( ku > kumax )  {
      usred=usred/(float)ku;
      ku=0;
      fprintf(outfile,"averaged squared velocity U  u**2=%10.3g\n",usred );
      usred = 0.;
    }
    /* ------------------------------------------------------------ */
        for(j=0;j<=n;j++)  {        /*  cycle on  j = 0,1,2,...,n  */
      for(i=1;i<=3;i++)  {
   ij=3*j+i-1;
   r1[ij]=r[ij]+u[ij]*ht;
        if( kmodel == 1 )
     u1[ij]=u[ij]+a[ij]*ht;
      }
```

```
    for (i=1; i<=3; i++)  {          /*   cycle on i=1,2,3;   */
      ij = 3*j+i-1;
      if( kmodel == 1 )
        rr = m*a[ij]+a[ij]+u[ij];
      else rr = u[ij];
  if (j == 0)  {
    uij = u[i-1]-u[2+i];
    aij = r[i-1]-r[i+2];
  } else
    if (j == n)  {
      uij = u[n*3+i-1]-u[(n-1)*3+i-1];
      aij = r[n*3+i-1]-r[(n-1)*3+i-1];
    } else {
      uij = 2.*u[j*3+i-1]-u[(j-1)*3+i-1]-u[(j+1)*3+i-1];
      aij = 2.*r[j*3+i-1]-r[(j-1)*3+i-1]-r[(j+1)*3+i-1];
    }
  rr = rr+uij*z+aij*z;
  deli2 = (i == 2) ? 1. : 0.;
  rr = rr+deli2*dpj_drj2*u[3*j+1];
  rr = rr+deli2*pj;
  rr = rr+(b+e)*u[ij];
   /* ---------------------------------------------------------------
  if ((j != 0) && (j != n))  {
    rrr = 0.;
    for (k=1; k<=3; k++)  {
      delik = (i == k) ? 1.: 0.;
      rrr = rrr+(eij(i,j,r) * eij(k,j,r) - delik/3) * u[3*j+k-1];
    }
    rr = rr-h*rrr;
  }
   ----------------------------------------------------------------*/
  /* cjk==0   rr=rr+sum() */
      stoh = gauss(1.,0.);
      stoh *= skvotk;
  stohas[ij] = stoh;
      if( kmodel == 1 ) {
    rr = rr*ht;
    if (ii%iprint == 0)
printf(" i=%d j=%d  a[ij]=%7.2f  rr*ht=%7.4f stoh=%7.4f  u*2[ij]=%7.4f\n\r",
          i,j,a[ij],rr,stoh,u[ij]*u[ij]);
    rr = stoh-rr;
    rr = rr/m;
    a1[ij] = a[ij]+rr;
        }
  else { u1[ij] = u[ij] - rr * ht + stoh ;

  if (ii%iprint == 0)
     printf(" i=%d j=%d u[ij]=%8.5f rr*ht=%7.4f stoh=%7.4f a[ij]=%8.3f\n\r",
       i,j,u[ij],rr*ht,stoh,a[ij]);
      }   /*   else   */
  }        /*   cicle on  i=1,2,3;         */
    }          /*   cicle on  j=0,1,2,...,n;      */
    /* ------------------------------------------ */

    t = t+ht;
    gotoxy(1,1);
    printf("Time = %f\n\r ", t);
            /* for slides  */
    if ((t-(int)(t/t_slides+EPSILON)*t_slides) <= ht) {
```

```
        fprintf(fs, "%d\n", n+1);
        for (j=0; j<=n; j++) fprintf(fs, "%f\n", r[3*j]);
        for (j=0; j<=n; j++) fprintf(fs, "%f\n", r[3*j-1]);
       }
       for (j=0; j<=n; j++)      /*  writing in r[], u[], a[] new value */
        for (i=1; i<=3; i++)   {
         ij=3*j+i-1;
         r[ij]=r1[ij];
         u[ij]=u1[ij];
         a[ij]=a1[ij];
        }

   /*    calculation  of corr. functions    */
        ls[0] = 0.;
        corfun( n, r, p, corrf, ls, p1, hcov, nvrcorrf, nvrls,hls,t, ht, tmin,
           nls, ncorrf,  nirrm, krrm, pcen, pcen1, rcen  );

      /*------------------------------------------------------------*/
       ii = ii+1;         /*  iprint - step of print  r[], u[], a[]  */
       if (ii%iprint == 0)  {
        print_matr(r,3,n+1);    fprintf(outfile,"\n");
        print_matr(u,3,n+1);    fprintf(outfile,"\n");
        print_matr(a,3,n+1);    fprintf(outfile,"\n");
        normro(n,r,p);
        fprintf(outfile," writing of normal coordinates s ,  t=%.3f \n",t);
        print_matr(p,3,n+1);
       }
      /*------------------------------------------------------------*/

 }          /*     while t<tmax        */
/*==================================================================*/

     fclose (frrm);
     fclose (frrm1);
     clrscr();

     //  correlation functions
     for (j = 1; j <= n; j++)
       for(k = 0; k <= ncorrf;k++)
        corrf[(ncorrf+1)*j+k] /= (float)nvrcorrf[k];

     //  displacement
      ls[0] = 0.;
     for( k = 1; k <= nls; k++) {
       ls[k] /= nvrls[k];
       ls[k] /= (float)(n+1);
     }
     /*   regression    */
      fmodes = fopen("modes.mlk","w");
      n_max = ncorrf;
      htt = hcov[1];
      fprintf (fmodes, "%-d\n", n);
      for( j=1;j <= n;j++)   {
        rrr=lin_yx( corrf + j * (ncorrf+1),hcov,ncorrf);
        fprintf (fmodes, "%-16.13f\n", htt);
        fprintf (fmodes, "%d\n", n_max+1);
        fprintf (fmodes, "%-16.13f\n", rrr);
```

```
        for (k=0; k<=n_max; k--) fprintf (fmodes, "%-16.13f\n",
        *(corrf + k+ j * (ncorrf+1))/(*(corrf + j * (ncorrf+1))));
        fprintf (outfile, " mode = %d, time of  relaxation = %f, R[0] = %g\n",
                  j, -0.5/rrr, *(corrf + j * (ncorrf+1)));
        printf ( "\n\r mode = %d, time of  relaxation = %f", j, -0.5/rrr);
    }  /* for j */
    /*   for plots   */
      fnul   = fopen("nulmode.mlk","w");
      for( k=1; k <= nls; k++)  hls[k] = log10( hls[k] );
      fprintf (fnul, "%-16.13f\n", hls[1]);
      fprintf (fnul, "%d\n", nls);
      for (k=1; k<=nls; k++)  ls[k] = log10(ls[k]);
      for (k=1; k<=nls; k++) fprintf (fnul, "%-16.13f\n", ls[k]);
      fclose (fnul);
      fclose (fmodes);

/*  coordinates keeping  */
   fdata = fopen("rua.mlk", "w");
   sprintf(p_str, "%d\n", n);      fputs(p_str, fp);
   for(j=0;j<=n;j++) {
    ij=3*j;
    sprintf(p_str, "%f %f %f\n", r[ij], r[ij+1], r[ij+2]);    fputs(p_str, fp)
    sprintf(p_str, "%f %f %f\n", u[ij], u[ij+1], u[ij+2]);    fputs(p_str, fp)
    sprintf(p_str, "%f %f %f\n", a[ij], a[ij+1], a[ij+2]);    fputs(p_str, fp)
      }
   fclose'(fdata);

free(r);
free(r1);
free(u);
free(u1);
free(a);
free(a1);
free(p);
free(stohas);
free(corrf);
free(p1);
free(ls);
free(hcov);
free(nvrls);
free(nvrcorrf);

}       /* end main */


/*================================================*/
float eij(int i, int j, float POINTER *r)  {
 /*  eij=(r[i,j+1]-r[i,j-1])/sqrt(sum k ((r[k,j+1]-r[k,j-1])**2)  */
 /*  i=1,2,3;  j=1,2,...,n-1;              */
float rrr,rs;
int k;
   rs=0.;
   for(k=1;k<=3;k++)   {
     rrr=r[3*(j+1)+k-1]-r[3*(j-1)+k-1];
     rrr=rrr*rrr;
     rs=rs+rrr;
   }
   rs = sqrt(rs);
   rrr = r[3*(j+1)+i-1]-r[3*(j-1)+i-1];
   if( fabs(rs) < EPSILON )  {
```

```
      rrr=0.;
       /*
    fprintf(outfile," r^[j+1]==r^[j-1]; º¡á¼Ñ¡áΓÑ½∞ eij=0\n");
    bioskey(0);
       */
   } else  rrr = rrr/rs;
  return(rrr);
}

/* ================================== */
void print_matr (float POINTER *x, int n, int m)  {
int i, j, ij;

  printf("\n\r");
  for(i=1; i<=n; i++)  {
    for (j=1; (j <= m) && (j <= 7); j++)  {
      ij = (j-1)*n-i-1;  /* ij=(i-1)*m+j-1; */
      fprintf (outfile, "%11.3g ", *(x+ij));
      printf ( "%11.3g ", *(x+ij));
    }
    fprintf(outfile,"\n");
    printf("\n\r");
  }
}

/*================================================================*/

void normro(int n,float POINTER *r,float POINTER *p)  {
/*  NORM. COORDINATIONS
    input - coordnates of particles r(i,j), i=1,2,3; j=0,1,...n
    output - normal coordinates  p(i,j)
    p(i,j)=sum(R(k,j)*r(i,k) */
    R(k,j)=sqrt((2-delta(j,0))/(n+1))*cos((j*(1+2*k)*pi)/(2*(n+1));j,k=0..n */

int i,j,ij,k;
float pi=M_PI;
float rr,rrr,delj0,uij,rik,ajk;
double drr,dcos;

  for (j=0; j<=n; j++)
    for (i=1; i<=3; i++)  {
      ij = 3*j+i-1;
      delj0 = (j == 0) ? 1. : 0.;
      rr = (2-delj0)/((float)n + 1.0);
      rr = sqrt(rr);
      rrr = 0.;
      uij = ((float)j)*pi/2.0;
      uij = uij/((float)n+1.0);
      for (k=0; k<=n; k++)  {
       rik = uij*((float)k*2.0+1.0);
       drr = (double) rik;
       dcos = cos(drr);
       rik = (float) dcos;
       ajk = rr*rik*r[3*k+i-1];
       rrr += ajk;
      }
      p[ij] = rrr;
    }
```

```
/*==============================================================*/
 float lin_yx( float POINTER *y, float POINTER *h, int T)  {
  /*  LINEAR REGRESSION
      y[h] = Kj(0) * exp(a*h);  ==>    log(y[h])=log(Kj(0)) - a*h ;
      y(j)[k]=corrf[j,k], h(k)=hcov[k], T=ncorrf;  k=0,1,...,ncorrf;
      a ={ sum(h[k]*log(y[k])-log(Kj(0))*sum(h[k]) } / sum(h[k])**2;  */

 float   s1 = 0.0, s2 = 0.0, s3 = 0.0, a;
 int  k;
   for( k = 1; k <= T; k++ )  {
     if( ( y[k] / y[0] ) > 1.e-30 )  {
       s1 += h[k] * log(y[k]/y[0]);
       s2 += h[k];
       s3 += h[k]*h[k];
      }
     else  printf("corrf[j,%d] / corrf[j,0] < 1.e-30 \n\r",k);
    }
   a = ( s1-s2*log(y[0]/y[0]) ) / s3;
   return(a);
 }
/*====================================================================*/

 void corfun( int n, float POINTER *r, float POINTER *p,
              float POINTER *corrf, float POINTER *ls,
         float POINTER *pl, float POINTER *hcov,
         int POINTER *nvrcorrf, int POINTER *nvrls,
         float POINTER *hls,
         float t, float ht, float tmin,
         int nls, int ncorrf,  int nirrm, int krrm,
         float  POINTER *pcen, float POINTER *pcenl,
         float POINTER *rcen )  {
  int  i, j, k , ij;
  static long k1, k2, kirrm=-1;
  static float t1, trrm, hh,trab;
  int flagp;                 /*  flag of normalization */
  float  tau, tk1, rr;
  static int flagcorrf=1;    /*     first call flag    */
  static int flagt=1;
  long curpos,kk,kkk;

  flagp = 1;
  if ( flagcorrf == 1 )  {
    trrm=0.;
    hh = ht*(float)krrm;
    flagcorrf=0;
    t1 = tmin + nirrm * hh;
    k1 = -1;
    k2 = -1;
   }

  /* writing in rrm, if  tmin <= t <= tmin + nirrm * krrm * ht
     nrrm = nirrm*3*(n+1) - dimension of rrm[]
     krrm - integrating number of steps between two notices in the   rrm */
  if( tmin <= t )  {
    k1++;
    if( (k1 % (long)krrm) == 0 )  {
      k2++;
      if(flagt == 1)  { trab=t; flagt=0; t1 = trab + nirrm * hh; }
```

```
    if( t <= t1 )  {
    p[ 3 * (n+1) ] = t;
    pcen[ 3 * (n+1) ] = t;
    normro(n,r,p);
    centrr( n, r, rcen) ;
    normro(n,rcen,pcen);
    flagp = 0;
    trrm = t;
    kirrm++;
    nvrcorrf[0] += 1;
    for( j = 0; j <= n; j++)
for( i = 1; i <= 3; i++)  {
  ij = 3*j+i-1;
  corrf[(ncorrf+1) * j] += pcen[ij] * pcen[ij];
 }
  fseek(frrm,0L,SEEK_END);
  fwrite( (void*)p, (3*(n+1)+1)*sizeof(float), 1, frrm);
  fseek(frrm1,0L,SEEK_END);
  fwrite( (void*)pcen, (3*(n+1)+1)*sizeof(float), 1, frrm1);
    }       /*          for  if( t <= t1 )            */

    /* Culculation of corr. function   */
    tau = t - trrm;
    for( k = 1; k <= ncorrf; k++)
     if( (trab + hcov[k] < t) && ( tau < hcov[k] ) )  {

   if( flagp == 1 )  {
     flagp = 0;
     normro(n,r,p);
     centrr( n, r, rcen) ;
     normro(n,rcen,pcen);
     if( k == 1 )  {
       nvrcorrf[0] += 1;
       for( j = 1; j <= n; j++)
         for( i = 1; i <= 3; i++)  {
           ij = 3*j+i-1;
      corrf[(ncorrf+1) * j] += pcen[ij] * pcen[ij];
     }
             }
    }   /*  if( flagp == 1 )       */

   kk = (long)( ( hcov[k] - tau ) / hh + 0.5 );
   kkk = kirrm - kk;
   if (kkk >= 0) {
    i = readzap( pcen1, kkk, (3*(n+1)+1)*sizeof(float), 1 );
    if ( i != 0 )  {
     if( fabs(t-pcen1[3*(n-1)]-hcov[k]) > 0.6*ht )   {
      printf("\n\rR (t-pcen1[3*(n+1)]-hcov[k] > 0.0001 t=%f pc1()=%f t--=%f"
        t,pcen1[3*(n+1)],t-pcen1[3*(n+1)]-hls[k]);
      printf("\n\rR  k=%d   hcov[k]=%g",k,hcov[k]);
      exit(1);
     }
     nvrcorrf[k] +=1;
     for( j = 1; j <= n; j++)
      for( i = 1; i <= 3; i++)  {
       ij = 3*j+i-1;
       corrf[(ncorrf-1)*j - k] += pcen[ij] * pcen1[ij];
      }
     }
```

```
        else   {
         curpos =kkk * 3*(n-1)*sizeof(float);
         printf("\n\r  not found  R(s) \n\r  kkk * 3*(n+1)*sizeof(float)=%ld "
                            curpos);
         exit(1);
        }
      }                  /*          for IF(kkk)                   */
      else { printf("K(s)  :  kkk<0 kk=%d  kirrm=%d kkk=%d k=%d t=%f\n\r",
              kk,kirrm,kkk,k,t);
             exit(1);    }
    }      /*   for IF( (tmin + hcov[k] < t) && ( tau <= hcov[k] ) )  */

   }       /*    for  if( k1 % krrm )      */

      /* Culculation of displacement  */
      tau = t - trrm;

    clrscr();
    for( k=1; k <= nls; k++)   {
    tk1 = trab + hls[k];
    if( ( tk1 < t ) && ( tau < hls[k] ) )  {
     rr = (t - hls[k] - trab + ht ) / hh;
     kk = (long)rr;
     rr = t - hls[k] - trab - hh * (float)kk;
//   printf("\n\r L k=%d kk=%ld delt=rr=%g t=%f h=%f",k,kk,rr,t,hls[k]);
     if( fabs(rr) < (ht/2.) )  {
      if( flagp == 1 )   {
        flagp = 0;
        normro(n,r,p);
       }
      kk = (long)( ( hls[k] - tau ) / hh + 0.5 );
      kkk =  kirrm - kk;
      if (kkk >= 0 ) {
       i = readzap( p1, kkk, (3*(n+1)+1)*sizeof(float), 0 );
       if ( i != 0 )   {
        if( fabs(t-p1[3*(n+1)]-hls[k] ) > 0.6*ht )   {
          printf("\n\rL (t-p1[3*(n+1)]-hls[k] > 0.0001 t=%f p1()=%f t--=%f",
              t,p1[3*(n+1)],t-p1[3*(n+1)]-hls[k]);
          exit(1);
         }
        nvrls[k] += 1;
        for( i = 1; i <= 3; i++)   {
          rr = p[i-1] - p1[i-1];
          ls[k] += rr * rr;
         }
        }
        else  { printf("\n\r not found L(s)  ");  exit(1); }

        }       /*          for IF(KKK)                    */
       else { printf("\n\rL(s)  :  kkk<0 kk=%ld  kirrm=%ld kkk=%ld k=%d t=%f\n\r
         kk,kirrm,kkk,k,t);
         exit(1); }
     }
    }
   }           /*      for     if( fabs(rr) < (ht/2.) )     */
  }    /* if( tmin <= t ) */
 }  /* end corrfun() */
/*==================================================================*/
```

```
float gauss( float s,float am )  {
float a, r, v;
int j;
 a = 0.0;
 v = slutch1();
for ( j = 0; j < 12; j++ )  {
  if( v < 0.2 )            r = slutch();
    else  if( v < 0.4 )  r = slutch2();
    else                 r = slutch1();
  a += r;
}
  a =  a - 6.0;
  v = a*s+am;
return( v );
} /* gauss */

/************************************************************/

float slutch()  {
static  long   is=7235591L;
   float    x;
is *= 125 ;
is -= (is/2796203L)*2796203L;
x  = (float)is/(float)2796202L;
return (x);
}
/*================================================================*/
  float slutch1()  {
   static double ksi = 0, zz = 0.011;
   float pi=M_PI,x;
   double dpi,rr;
   long kk;
  dpi = pi;
  rr = ksi / zz + dpi;
  zz += 1.e-8;
  kk = rr;
  ksi = rr - kk;
  x = ksi;
  return(x);
 }
/*======================================================================*/

   float slutch2()  {
     static float rr = 1.;
     float  dm, c, b, d;
     int  l;
    dm = 8192.;
    d = 67101323.;
    b = dm * dm - d;
    c = rr / dm;
    while( c >= 10000.)  c /= 10000.; .
    l = c;
    rr = b * (float)l + dm * (rr - dm*(float)l);
    if( rr > d )  rr -= d;
    return( rr / d );
  }
/*======================================================================*/
  void centrr( int n, float POINTER *r, float POINTER *r1)  {
    int i,j,ij;
```

```
     float rr;
       for(i=1;i <= 3;i++)  {
         rr = 0.;
         for(j=0;j <= n;j+-)  {
           ij = i-1 + 3*j;
           rr += r[ij];
           }
         rr /= (float)(n+1);
         for(j=0;j <= n;j++)  {
           ij = i-1 + 3*j;
      r1[ij] = r[ij] - rr;
           }
         }          /*   for i    */
       }            /*   end      */
/*=================================================================*/
void print_e (float POINTER *x, int n, int m)  {
int i, j, ij;

  printf("\n\r");
  for (j=0; (j <= n) && (j <= 6); j++)  {
   for(i=1; i<=m; i++)  {
      ij = m*j + i-1;
      printf ( "%10.3g ", *(x+ij));
     }
    printf("\n\r");
  }
}

/*=============================================================*/
int readzap( float *p, long n, int len, int pcenn )  {
long nl;
int  cod=1;
   nl = n*len;
   if (pcenn == 0 )  {
     fseek(frrm, nl, SEEK_SET);
     if ( fread( (void*)p,len,1,frrm ) != 1 )    cod = 0;
     }
    else  {
     fseek(frrm1, nl, SEEK_SET);
     if ( fread( (void*)p,len,1,frrm1 ) != 1 )    cod = 0;
     }
//      ii = ftell(frrm);
// printf("\n\r READZAP  curpos = %ld ii=%ld cod=%d",curpos,ii,cod);
// print_e (p, 4,3);
// bioskey(0);
   return cod;
}
/*=================================================================*/
```

APPENDIX 2
PROTOTYPE PROGRAM OF THE FINITE ELEMENT METHOD

```
                                PROGRAM DEFORM
C
C*********************************************************
C
C  DEFORM - DEFORM CALCULATION PROGRAM BY FINITE
C           ELEMENT METHOD.
C
C  ENTRY PARAMETERS : NE,NUZ,NBW,T,TEMP;
C           (SEE INPUT-OUTPUT COMMENTARY)
C
C  OUTPUT PARAMETERS :
C           (SEE OUTPUT COMMENTARY)
C
C  USED SUBROUTINES : WURZEL , MATPR
C
C  NOTE : ENTRY PARAMETERS INPUT IS PRODUCED FROM FILE
C         indef.dat; JUNCTION COORDINATES GRID IS PRODUCED
C         FROM FILE netz.dat FORMED BY MEANS OF PROGRAM
C         netz.
C*********************************************************
C
      DIMENSION NS(6),ND(3),ESM(6,6),EF(6),B(3,6),C(6,3),D(3,3)
      DIMENSION ET(3),A(50000),EMM(10),PRM(10),ALPHAM(10)
      DIMENSION UGRAN(1000),IGRAN(1000),F(1000),U(1000)
      DIMENSION NUMUZ(3000),XCOOR(1000),YCOOR(1000),MAT(1000)
      OPEN (UNIT=10,FILE='G:\GAVR\indef.dat')
      OPEN (UNIT=11,FILE='G:\GAVR\netz.dat')
      OPEN (UNIT=20,FILE='G:\GAVR\outdef.dat')
      OPEN (UNIT=21,FILE='G:\GAVR\outnetz.dat')
      OPEN (UNIT=22,FILE='G:\GAVR\defnetz.dat')
C
C INPUT-OUTPUT OF MAIN DETAIL PARAMETERS :
C      NE      - TOTAL ELEMENTS NUMBER
C      NUZ     - TOTAL JUNCTIONS NUMBER
C      NBW     - RIGID MATRIX TAPE WIDTH
C      NMAT    - NUMBER OF COMPOSITION MATERIALS
C      IND     - INDICATOR OF MATERIAL FORM TYPE
C      ITEMP   - TEMPERATURE PARAMETER ( 0 ; 1 )
C      T       - DETAIL THICKNESS
C      TEMP    - DETAIL TEMPERATURE
C
      READ (10,100) NE,NUZ,NBW,NMAT,IND,ITEMP,T,TEMP
      WRITE (20,200) NE,NUZ,NBW,NMAT,IND,ITEMP,T,TEMP
      NP=NUZ*2
100   FORMAT (///30X,I5/30X,I5/30X,I5/30X,I5/30X,I5/30X,I5/
     *       30X,F10.4/30X,F10.4)
200   FORMAT (/1X,'*** INITIAL PARAMETERS OF PROG. DEFORM  ***'//
     *  2X,'TOTAL ELEMENTS NUMBER             : ',I5/
     *  2X,'TOTAL JUNCTIONS NUMBER            : ',I5/
     *  2X,'RIGID MATRIX TAPE WIDTH           : ',I5/
     *  2X,'NUMBER OF COMPOSITION MATERIALS : ',I5/
     *  2X,'INDICATOR OF MATERIAL FORM TYPE : ',I5/
     *  2X,'TEMPERATURE PARAMETER             : ',I5/
     *  2X,'DETAIL THICKNESS                  : ',F10.5/
     *  2X,'DETAIL TEMPERATURE                : ',F10.5)
```

```
C
C  INPUT-OUTPUT OF DETAIL MATERIAL (MATERIALS I=1,4) PARAMETERS :
C      EMM    - ELASTIC MODULUS
C      PRM    - POISSON'S COEFFICIENT
C      ALPHAM- COEFFICIENT OF THERMAL EXPANSION
C
      READ (10,110)
      WRITE (20,210)
      DO 90 I=1,4
      READ (10,111) EMM(I),PRM(I),ALPHAM(I)
90    WRITE (20,211) I,EMM(I),PRM(I),ALPHAM(I)
110   FORMAT(//)
210   FORMAT (/'MATERIAL PARAMETERS : '/
     *    2X,'MATER.   EL.MOD.  POISSON'S COEFF.  COEFF.THERM.EXP.')
111   FORMAT (7X,3E10.3)
211   FORMAT (2X,I5,3E10.3)
C
C INPUT-OUTPUT OF LIMIT JUNCTION COORDINATES
C
      DO 13 I=1,NP
13    UGRAN(I)=0.
      READ (10,114)
      WRITE (20,214)
      K=0
2     K=K+1
      READ (10,113) I,XI
      WRITE (20,113) I,XI
      IF (I.LE.0) GO TO 1
      IGRAN(K)=I
      UGRAN(I)=XI
      GO TO 2
1.    NGRAN=K-1
114   FORMAT (//)
113   FORMAT (7X,I5,2F10.3)
214   FORMAT (/'LIMIT JUNCTION COORDINATES :'/
     *     2X,'JUNC. NUMB.  X-COORD. Y-COORD.')
C
C  INPUT-OUTPUT OF JUNCTION FORCES
C
      DO 4 I=1,NP
4     F(I)=0.
      READ (10,115)
      WRITE (20,215)
115   FORMAT (//)
215   FORMAT (/'JUNCTION FORCES :'/
     *     2X,'JUNC. NUMB.  X-COORD. Y-COORD.')
      K=0
6     K=K+1
      READ (10,113) I,XI
      WRITE (20,113) I,XI
      IF (I.LE.0) GO TO 5
      F(I)=XI
      GO TO 6
5     CONTINUE
```

```
C
      NEND=NP*NBW
      DO 8 I=1,NEND
8     A (I)=0.0
      R0=1./(1.0-PR*PR)
C
C  INPUT-OUTPUT OF ELEMENT JUNCTION COORDINATES
C
      DO 7 KK=1,NE
      READ (11,112) NEL,ND,X1,Y1,X2,Y2,X3,Y3
      NM1=3*NEL-2
      NM2=3*NEL-1
      NM3=3*NEL
      N1=ND(1)
      N2=ND(2)
      N3=ND(3)
      NUMUZ(NM1)=N1
      NUMUZ(NM2)=N2
      NUMUZ(NM3)=N3
      XCOOR(N1)=X1
      XCOOR(N2)=X2
      XCOOR(N3)=X3
      YCOOR(N1)=Y1
      YCOOR(N2)=Y2
      YCOOR(N3)=Y3
C*      WRITE (21,212) NEL,ND,X1,Y1,X2,Y2,X3,Y3
      WRITE (21,212) NEL,NUMUZ(NM1),NUMUZ(NM2),NUMUZ(NM3),
     *    XCOOR(N1),YCOOR(N1),XCOOR(N2),YCOOR(N2),
     *    XCOOR(N3),YCOOR(N3)
112   FORMAT (4I5,6F10.5)
212   FORMAT (1X,I3,2X,3I4,3X,6(2X,F8.4))
      DO 18 I=1,3
      NS(2*I-1)=ND(I)*2-1
18    NS(2*I)=ND(I)*2
C
C CALCULATION OF ELEMENT <NEL> STIFF MATRIX
C
      IF (NMAT.EQ.1) IMAT=1
      IF (NMAT.EQ.1) GO TO 289
      CALL MATPR (IND,X1,Y1,X2,Y2,X3,Y3,IMAT)
289   MAT(NEL)=IMAT
      PRINT *,'ELEMENT CALCULATION  ',NEL,' IMAT = ',IMAT
      EM=EMM(IMAT)
      PR=PRM(IMAT)
      ALPHA=ALPHAM(IMAT)
      R=EM*R0
      D(1,1)=R
      D(2,2)=D(1,1)
      D(3,3)=R*(1.0-PR)/2.
      D(1,2)=PR*R
      D(2,1)=D(1,2)
      D(1,3)=0.0
      D(3,1)=0.0
      D(2,3)=0.0
```

```
      D(3,2)=0.0
      DO 20 I=1,3
      DO 20 J=1,6
20    B(I,J)=0.0
      B(1,1)=Y2-Y3
      B(1,3)=Y3-Y1
      B(1,5)=Y1-Y2
      B(2,2)=X3-X2
      B(2,4)=X1-X3
      B(2,6)=X2-X1
      B(3,1)=B(2,2)
      B(3,2)=B(1,1)
      B(3,3)=B(2,4)
      B(3,4)=B(1,3)
      B(3,5)=B(2,6)
      B(3,6)=B(1,5)
      ET(1)=0.0
      ET(2)=0.0
      ET(3)=0.0
      AR2=X2*Y3+X3*Y1+X1*Y2-X2*Y1-X3*Y2-X1*Y3
C*      DO 700 III=1,3
C*700    WRITE (20,750) (B(III,I17), I17=1,6)
C*750    FORMAT (2X,'B: ',6F10.5)
C
C  MATRIX CALCULATION  C = (BT) (D)
C
      DO 22 I=1,6
      DO 22 J=1,3
      C(I,J)=0.0
      DO 22 K=1,3
22    C(I,J)=C(I,J)+B(K,I)*D(K,J)
C
C  MATRIX CALCULATION OF ELEMENTS <ESM> AND <EF>
C     ESM = (BT) (D) (B)  = (C) (B)
C     EF  = (BT) (D) (ET) = (C) (ET)
C
      DO 27 I=1,6
      SUM1=0.0
      DO 29 K=1,3
29    SUM1=SUM1+C(I,K)*ET(K)
      DO 27 J=1,6
      SUM=0.0
      DO 28 K=1,3
28    SUM=SUM+C(I,K)*B(K,J)
      ESM(I,J)=SUM*T/(2.*AR2)
27    EF(I)=SUM1*T/2.
C*      DO 152 III=1,6
C*152    WRITE (20,780) (ESM(III,JJJ), JJJ=1,6)
C*780    FORMAT (1X,'ESM  ',6F8.4)
```

```
C
C REPEATED COUNTING OF MATRIX <ESM> INTO MATRIX <A>
C               BY DIRECT STIFF METHOD
C
      DO 7 I=1,6
      II=NS(I)
      IF (ITEMP.EQ.1) F(II)=F(II)+ET(I)
      DO 17 J=1,6
      JJ=NS(J)
      JJ=JJ+1-II
      IF (JJ.GT.NBW) WRITE (20,220)
      IF (JJ.GT.NBW) STOP 111
220   FORMAT (2X,'MATRIX <A> BAND WIDTH IS WRONG !')
      IF (JJ) 17,17,16
16    J5=(II-1)*NBW+JJ
      A(J5)=A(J5)+ESM(I,J)
C*      WRITE (20,790) I,J,NS(I),NS(J),J5
C*790   FORMAT (' I,J,NS(I),NS(J),J5 ',5I7)
17    CONTINUE
7     CONTINUE
      WRITE (20,221)
221   FORMAT (/1X,'*** RESULTS OF PROG. DEFORM CALCULATION *** ')
C
C  MODIFICATION OF MATRIX <A> WITH LIMIT MEANINGS CALCULATION
C
C*      DO 151 III=1,NP
C*      I1=(III-1)*NBW+1
C*      I2=I1+NBW-1
C*151   WRITE (20,771) I1,I2,(A(IAI), IAI=I1,I2)
C*      WRITE (20,800) NGRAN,(IGRAN(I),I=1,4)
C*800   FORMAT (2X,'NGRAN, IGRAN() ',5I5)
      IF (NGRAN.EQ.0) GO TO 31
      DO 30 I=1,NGRAN
      J=IGRAN(I)
      JA=(J-1)*NBW+1
      A(JA)=A(JA)**4
C*      DO 32 J1=2,NBW
C*32    A(JA+J1-1)=0.0
30    F(J)=UGRAN(J)*A(JA)
31    CONTINUE
C*      WRITE (20,801) (F(III),III=1,10)
C*801   FORMAT(1X,'F ',10F7.3)
C*      DO 150 III=1,NP
C*      I1=(III-1)*NBW+1
C*      I2=I1+NBW-1
C*150   WRITE (20,771) I1,I2,(A(IAI), IAI=I1,I2)
C*771   FORMAT (1X,2I3,' A ',8F7.3)
C
C  LINEAR EQUATIONS SYSTEM DECISION : (A) (U) = (F)
C
      CALL WURZEL (A,F,NP,NBW,U,IERR,ENEW)
      WRITE (20,224) IERR,ENEW
224   FORMAT (2X,'IERR= ',I3,'  ENEW= ',E12.4)
```

```
      WRITE (20,223)
223   FORMAT (//2X,'JUNCTION DEFORMATIONS MEANINGS :'/
     *   2X,'JUNC.NUMBER  X-COORD.  Y-COORD. ')
      DO 40 I=1,NUZ
40    WRITE (20,222) I, U(2*I-1), U(2*I)
222   FORMAT (2X,I5,2X,2E14.5)
      DO 42 I=1,NP
      F(I)=XCOOR(I)
      UGRAN(I)=YCOOR(I)
      XCOOR(I)=XCOOR(I)-U(2*I-1)
42    YCOOR(I)=YCOOR(I)+U(2*I)
      DO 41 I=1,NE
      NM1=3*I-2
      NM2=3*I-1
      NM3=3*I
      N1=NUMUZ(NM1)
      X1=XCOOR(N1)
      Y1=YCOOR(N1)
      N2=NUMUZ(NM2)
      X2=XCOOR(N2)
      Y2=YCOOR(N2)
      N3=NUMUZ(NM3)
      X3=XCOOR(N3)
      Y3=YCOOR(N3)
C
C  CALCULATION OF TENSIONS
C
      X1S=F(N1)
      X2S=F(N2)
      X3S=F(N3)
      Y1S=UGRAN(N1)
      Y2S=UGRAN(N2)
      Y3S=UGRAN(N3)
      CX=X1-X2
      CXS=X1S-X2S
      IF (ABS(CXS).LT.ABS(X1S-X3S)) CX=X1-X3
      IF (ABS(CXS).LT.ABS(X1S-X3S)) CXS=X1S-X3S
      IF (ABS(CXS).LT.ABS(X2S-X3S)) CX=X2-X3
      IF (ABS(CXS).LT.ABS(X2S-X3S)) CXS=X2S-X3S
      EPSX=(CXS-CX)/CXS
      CY=Y1-Y2
      CYS=Y1S-Y2S
      IF (ABS(CYS).LT.ABS(Y1S-Y3S)) CY=Y1-Y3
      IF (ABS(CYS).LT.ABS(Y1S-Y3S)) CYS=Y1S-Y3S
      IF (ABS(CYS).LT.ABS(Y2S-Y3S)) CY=Y2-Y3
      IF (ABS(CYS).LT.ABS(Y2S-Y3S)) CYS=Y2S-Y3S
      EPSY=(CYS-CY)/CYS
      EM=EMM(MAT(I))
      GXY=(CXS-CX)/CYS+(CYS-CY)/CXS
      SIGX=ABS(EM*R0*(EPSX+EPSY*PR))
      SIGY=ABS(EM*R0*(EPSX*PR+EPSY))
      TAUXY=ABS(EM*0.5*(1.-PR)*R0*GXY)
41    WRITE (22,117) I,N1,N2,N3,X1,Y1,X2,Y2,X3,Y3,
     *                 SIGX,SIGY,TAUXY
117   FORMAT (4I5,6F10.5,3E14.5)
      STOP
      END
```

Index

GPSR Compliance

The European Union's (EU) General Product Safety Regulation (GPSR) is a set of rules that requires consumer products to be safe and our obligations to ensure this.

If you have any concerns about our products, you can contact us on ProductSafety@springernature.com

In case Publisher is established outside the EU, the EU authorized representative is:

Springer Nature Customer Service Center GmbH
Europaplatz 3
69115 Heidelberg, Germany

Batch number: 10815964

Printed by Printforce, the Netherlands